Differences, Similarities and Meanings

Semiotics, Communication and Cognition

Edited by
Paul Cobley and Kalevi Kull

Volume 30

Differences, Similarities and Meanings

—

Semiotic Investigations of Contemporary
Communication Phenomena

Edited by
Nicolae-Sorin Drăgan

DE GRUYTER
MOUTON

ISBN 978-3-11-125787-7
e-ISBN (PDF) 978-3-11-066290-0
e-ISBN (EPUB) 978-3-11-065923-8
ISSN 1867-0873

Library of Congress Control Number: 2021940045

Bibliographic information published by the Deutsche Nationalbibliothek
The Deutsche Nationalbibliothek lists this publication in the Deutsche Nationalbibliografie;
detailed bibliographic data are available on the Internet at http://dnb.dnb.de.

© 2023 Walter de Gruyter GmbH, Berlin/Boston
This volume is text- and page-identical with the hardback published in 2021.
Cover design based on a design by Martin Zech, Bremen
Typesetting: Integra Software Services Pvt. Ltd.
Printing and binding: CPI books GmbH, Leck

www.degruyter.com

Contents

Nicolae-Sorin Drăgan
Introduction —— 1

Part I: Theoretical advances

Augusto Ponzio
Communication, listening, non-functionality —— 13

Susan Petrilli, Augusto Ponzio
Difference and similarity in the I-Other relation: between two individuals or two singularities? A semioethic approach —— 29

Susan Petrilli
Global semiotics and its developments in the direction of semioethics —— 59

Göran Sonesson
The relevance of the encyclopaedia. From semiosis to sedimentation and back again —— 97

Luis Emilio Bruni
Cultural narrative identities and the entanglement of value systems —— 121

Part II: Applied Semiotics

A The digital age in semiotics and communication

Kristian Bankov
Lying as a transaction of value: explorations in semiosis and communication from a new perspective —— 151

Loredana Ivan, Corina Daba-Buzoianu, Ioana Bird
Effective affective campaigns? An analysis of campaigns centered on Roma —— 163

B Political semiotics and communication

Massimo Leone
The semiotics of extremism —— 181

Nicolae-Sorin Drăgan
The dynamic aspect of the semiotic behavior of political actors in TV debates —— 203

C Communication, semiotics and multimodality

Dario Martinelli
The different rifles of audiovisual communication: a semiotics of foreshadowing and the case of Roberto Benigni —— 227

Evripides Zantides
Differences, similarities and changes of national identity signs in print advertisements. The advertising discourse as a mirror of locality and vice-versa —— 245

Elena Negrea-Busuioc, Diana Luiza Simion
What's in a nickname? Form and function of sports' team nicknames —— 275

Constantin Popescu
Architecture and painting codes in *the Annunciation*. Oltenia (XVIIIth–XIXth centuries) —— 289

Index —— 305

Nicolae-Sorin Drăgan
Introduction

This volume focuses on the interdisciplinary aspect of semiotics and communication studies. It features thirteen articles on topics ranging from semioethics to lying and focusing on case studies which range from Roma communities to sports nicknames. The chapters included in this volume – expanded and refined from papers and keynote lectures delivered atthe 2nd International Conference *Semiosis in Communication* held at the National University of Political Studies and Public Administration, Bucharest, Romania, in June 2018 – explore the various forms of manifestation of *differences and similarities* in contemporary communication phenomena and examines how this dichotomy generates meanings in different communication situations, both from the theoretical and applied semiotics perspectives.

The dichotomy *differences – similarities* is fundamental to understanding meaning-making mechanisms in language (Saussure 1966; Derrida 1978; Deleuze 1995), as well as in other sign systems (Ponzio 1995; Sebeok & Danesi 2000; Deely 2009; Petrilli 2014). While in science, "the oppositions, the differences are relevant", in history "social facts have two aspects: one of *coexistence*, which is described by *similarities*, and another of *succession*, which is described by *differences*" (Marcus 2011: 351). Meaning always appears in the "play of *differences*" (Derrida 1978: 220) and *similarities*. Jacques Derrida (1978) seems to confirm the intuition of Jean-Jacques Rousseau (1998: 305), according to which "one must first observe the differences in order to discover the properties". An experience related by Eco (1976), which shows how his four-year-old son imitates the movements of a helicopter, is relevant in this regard. In his effort to understand and describe the fundamental feature of a helicopter, which makes it different from other machines, the little one selects an essential property of the movement of helicopter blades, namely the rotational motion. He had discovered "the basic geometric relation between the two blades, a straight line pivoting upon its center and rotating through 360 degrees", a movement he was able to reproduce "it in and with his own body" (Eco 1976: 206). Basically, he described – using his own body – a certain type of similarity that he had discovered by observing the pertinent differences from other objects. He used, intuitively, the interplay of differences and similarities to give meaning to a certain communication experience.

Nicolae-Sorin Drăgan, National University of Political Studies and Public Administration (SNSPA), Bucharest, Romania

https://doi.org/10.1515/9783110662900-001

In cognitive mechanisms that generate meaning, the two *modus operandi* have a permanent tendency to act and be used together (Krampen et al. 1987). While similarities are the basis for categorization and therefore for understanding, "only differences can resemble each other" (Deleuze 1990: 261), triggering various processes of semiosis. Therefore, the phenomena of similarities and differences must be considered complementary (Marcus 2011). The expression "the interplay of differences and similarities" captures this type of complementarity. The interplay of differences and similarities has a certain complexity that is due on the one hand to the new modes of sign production, and on the other hand to the increasingly varied perceptual experiences. In a fairly general formula, the dichotomy *differences – similarities* is one of the primary semiotic mode of operation of the human mind, one of the first ways to understand and signify the experience of reality. We thus gain a certain interpretative experience of reality that helps us to build meanings. Each of these previous interpretative experiences shape our semiosic processes, an idea that will be reformulated in an original way by Göran Sonesson in this volume. Inspired by Husserlian phenomenology, the author explains how previous interpretive experiences shape the most recent meanings starting from the concept of *sedimentation*, understood as "the passive mnemonic remnants of earlier semiotic acts, which form the background to the interpretation of any current act". In the same vein, in another article of this volume, Constantin Popescu also captures in his analysis a particular way in which previous experiences shape the way in which certain cultural artifacts are created. By examining the pictorial representations of the scene of the Annunciation on the eastern exterior walls of some churches of the Vâlcea county, the author shows how the builders and painters of these places found new ways of semic production for the creation of this type of artifact, which highlights their own experience of interpreting well-known biblical scenes and renews the traditional canons of Orthodox painting.

Our readers have already noticed that the proper understanding of the variety of forms of manifestation of the interplay of differences and similarities in contemporary communication phenomena and of the way in which they generate meanings imposed an eclectic choice. We hope that our preference for such an apparent heterogeneity will give readers the chance to have fertile encounters with the communication experiences analyzed by contributors, maintain their interrogation and open new possibilities for dialogue on the subject under debate. The investigation models of this type of dichotomy inevitably present a certain heterogeneity, representative for all major schools and research paradigms in the area of semiotics and communication studies, from cognitive semiotics to cultural-semiotic approaches, or from visual analyses of cultural artifacts to multimodal approaches to communication. All this combination of methods and levels of study of the meanings determined by this type of dichotomy reveals the interdisciplinary nature of this volume. For example, Jordan Zlatev (2011) discusses this type of interdisciplinary study of meaning when defining cognitive semiotics as an:

> interdisciplinary matrix of disciplines and methods, focused on the multifaceted phenomenon of *meaning* or as an emerging field with the ambition of " . . . integrating methods and theories developed in the disciplines of cognitive science with methods and theories developed in semiotics and the humanities, with the ultimate aim of providing new insights into the realm of human signification and its manifestation in cultural practices.

Therefore, we tried to bring close to the readers those common aspects and features that best capture the interplay of differences and similarities in contemporary communication phenomena, to integrate them in a complementary way and to create a coherent and consistent story about how this type of dichotomy generates meaning. Let us see in the following, some of the hypostases of the interplay of differences and similarities in various communication situations in the contemporary world.

Perhaps the most familiar perspective from which we can understand this type of dichotomy, particularly sensitive in our times, is the problem of *otherness*. "The difference exists, the otherness is built", states strongly Victor Ieronim Stoichiță (2017: 5). The image historian is interested in how the "Different", the "Non-Similar", the *Other* in a word, is "revealed, built, sometimes invented" in visual representations (Stoichiță 2017: 7). The differences that "exist" require certain efforts to recognize, represent and of course interpretive efforts, semiosic processes which will not be innocent at all. Even here, in the space of representation designated as the "iconosphere of difference", the "meeting with the Other" updates the dichotomy of "attraction/fear of the Other", a dilemma that still dominates the *imaginary of otherness* in modern times (Stoichiță 2017: 213). In their article for this volume, Loredana Ivan, Corina Daba Buzoianu and Ioana Bird emphasize the mediating role of affective communication campaigns between the social reality and the personal experience of each of us vis-à-vis the *imaginary of otherness*.

From the anthropologist's point of view, the *invention of the Other* may mean "mediating between identity and difference" (Kilan, 2000: 14). Mediation that is not done through "reconfiguring the other from the same, because it would certainly be his decline, but [through] the fascinated recognition of the distance" suggests Mondher Kilani (2000: 12). The other is already vulnerable, simply because it is different. Therefore, such an effort to recognize the difference is possible if we are interested in the "inner stranger", rather than "outer different". The Apostle Matthew had drawn our attention to the fact that "first take the plank out of your own eye, and then you will see clearly to remove the speck from your brother's eye" (Matthew 7: 1–12). Understanding the difference – in terms of otherness –begins with an inner look that helps you meet the other within the space of your own identity. Augusto Ponzio draws our attention in his article to the fact that in order for such an understanding to take place we need "listening" and "responsive understanding" of the other. It takes effort to get out of the selfishness of your own comfort and open up to the other. That involves *responsibility*, "'response-ability', 'answer-ability', in the double sense of the ability to respond to the other and to take responsibility for the other". In the same vein, Susan Petrilli states in another article that listening is "an interpreter of responsive understanding, a disposition for hospitality, for welcoming the signs of the other". How can we cultivate such an understanding of otherness today? The authors mentioned earlier offer readers a fertile ground for reflection and debate on this topic.

Let us remember, for example, that despite its imperfections, the Jew Shylock, one of the most complex characters of *The Merchant of Venice*, seems to have the intuition of what Quine (1969: 134) called "sense of similarity" when using the interplay of differences and similarities to express his own humanity:

> I am a Jew. Hath not a Jew eyes? Hath not a Jew hands, organs, dimensions, senses, affections, passions; fed with the same food, hurt with the same weapons, subject to the same diseases, healed by the same means, warmed and cooled by the same winter and summer as a Christian is? If you prick us, do we not bleed? If you tickle us, do we not laugh? If you poison us, do we not die? And if you wrong us, shall we not revenge? If we are like you in the rest, we will resemble you in that.
>
> (Shakespeare, *The Merchant of Venice*, Act 3 Scene 1)

Shylock proposes here a "perceptual judgment", skilfully constructed through an "inferential process" based on similarities that have a perceptual basis (Eco 2000: 125). Quine (1995: 253) calls them "perceptual similarities" and claims that they "is the basis of all learning, all habit formation, all expectation by induction from past experience; for we are innately disposed to expect similar events to have sequels that are similar to each other" (Quine and Gibson 2004: 277). Such a "sense of similarity", or the ability to perceive similarities and differences, is one of the fundamental aspects for human cognition (Quine 1969; Tversky and Gati 1978; Vosniadou and Ortony 1989; Eco 2000). In order to develop and refine such an ability, the *sense of similarity*, we saw that we need a responsive understanding of otherness, of the different. However, in the public space, we are witnessing a *mystique of dialogue* instead of an *authentic dialogue*, as we show in our article. Politicians are concerned with how to perform better in the role of president, as a *political persona* and not as an *individual* in front of the public, through various strategies of complementarity of the semiotic resources.

In a world of global communication, where each one's life depends increasingly on signs, language and communication, understanding how we relate and opening ourselves to otherness, to *differences* in all their forms and aspects is becoming more and more relevant. Today, we often understand the *differences* in terms of adversity or opposition and forget the value of the *similarities*. According to Umberto Eco (1995), the radicalization of these concepts can lead to problematic situations such as "fear of difference", a typical feature of *Ur-Fascism*, or *Eternal Fascism*. Massimo Leone takes things further in his contribution to this volume and introduces a reflection leading to understanding some of the linguistic and semiotic mechanisms through which a logic of extremism is introduced in the semiosphere of a society. Against this background, the plea of Susan Petrilli and Augusto Ponzio for understanding dialogue as openness to the other, where "otherness is not only the otherness of others, but also one's own otherness" is more relevant in the context of globalization.

The discourse on certain aspects of the phenomenon of globalization seems to alleviate this fear of difference, capturing new aspects that allow us to understand how the interplay of differences and similarities operates in a multicultural context. While Tomlinson (2007: 364), concerned with the cultural conditions of the phenomenon of globalization, considers such a phenomenon "the most significant force in creating and proliferating cultural identity", Jonathan Friedman (1994) is rather concerned with the social conditions of construction and attribution of meanings in the context of globalization. Friedman argues that cultural globalization is a product of the system of globalization, which involves transforming *difference* into *essence* (Friedman 1994: 209). At the beginning we are aware of what is specific, the differences from the way we do things similar to others. When these differences can be attributed to delimited communities, we are talking about culture or cultures, says Friedman. From this moment, the transformation of differences into essences comes naturally. When he speaks of essences, Friedman refers to concepts such as race, text, paradigm, code, or structure, in other words, the symbolic universe built by that community. In this way, culture generates an essentialization of the world, configurations of different cultures, ethnic groups or races. Therefore, "a difference can be recognized only on the basis of similarity (and vice-versa)" (Krampen et al. 1987: 220), even in debates about cultural diversity and globalization. Concerned with what he calls "processual cultural narrative identities", Luis Emilio Bruni has a similar perspective in his contribution for this volume, highlighting the centrality of "values" in the determination of identities. Also, in his contribution to this volume, Evripides Zantides also sits on the common ground of investigating the issue of identity. The author seeks to identify specific categories of national identity signs, and define their particular cultural function and significance.

With the development of new technologies, new challenges arise for understanding how we operate with differences and similarities in the network society. In the world where artificial intelligence algorithms can recreate paintings in the style and manner of great artists (Gatys, Ecker and Bethge 2015; Ruder, Dosovitskiy & Brox 2016; Bourached and Cann 2019), in which they can identify distinctive patterns in the way authors write (Plecháč 2019), or learning how to write sonnets, compose classical music (Hadjeres, Pachet and Nielsen 2017), the interplay of differences and similarities it acquires new meanings. From a semiotic perspective, a painting "is a complex text resulting from the network of many modes of production" (Eco 1976: 259). As any aesthetic text can be falsified, a "similar copy" can be produced. What seemed to Umberto Eco (1976: 258) a difficult task to achieve, a "para-artistic achievement", now seems to have become just a matter of programming intelligent algorithms. Gatys, Ecker, and Bethge (2015: 1) use a certain type of similarity between "performance-optimised artificial neural networks and biological vision" in the effort to "separate and recombine content and style of arbitrary images, providing a neural algorithm for the creation of artistic images". What are the new modes of sign production generated by computational algorithms and how do they differ from those known so far? How are the similitude rules established in such situations? Are we talking about an *impression of similarity* or it is an act of *invention* that makes it possible to paint *à la manière de*? From the perspective of differences, which are the criteria that machine-learning algorithms use to identify distinctive patterns in the way authors write?

The examples presented earlier and new modes of sign production that they can generate capture only a few aspects of the mode of operation with differences and similarities in various communication situations. The variety and complexity of the forms of manifestation of the interplay of differences and similarities in contemporary communication phenomena has allowed the authors to initiate a critical, insightful and inspiring debate on this aspect of communication from the perspective of semiotic-type approaches and communication studies. While the first part of the book contains cutting-edge theoretical developments, examining fundamental concepts in the semiotic investigation of various types of *differences* and *similarities* in contemporary communication phenomena, the second part of the book offers practical analyses and examples of applied semiotic research of the topic, in which semiotic theories are put into practice. As we have shown earlier, the authors' contributions for this volume are viewed in a complementary way. The ideas and concepts discussed in the theoretical chapters find their fertility in the analyses and studies presented in the application section, which ensures the conceptual unity of this volume.

In the first paper of this volume, **Augusto Ponzio** argues that the right to non-functionality, the fundamental human right, is denied in today's communication-production world where development, efficiency, competitiveness are driving values; consequently, to assert this right becomes an act of subversion. Reflections on understanding otherness from a semiotic perspective continue in the following paper written by **Augusto Ponzio** and **Susan Petrilli**, dedicated to the memory of **Solomon Marcus**, honorary president of the Bucharest conference. The I-other relationship is not a relationship that renders them elements of an assemblage, of a collective of some sort, but instead the I and the other remain mutually other with respect to each other, different, unique. In the same vein, in the third paper, **Susan Petrilli** details this new way of understanding the life of signs in society, *semioethics*. Semioethics, according to **Susan Petrilli** is the specific vocation of semiotics when it recovers the goal of semeiotics as practiced by Hippocrates and Galen, that is, the health of life.

Taking its inspiration from Husserlian phenomenology and the Prague School, **Göran Sonesson**'s paper reinterprets communication as the co-creation of meaning on the part of addresser and addressee. On the basis of Schütz's idea of relevancy and Eco's notion of encyclopaedia, it then proceeds to resolve the paradoxes of the extended cognition. In the last (five) paper of the first part of this volume **Luis Emilio Bruni** brings together the notions of "narrative identity" and "heterarchy of values" in order to aid our understanding of the paradoxes and contradictions that arise in global digital culture. In this perspective, the sense of *cultural narrative identity* is continuously challenged by the competition and coexistence of overlapping value systems with multiple "regimes of worth", which give place to *dynamic systems of heterarchical belonging*.

In the first paper – sixth in this volume – of the second part, **Kristian Bankov** introduces a new semiotic approach in accounting for one of the most human features – lying. He is conceiving the act of lying as a semiotic transaction of value similar to theft. The classical dichotomy of instrumental and existential values (Floch) opens a direction for an interesting typology of practical and ego lies. In the next paper, **Loredana Ivan, Corina Daba Buzoianu** and **Ioana Bird** uses a qualitative textual analysis of the Romanian campaigns centered on Roma communities to investigate the presence of cognitively complex self-conscious/moral emotions: shame, guilt, embarrassment, and pride. The textual analysis focuses on the messages in the communication outputs and the way the messages relate to self-conscious moral emotions.

Massimo Leone's paper focuses on a possible semiotic definition of extremism and extremist discourse in the frame of an understanding of social ideologies and public opinion shaped on the model of Jurij Lotman's cultural semiotics. Extremism is therefore defined as a particular dynamic in the semiosphere, a dynamic that, independently from its specific content, leads some marginal cultural positions to gain momentum through their massive exposure and virality in relation to the mainstream. **Nicolae-Sorin Drăgan**'s paper proposes a formalized mathematical model to analyze the dynamics of the political actors' positioning game in a specific communication situations, such as the TV presidential debates.

The paper by **Dario Martinelli** offers a semiotic view on the notion of "foreshadowing" in audiovisual narration, introducing the concepts of preparation, reinforcement, variation and delivery, and using Roberto Benigni's filmography as case study. Through quantitative content, and semiotic analysis **Evripides Zantides** identifies the kinds of national identity signs in images of commercial print advertisements in the Republic of Cyprus from state independence in 1960 to 2010. While he investigates which ones prevail in the overall corpus, he examines how these signs fall under the rhetoric of political or economic nationalism, as well as he aligns, and critically assesses them in respect of local socio-political changes.

The paper belonging to **Elena Negrea-Busuioc** and **Diana Luiza Simion** seeks to discuss the function of team nicknames as mechanisms for community identification of both team members and fans. We argue that the nicknames assigned to sports teams fulfill a referential, descriptive function rather than an evaluative one, and that the nicknaming mapping practices reveal the dynamic nature of sports culture.

The paper of **Constantin Popescu**, the last in this volume, examines churches from the hilly areas of Oltenia, Romania (18th and 19th centuries). The constructors and painters, most of them peasants, tried different solutions to harmonise codes of architecture and fresco painting, often – the Annunciation proves it – without notable results. But their searches and some touching failures are full of lessons.

Each of the articles in this book captures certain aspects of what we have called here the *interplay of differences and similarities* in contemporary communication phenomena. The interdisciplinary nature of this volume has allowed us to integrate all these perspectives in a coherent and accessible way, we hope, for the reader. We all have, without a doubt, a certain sense of perceiving differences and similarities, a certain familiarity with this type of dichotomy through which we give meaning to our communication experiences. We are trying to open here a platform for further discussion and research on this topic, on the way in which the different aspects of the interplay of differences and similarities shape semiosic processes.

Nicolae-Sorin Drăgan
National University of Political Studies and
Public Administration (SNSPA), Bucharest, Romania
14 April 2021

References

Bourached, Anthony & George H. Cann. 2019. Raiders of the Lost Art. *ArXiv:1909.05677v1*. 1–4.
Deely, John. 2009. *Purely Objective Reality*. Berlin: Mouton de Gruyter.
Deleuze, Gilles. 1990. *The Logic of Sense* (trans. Mark Lester with Charles Stivale, ed. by Constantin V. Boundas). New York: Columbia University Press.
Deleuze, Gilles. 1995. *Negotiations* (trans. by M. Joughin). New York: Columbia University Press.
Derrida, Jacques. 1978 [1967]. *Writing and Difference* (trans. by A. Bass). Chicago: The University of Chicago Press.
De Saussure, Ferdinand. 1966. *Course in general linguistics* (trans. by B. Wade). New York: McGraw-Hill.
Eco, Umberto. 1976. *A Theory of Semiotics*. Bloomington: Indiana University Press.
Eco, Umberto. 1995. Ur-Fascism. *The New York Review of Books*, June 22, 1995. http://www.nybooks.com/articles/1995/06/22/ur-fascism/.
Eco, Umberto. 2000 [1997]. *Kant and the Platypus: Essays on Language and Cognition*. New York: Harcourt Brace & Company.
Eco, Umberto. 2011 [2000]. Corectitudine sau intoleranță politică? In Umberto Eco, *Pliculețul Minervei*, 28–31. București: Humanitas.
Friedman, Jonathan. 1994. *Cultural Identity and Global Process*. London: Sage Publications Ltd.
Gatys, Leon A., Alexander S. Ecker & Matthias Bethge. 2015. *A Neural Algorithm of Artistic Style. ArXiv, abs/1508.06576*. https://doi.org/10.1167/16.12.326.
Goldstone, Robert L. & Ji Yun Son. 2005. Similarity. In Keith J. Holyoak & Robert G. Morrison (eds.), *The Cambridge handbook of thinking and reasoning*, 13–36. Cambridge: Cambridge University Press.
Goodman, Nelson. 1972. Seven strictures on similarity. In Nelson Goodman (ed.), *Problems and Projects*, 23–32. New York: Bobbs-Merrill.
Hadjeres, Gaëtan, François Pachet & Frank Nielsen. 2017. DeepBach: a Steerable Model for Bach Chorales Generation. In D. Precup & Y. W. Teh, (eds.), *Proceedings of the 34th International Conference on Machine Learning, ICML 2017, Sydney, NSW, Australia, 6–11 August 2017*, volume 70 of *Proceedings of Machine Learning Research*, 1362–1371. PMLR.
Kilani, Mondher. 2000. *L'invention de l'autre. Essais sur le discours anthropologique*. Paris: Payot Lausanne.
Kilani, Mondher. 2012. *Anthropologie: du local au global*. Paris: Armand Colin.
Krampen, Martin, Klaus Oehler, Roland Posner & Thure von Uexküll (eds.). 1987. *Clasics of Semiotics*. New York: Springer Science + Business Media, LLC.
Marcus, Solomon. 2011. *Paradigme Universale*. Pitești: Paralela 45.
Medin, Douglas L., Robert L. Goldstone & Dedre Gentner. 1993. Respects for similarity. *Psychological Review*, *100*(2). 254–278. https://doi.org/10.1037/0033-295X.100.2.254.
Plecháč, Petr. 2019. *Relative contributions of Shakespeare and Fletcher in Henry VIII: An Analysis Based on Most Frequent Words and Most Frequent Rhythmic Patterns. ArXiv:1911.05652v1*. 1–11. https://doi.org/10.1093/llc/fqaa032.
Petrilli, Susan. 2014. *Sign Studies and Semioethics: Communication, Translation and Values*. Berlin: Mouton de Gruyter.

Petrilli, Susan & Augusto Ponzio. 2005. *Semiotics Unbounded: Interpretive Routes through the Open Network of Signs*. Toronto: University of Toronto Press.

Ponzio, Augusto. 1995. *La differenza non indifferente. Comunicazione, migrazione, guerra*. Milan: Mimesis.

Quine, Willard Van Orman. 1969. *Ontological relativity and other essays*. New York: Columbia University Press.

Quine, Willard Van Orman. 1995. Naturalism; Or, Living Within One's Means. *Dialectica* 49 (2/4). 251–261. Reprinted in W. V. Quine & Roger F. Gibson (ed.), *Quintessence: Basic Readings from the Philosophy of W.V. Quine, 2004*, 275–286. Cambridge, Massachusetts: The Belknap Press of Harvard University Press.

Rousseau, Jean-Jacques. 1998 [1781]. *Essay on the origin of languages and writings related to music*. The collected writings of Rousseau, vol. 7(translated and edited by John T. Scott). London: Dartmouth College Press.

Ruder, Manuel, Alexey Dosovitskiy & Thomas Brox. 2016. Artistic Style Transfer for Videos. In Bodo Rosenhahn & Bjoern Andres (eds.), *Pattern Recognition. GCPR 2016. Lecture Notes in Computer Science*, vol. 9796, 26–36. Springer, Cham. https://doi.org/10.1007/978-3-319-45886-1_3.

Sebeok, Thomas A. & Marcel Danesi. 2000. *The Forms of Meanings. Modeling Systems Theory and Semiotic Analysis*. Berlin: Mouton de Gruyter.

Stoichiță, Victor Ieronim. 2017. *Imaginea Celuilalt. Negri, evrei, musulmani și țigani în arta occidentală în zorii epocii moderne 1453-1800*. București: Humanitas.

Tomlinson, John. 2007. Cultural Globalization. In George Ritzer (ed.), *The Blackwell Companion to Globalization*, 352–366. Blackwell Publishing Ltd.

Tversky, Amos. 1977. Features of similarity. *Psychological Review* 84(4). 327–352. https://doi.org/10.1037/0033-295X.84.4.327.

Tversky, Amos & Itamar Gati. 1978. Studies of similarity. In Eleanor Rosch & Barbara Lloyd (eds.), *Cognition and Categorization*, 79–98. Hillsdale, NJ: Wiley.

Vosniadou, Stella & Andrew Ortony. 1989. Similarity and analogical reasoning: A synthesis. In Stella Vosniadou & Andrew Ortony (eds.), *Similarity and Analogical Reasoning*, 1–18. Cambridge: Cambridge University Press. https://doi.org/10.1017/CBO9780511529863.

Zlatev, Jordan. 2011. What is cognitive semiotics? *Semiotix: a global information bulletin*, Vol. 6. http://www.semioticon.com/semiotix/2011/10/what-is-cognitive-semiotics/.

Part I: **Theoretical advances**

Augusto Ponzio
Communication, listening, non-functionality

Abstract: Communication has generally been understood in terms of messages transmitted from a sender to a receiver on the basis of a code. But since Saussure and all those who have contributed to broadening semiotics as a field of enquiry (Husserl, Jakobson, Bakhtin, Barthes, Peirce, Morris, Rossi-Landi to name the authors more closely connected with our theme), communication has acquired far greater consistency. A noteworthy contribution comes from philosophy of language understood in terms of the "art of listening" as distinguished from "wanting to hear". Interpretation requires "listening" and "responsive understanding". Communication is "dialogue", where "otherness" is essential. In the communication-production world today where development, efficiency, competitiveness are driving values, we affirm the right to non-functionality which as such takes on a subversive character. With liberation from indifferent work in the form of unemployment, with de-commodification of traditional emigration in the form of migration, and with increase in the need for occupations dedicated to non-functional otherness, communication-production opens ever growing spaces to the non-functional. The non-functional is the human. Yet the "rights of man" do not contemplate the right to non-functionality.

Keywords: Communication-production, dialogue, humanism of otherness, human rights, identity, listening v. hearing, non-functionality, responsibility/responsiveness, semioethics

1 Listening and non-functionality as conditions of communication

Communication has generally been understood in terms of messages transmitted from a sender to a receiver on the basis of a code. Since Saussure and his *sémiologie* and all those who have contributed to broadening semiotics as a field of enquiry endowing it with its current configuration (from Husserl, Jakobson, Bakhtin, Barthes, Peirce, Morris, Rossi-Landi to name only those authors more

---
Translation from Italian by Susan Petrilli

Augusto Ponzio, University of Bari "Aldo Moro", Italy

closely connected with our theme), communication has acquired far greater consistency. A noteworthy contribution has come to semiotics from philosophy of language. We have described the philosophy of language in terms of the "art of listening". To "want to hear" is one thing, to "listen" is another. For what concerns human signs this distinction is important. Interpreting signs requires "responsive understanding": *responsive understanding* implies *responsibility*, "response-ability," "answer-ability," in the double sense of the ability to respond to the other and to take responsibility for the other. Communication thus acquires the character of *dialogue*, a dialogue in which the relation of otherness is essential. This is dialogue where *otherness* is not only the otherness of others, but also one's own otherness; where otherness is no longer connected with belonging, with identity, is not limited internally to a group, a community to the exclusion of another group or community.

Listening is also *hospitality* without conditions and hospitality without conditions is the most important of human rights, what we have specified as the right to *non-functionality*. In the love relationship, in all its forms, in relations of friendship, if a question of true love, of true friendship, the point is that there is no self-interest, no profit, no gains. Each one of us wants to be loved for nothing, in one's complete non-functionality. This is something we all know well. Otherwise, we speak of false love and self-interested friendship.

But un-self-interested love only occurs in the so-called "private sphere". Instead, in the "public" sphere what counts is functionality, productivity, self-interest. Hospitality for the "foreigner" is conditioned by one's capacity to be functional, productive, useful. From this point of view, there is no difference between the foreigner's condition today and the situation portrayed by Spielberg in the film *Schindler's List* where even the Jew was saved from the lager if that Jew was useful, for the case in point if included as a worker in Schindler's list.

In 1948 the semiotician Charles Morris published a book titled *The Open Self* in which he expresses his hope to see the formation of open *communities*. Semiotics must proceed in this direction and contribute to recognizing the right to *non-functionality* well beyond the private sphere, the right to promoting non-functionality as a *fundamental right* in the sphere of the human. Even the right to life becomes the right to mere survival, to life that is not life, life that can no longer be characterized as human, if it is not accompanied by the right to non-functionality.

Nor can there be real communication without listening, without hospitality, without recognition of this fundamental right. Thus understood, semiotics and philosophy of language take on the form of a well-motivated dissidence towards communication as it is generally understood and as it presents itself essentially on the international scene today under the denomination of "world communication" or "global communication".

The problem of alterity and the critique of identity are of pivotal importance in Occidental Reason. In this sense an important contribution, as anticipated, comes from philosophy of language. I have taught philosophy of language since 1970 and my interest in this discipline comes from my readings of Emmanuel Levinas. I graduated in 1966 with a dissertation on Levinas under the title *La relazione interpersonale*. I can say that the whole course of my research has developed from Levinas to Levinas. I will here recall some of my works:

Subjectivité et alterité dans la philosophie de Emmanuel Lévinas (1996), *Emmanuel Levinas, Globalisation, and Preventive Peace* (2009), *L'Écoute de l'autre* (2009), *Rencontre de paroles* (2010), and *The I questioned. Emmanel Levinas and the critique of occidental reason* (2006), published as a monographic issue of the journal *Subjects Matters*, directed by Paul Cobley, with contributions from other Levinas specialists commenting my text.

We may begin with the undisputable observation that the self forever needs justification in front of the other forever. This is the inevitable situation of the self. And identity – that is, belonging to an identity – is a prevalent means of justification of the self. The problem of identity and its expedients for justification in front of the other is central in the writings of Emmanuel Levians.

The first case of the self is not the nominative, as Levinas says, but the accusative. The real question, as Levinas says, is not the question posed by Martin Heidegger: "Why being and not nothing?", but "Why me here in this situation, in this locality, in this dwelling place, in this favourable condition, and not the other?".

One common possibility in globalized communication to find justification in front of the other is for the self to claim its specific identity. And identity may be asserted in terms of nation, ethnic group, race, skin colour, religion, language, territory, gender, profession, social status.

The reason of identity is the reason of the self and not the reason of the other.

All Western culture is a justification of identity in the face of the other.

Identity is a real abstraction, a constitutive abstraction, of reality. It has an ontological value. Identity is an essential aspect of being. And violence, in all possible forms – war included – is one of its consequences.

War is connected to the justification of identity, to the defense of identity, of the reality of being. Levinas opens his Preface to his principal work, *Totalité et infini*, with reflections on what he considers as the recurrent aspect of the real, of being: that is, war. The state of war suspends *ad interim* the moral imperatives, renders them illusory, as Levinas says.

The face of being, that manifests itself in war, is the face of reason. War evidences the connection between politics and ontology and the subordination of individual singularity, in its uniqueness to the totality, to an ontological order without escape.

In the conception of the reason of identity, in the rhetoric of identity, in the reason of Western culture, in its logic the sense of individual singularity with its uniqueness, is derived only from the totality.

Singularities, uniqueness are sacrificed to objective sense, the totality of ontology, the sense, the judgement of history.

In the logic of reason, identity, reality, history, Being, war, peace is only the peace of war, the end of war, a truce in preparation for war.

In contrast to combatting war through "preventive war", that is, violence through violence, war against war, terrorism against terrorism, Levinas considers the possibility of preventive peace.

Preventive peace is connected, it too, with the necessity of justification of the self in front of the other. But in this situation the need for justification in front of the other is not achieved through the expedient of difference between two identities, identity of the self and identity of the other; it is not achieved on the basis of difference between human rights and the rights of the other.

"Human rights and the rights of the other" is the title of an essay by Emmanuel Levinas. This title implies and subtends the claim that human rights are not the rights of the other.

The common expedient for justification of the rights of the self in front of the other achieves the passage of the self from a bad conscience to a good conscience, a clean conscience. Instead, preventive peace consists in maintaining the self in the condition of a bad, a dirty conscience without resorting to the alibis of a good, clean conscience. Preventive peace consists in the manifestation of difference among individuals in their uniqueness, as singularities and not as representatives of a collectivity, as individuals belonging to indifferent collectivities, abstractions, or ontological entities.

Preventive peace is non-justification, a dirty conscience, non-indifference. Non-indifference to the other, non-indifference as responsibility for the other is the very difference between me and the other, as Levinas says, because my responsibility for the other is not reciprocal, reversible, or exchangeable. I am responsible of the other, responsible for all others, for the guilt of another. The condition of hostage, as Levinas says, is an authentic figure of responsibility for the other.

Peace otherwise than the peace of war is otherwise than being, *autrement qu'être*. This is the title of Levinas's 1974 book. Peace otherwise than peace is non-indifference toward the other, responsibility without alibis in the face of the other, "*dis-inter-essement*".

Preventive peace is openness to the other, without closure in identity, or illusory walls, preventive peace is the face to face relation with the other, one-to-one, uniqueness to uniqueness.

Original peace is what Levinas calls asymmetry of I and other. Levinas recalls Dostoevsky on this aspect. In *Brothers Karamazov* of the brothers says: "we are all guilty for everything and everybody, and me more than anybody else".

I am responsible for the other and nobody can substitute me. In this sense I am chosen by the other and consequently I realize my uniqueness and also my freedom. I am responsible for the other even when this other commits crimes, or suffers crimes and persecutions.

According to Levinas, the state emerges from the limitation and regulation of non-indifference to the other, and not, as in Hobbes's vision, from the limitation of violence and fear of the other (*homo homini lupus*). The original fear of the I, in bad conscience, is not fear of the other but fear for the other. Consequently, the problem of law presents itself in terms of justice and defence of the other and in terms of perfectibility, of justice, in the sense of non-indifference for the other.

Can human rights also be the rights of the other? This is the problem of communication today, one that requires an urgent response. This response no doubt concerns semiosis, but it cannot be limited to semiotics. It requires a shift in semiotics, a special orientation in semiotics, what we have denominated as "semioethics".

Semiotics recovers an ancient branch of semiotics, medical semeiotics (Hippocrates, Galen), symptomatology, whose interest is to interpret symptoms to the end of keeping life healthy. This too is part of the special bend of semiotics today: concern with present day communication at the global level and identify the symptoms that not only contradict sustainable conditions of life, but that are lethal for life – human life and non-human, for life over the entire planet.

The realization of human rights, not only as the rights of the self but also as the rights of the other, requires that "human" should refer not to a humanism of identity, but to humanism of otherness, alterity. *L'humanisme de l'autre homme*, is the title of another important book by Levinas (1972). In the perspective of the humanism of alterity at the basis of human rights there must be a right that has not been mentioned so far. We named it earlier: the right to non-functionality.

Returning to the topic of my book of 1997 (2nd edn. 2004), *Elogio dell'infunzionale. Critica dell'ideologia della produttività* (*Praise of the non-functional. Critique of the ideology of productiveness*), this paper addresses the concept of the "right to non-functionality". The importance of such a right for a social system characterised by productiveness, as is our own, is enormous.

In contrast to this system, the "right to non-functionality" is connected with a system open to humanism. But this is not a matter of humanism centred on identity self-interest (whether individual or collective), but of the humanism of otherness. "Humanism of the rights of the other" is something different from the proclamation of "human rights" which inevitably end up serving the "humanism of identity". In fact, so-called human rights are functional to the humanism of identity, they refer to the rights of the self, and not of the *other*. It is eloquent, meaningful and revealing that Emmanuel Levinas should have entitled one of his essays "Human Rights and the Rights of the Other" (in Levinas 1987). This expression underlines that the other is excluded from human rights.

We may say that non-functionality subtends all human rights understood as the right to alterity.

By contrast with "productivity", the right of the other and of the self as other is specified as "the right to non-functionality". This is the right to value not as the other *relatively to* . . ., not as one of us or one of them, not as an individual representative of a genre, group, or community of some sort, but rather as a *unique other*, an *absolute other*, an *irreplaceable other*, as occurs in a relation of friendship or of love. In friendship and in love we do not expect the other to be interested in us simply because we are useful. In this case we would speak of cupboard love, false friendship, a marriage of convenience. Such examples testify to the difference, indeed the separation as established in our society between private life and public life. In private life not only are we familiar with the value of non-functionality, but we also make claims to such a right.

Implicitly, tacitly we have the right to non-functionality at heart and we know that ultimately the human is the non-functional. But "human rights" do not contemplate the right to non-functionality. And yet we use such expressions as "this life is not life!" (from the Italian saying "questa vita non è vita!"). And generally when we make such claims we are referring to a life in which the right to non-functionality is neglected, eliminated, excluded.

In relationships of friendship and love we expect, indeed we make claims to an un-self-interested interest in the other. We claim that we are not interested in the other's functionality. If this were not the case, even if someone declares they love you and want to be with you, the perception is that the friendship is self-interested and that the love is untrue. But all this importance that we attribute to friendship and love in private life (private life is ever more deprived of its real rights, in spite of all the trumpeting in today's world about the "right to privacy") is altogether absent in public life, in public social relationships, in public relations.

Until the *right to life* is not firmly connected to the right to non-functionality, it remains bonded to a vision of man reduced to the status of means, to "capital", to the status of a "human resource for production". The value of this man is recognised "for the whole course of active life" (as recited in programmatic documents produced by the European Commission), that is, for as long as he is productive and contributes to reproduction of the production system.

But the human is not a resource, because the human is not a means, the human is not endowed with instrumental value: the human is an end. To invest the human with an instrumental function for the increase of "global competitiveness" on the world market is already demeaning in itself, and even more so when such a value becomes the goal of education and professional training.

The right to non-functionality takes the character of another form of humanism, another way of understanding and promoting human rights, recognition of the other – including myself as other and all my own possible identities – as a value in itself.

2 The right to non-functionality in the ideo-logic of communication-production

In *Elogio dell'infunzionale. Critica della ideologia della produttività* (1997), I work on the concept of "the right to non-functionality". I underline the importance of this concept for today's society (now characterized in terms of worldwide communication-production) in opening the way towards a new form of humanism. This new humanism is no longer centred on the interests "of life lived from the inside", that is, on the interests of (individual and collective) identity, but rather it presents itself in terms of the humanism of alterity, otherness.

The expression "humanism of the rights of alterity" can be reconducted to the sense of the humanism of identity which is wholly internal to today's dominant social system, based on capitalist, mercantile production. In fact, the humanism of the rights of alterity has developed with the capitalist social system and is complementary to it, an organic part of it. "Humanism of the rights of alterity, humanism of the rights of others" are expressions that can resound in the same sense of "humanism of identity" which in fact proclaims the "rights of man", "human rights". Emmanuel Levinas titled one of his essays "The Rights of Man and the Rights of Others" (this essay appeared in Levinas 1985, also in Levinas 1987, Eng. trans., 116–125). This was a way of underlining that the second term in this expression cannot be reduced to the first, that alterity cannot be reduced to identity. All the same, interest in identity, support for identity are so strong that the claim to the "rights of alterity," similarly to the claim to the "rights of difference" ends by referring to the rights of identity. For this reason, it is preferable to speak of the humanism of alterity as humanism centred on the *right to alterity*.

Misunderstandings relative to identification of "others" with "us", to identification of the "other" with the "other I" and therefore with the "I" are part of daily life, and are fostered by publicity experts, journalists and "writer-copyists" ready to do anything to keep the "public" ear happy. This leads to extraordinary conclusions of the type: "the true other is me", "others are us". *Italians, the others are us too* is the title of an article by Alberto Arbasino, which appeared in "la Repubblica", 12 January 1999, after protests and declarations of intolerance towards "extracommunitarians", that is, the illegals, the aliens, occasioned (every occasion is useful) by increase in criminality and violence in Milan. The article begins with a reference to Rimbaud as a way of immediately stating that "'the I *and* the other', with multiple variants is a formula for success for the pious followers of Emmanuel Levinas and Jacques Derrida"; (in addition to a photograph of Derrida who is never mentioned again in this article which consists of two pages of contorsions and pirouettes – with the sole aim of earning the public's applause – in which the categories of 'I' and 'other' are repeatedly relaunched and recaptured with nonchalance), this article is furnished with titles of the series "The complexes of the West", "When it enters in all its pompousness, the 'politically correct' of academics prescribes that the 'I' owes maximum reverence and also welcome to the values and needs of others": well and truly an "*arb-asinata*"! (that is, the judgement of a donkey playing on the ambiguity of the name Arbasino, that is, *arb-asino* where "asino" means donkey).

By contrast to "productivity" which moves and orients the entire communication-production system, the right to alterity is the "right to non-functionality". Here by "alterity" is not understood relative alterity, that is, alterity connected to roles, social, professional position, etc., on the basis of which one is other "relatively to" – professors with respect to "students", "fathers" with respect to "sons", "work force" with respect to "capital", "citizens" with respect to "illegals", that is, those who belong to the community with respect to those who do not belong to the community, extracommunitarians, etc. Relative alterity is that which forms our identity. But if, following a "reduction" hypothesis, we free ourselves of all the relative alterities that constitute our identity, does nothing remain, or does a residue independent from these relative alterities persist? In truth, contrary to that which this social system wants to make us believe, a residue persists, a non-relative alterity which allows each one of us to exist non simply as an *individual* and therefore as the representative of a genre, a class, a set, as other-relatively-to . . ., nor as a *person* (a term of reference for that which is "personal", which "belongs", is "one's own"), but as a *unique single individual*, as *absolutely other*, that cannot be replaced, interchanged, a genre in itself, *sui generis*.

The right to non-functionality is the right to be of value on one's own account, to be an end in oneself, as non-relative otherness (see chapter 13, "The Otherwise of Communication-Production: Levinas and the problem of Evasion", 119–123, in Ponzio 1997b).

In the present day communication-production world in which development, efficiency, competitiveness (to the point of acknowledging the *extrema ratio* of war) are the fundamental values, the right to non-functionality takes on a subversive character (non suspect subversion?). And yet, with liberation from indifferent work in the form of spreading unemployment, with the de-commodification of traditional emigration in the form of the irreducible phenomenon of migration, and with the increase in the need for occupations dedicated to non-functional otherness, communication-production itself, indeed in spite of itself, opens spaces to the non-functional that are becoming larger and larger.

The non-functional is the human. Yet the "rights of man" do not contemplate the right to non-functionality. The right to non-functionality exceeds and transcends the humanism of identity. It is at the foundation of all the rights of alterity.

In his notes of 1950s (in Jachia and Ponzio 1993), Bakhtin distinguishes between the "small experience" and the "great experience". The small experience is reduced and partial and remains attached to the concrete, to the effective world, the world as it is, the small experience responds to contemporaneity, it is connected with interest, with utility, with knowledge functional to practical action, with the economy of memory that excludes, through oblivion, all that which results as distracting and dispersive, inconclusive with respect to logicality, simplicity and uniformity in programming, to the univocality of sense. Instead, in the "great experience", the world does not converge with itself (it is *not that which is)*, it is not closed nor finalized. In the world, memory flows and loses itself in the human depths of matter and unlimited life, the life experience of worlds and atoms. And for such memory, the history of each one of us begins much earlier than its cognitive acts (its knowable "self") ((in Jachia and Ponzio 1993: 195–196).

The knowable self of the small experience is the self-produced by the "technologies of self", whose process of development Foucault in particular aimed to reconstruct.

The problem of the "technologies of self" is the problem of the formation of individual identity, which is complementary to the assertion of belonging to a given social entity, to a community, a nation, a State, ethnic group, genre. The technology of self and the technology of individuals are connected, both social technologies in a broad sense, political technologies. The formation of individual identities and of collective identities are part of a unitary process. In any case it is a question of a process that determines the consciousness awareness of autonomy, whether this is the autonomy of the individual or of the State.

Here we are not concerned with the study of the genesis of this autonomy and of relative individuality, separation, belonging, that is, we are not concerned here with the study of the historico-social genesis of the assertion of identity and the exclusion of alterity which is connected to this process. Instead, we wish to investigate that which *questions* this assumed autonomy, that which renders it delusory, even ridiculous. And we can immediately indicate the body, in its constitutive intercorporeity, as the central term of this questioning.

De-possession and extralocality: the body that flourishes in intercorporeity is shifted, extralocalized, external with respect to the chronotopic coordinates of consciousness. The body has a life that is other, unique, unrepeatable with respect to that which is circumscribed to the boundaries of the individual, with respect to the individual *localization* of human bodies. Moreover, insofar as the body is individual and localized, it is reciprocally replaceable, interchangeable to the extent that we are each defined on the basis of the place we occupy by comparison to others as much as on the basis of the gap, the difference that separates us from others. This is a question of the body's extralocalization with respect to the structures, mechanisms and techniques used for its submission, as an individual body, as body-identity, to the knowledge-power of *bio-politics* (Foucault); extralocalization which emerges in its "escape without rest" (above all its "persistence in dying") from the techniques that intend to dominate and control the body.

The worldwide expansion of capitalist production and of bio-power has led to the controlled insertion of bodies into the production system and to spreading the idea of the individual considered as a separate and self-sufficient entity. This has led to the almost total extinction of cultural practices and worldviews that are based instead on intercorporeity, interdependency, exposition and opening of the body. All those forms of perception in popular culture, discussed by Bakhtin in *Dostoevskij* (1963) and in *Rabelais* (1965) are now almost completely extinct. Our allusion is to all those forms of "grotesque realism" that present the body as a non defined body, as a body not confined to itself, a body that flourishes in the relation of symbiosis with other bodies, in relations of transformation and renewal that transgress and cross over the limits of individual life.

And yet, the technologies of separation among human bodies, among interests, among individual and collective subjects, functional to production and to the connection between production and communication which is getting ever closer to the very point of identification (typically of today's capitalist production system) – such technologies of separation cannot cancel the signs of compromission of every instant in our life as individuals with all of life over the entire planet. Recognition of such compromission is ever more urgent the more the reasons of production and communication functional to production impose ecological conditions in which communication between our body and the environment is rendered ever more difficult and distorted.

Therefore, we can immediately indicate the body, the body in its constitutive intercorporeity, as the central term for interrogation and questioning of the delusory autonomy of identity. Reference here is to the body in its singularity, unrepeatability, non-functionality, to the body which finds in death, an unconclusive end, the expression of excess with respect to a project, story, "authentic" choice: the living body that knows before it is known, that feels before it is felt, that lives before it is lived. This body is connected without interruption to other bodies, it is implied, involved in the life of the entire ecosystem on the planet Earth, in a web of relations that no technology of self can ever exit.

The body is refractory to the "technologies of self" and to the "political technology of the individual". The body is *other* with respect to the subject, to consciousness, to memory understood as addomesticated, selected, filtered, accomodated memory; the body is other with respect to the narration that the individual or collective subject constructs for itself and through which it delineates its identity, the image of itself to exhibit, the self in which to take an interest, the physiognomy through which it can be recognized, the role it must perform.

Production-reproduction in today's world does not only destroy products, or the means of work now achieved through the automatic machine, it does not only destroy trades, crafts, and professions, jobs, employment, but rather production-reproduction today destroys the environment, the body, the quality of life which is made to depend on indifferent work, work reduced to the alternation between work-time and free-time (free-time is that which work requires for rest, encouragement and regeneration; that which work allows or concedes, whose availability and use is always decided by work itself), or it is emptied and impoverished by the lack of work understood in terms of having a job, as employment. Work occupies, pre-occupies daily life, even as non-work whether in the form of free-time or of unemployment.

From the perspective of dominant ideo-logic today, insistence on the fundamental character of labour as such, of generic, undifferentiated labour is such that even a project for an alternative social system generally does not succeed in imagining another source of social wealth that is not labour, another optimal solution if not "work for all." Walter Benjamin observes that the German labourer's *Gotha programme* (Marx 1970 [1875]),which defines labour as "the source of all wealth and culture," already bears traces of the misunderstanding we are signalling (Benjamin 1986 [1931]) – a misunderstanding that ends by favouring the transition from socialism to fascism and nazism.

In his *Critique of the Gotha Programme* (1875), Marx clarified that "labour is not the source of all wealth", adding that: "the bourgeousie has its good reasons for investing work with a supernatural creative force" (Marx 1970 [1875]: 15). In his manuscripts of 1844, Marx criticizes "vulgar and material communism" (as well as *ante litteram, ante factum,* "real socialism") (Marx 1959 [1844]: point 1) which suppresses private property by generalizing it and which to private property opposes general private property, physical possession, ownership extended to all. Marx critiques the misunderstanding that subtends the project for a new society insofar as it continues to consider work-in-general as the source of wealth, as in capitalist society, so that "the category of the worker is not done away with, but is extended to all men" (Marx 1959 [1844]: point 1). For crude and vulgar communism understood in such terms, the community is no more than "a community of labour, and equality of wages paid out by communal capital – by the community as the universal capital" (Marx 1959 [1844]: point 1).

We cannot understand the process of total identification with the Community – "*Gemeinschaft*" – (the entire lexicon of Nazi Germany was functional to this identification), if we don't start from work-in-general, indifferent work, which produces value as a function of exchange value, and which is a structural, constitutive part of our social system. In the lexicon of Nazi Germany the term "Arbeit" not only signifies "abstract labor," "undifferentiated labour," which is quantified and paid by the hour, but also undifferentiated labour in the interclassist sense, in other words, labour freed from any association with "class," "alienation," "exploitation." In Italy, during fascism, the generically interclassist connotation assumed by the term "work" is considered as a sign among others of innovation contributed by fascism to the Italian language. In 1934, Giuseppe Bottai wrote with tones of satisfaction that the meaning of the term "Work" ("*Lavoro*") had become far broader to include all organizational and executive, intellectual, technical, manual forms, and it did not necessarily refer to the particular labours of a given class. Moreover, Bottai praised introduction of the expression '*datore di lavoro*' (literally, he who gives work, that is, employer) to replace the outdated term '*padrone*' (owner, master). He considered this an expression, consecrated by revolutionary laws, of the singular identification in popular consciousness of giuridical equality subtending our social order (referred in Foresti 1977: 11).

This social system, the capitalist, is based on the reality of free work and free exchange ideology and activates mechanisms for the specification and assertion of identity, such as self-belonging, capacity for decision-making, freedom, responsibility, the possibility of building one's own destiny. Subjectivity, at the individual and collective level is exalted, encouraged and deluded. But free work is abstract, quantified, indifferent work, it is subject to the production of exchange value and it is functional to the amplified reproduction of the production mechanism itself. As such the social system upon which free work is based mortifies subjectivity. And this is ever more the case the more the indifference characterizing the social relations typical of the capitalist social system spreads and activates processes of alienation, de-identification, uprooting, expropriation, homologation which increase ever more with the development of capital. Such a situation generates a paroxysmal search for identity which involves negating the other from self and the other of self.

Together with these mechanisms which create frustration and delusory identifications insofar as they are based on the demand that differences be indifferent to otherness, communication-production (an expression we propose for the present phase in capitalist development) produces means that are exorbitant with respect to the ends proposed, such as profit and increased competitiveness of the product. The communication-production phase is in fact characterized by the extraordinary extension of cognitive activities, above all through computational media. Consequently we are not witnesses to the continuous and speedy superceding of competencies, trades, crafts, professional capacities, specializations. As a result of present day development in communication, the sciences are not in a condition to maintain the separate and fragmented character of the traditional capitalist division of labour. The need for interdisciplinarity and for overcoming specialisms have become concrete requisites for communication in today's globalized world.

All this implies encounter among different languages. Dialogism is no longer just a theoretical or ethical requirement, it is a necessity in today's social production system.

The term "dialogue" is abused in our society where "democracy" is a value recognized by public opinion, to the point of becoming a common place to count on as a premise for immediate consensus. Indeed, it is a point of honour for any associative organism of a professional order, a political order, etc. to be able to boast a democratic character and orientation in the sense of being "open" and available for "dialogue". All the same, as regards the combination "democracy"–"dialogue", it must be remembered that the first and foremost master of dialogue, Socrates, was condemned to death by the democratic government of Athens. With the term "dialogism" is understood a relation that exists behind the subject's back, so to say, without his knowing it, dialogism not as a result of his "respect" for others, but in spite of himself, against his own will. Dialogue understood in such a framework is different from dialogue generally understood as an orienation adopted by the subject, as an initiative taken by the subject, a disposition of the subject, a concession made by the subject. Dialogism is involvement, compromise, assumption of responsibility by the subject, difference that opens to unindifference, according to an orientation that is not at all chosen and willed by the subject, but experienced as inevitable and irreducible – the more so the more attempts at extricating oneself from the other through alibis, excuses, ways out, limits and defensive forms of closure, prove to be vain.

Paradoxically, the *monologism* of worldwide and global communication on which the capitalist communication-production social system stands, requires encounter and interaction among different languages, it requires dialogue among different logics, visions and linguistic practices for the construction of worldviews. In other words, this particular phase in the development of the capitalist social system cannot avoid standing on an architectonics that is *plurilinguistic* and *polylogic*. The risk of "cultural uniforming" (which the monologism of world communication necessarily involves) passes across dialogic encounter among the most diverse historical languages and special languages – *and we are exactly in this transitional phase*. We ought to take advantage of this transition and avoid confirming the situation of monologism and cultural uniforming.

It should also be remembered that while automation produces unemployment, with the reduction of overall work-time (promoted as a function of profit and competitiveness) automation also creates at the same time the conditions for increase in available time for the full personal development of each one of us. *Available time* and not *work-time* may be envisaged as the real *social wealth*.

But work itself is undergoing a metamorphosis, *linearity* is in the process of being replaced by *interactivity*. This is particularly obvious in the field of multimedia, a symbol of the current phase in the development of communication-production. The linear and hierarchized organization of work is now yielding to co-participation, interactivity, interfunctionality, modularity, and flexibility in structures that favour innovation and inventiveness. The division of labour connected to separation between manual labour and intellectual labour no longer holds as digital technologies take over. Until not long ago "linguistic work" and "non-linguistic work" represented two distinct and separate realities for researches looking for connections and homologies, like Ferruccio Rossi-Landi towards the end of the 1960s. Now, instead, thanks to progress in technology and artificial intelligence, linguistic work and non-linguistic work have at last come together in computers.

"Productivity", "competitiveness", "employment": in the last analysis these goals are rather miserable by comparison with the wealth of means that have been employed to attain them. Only if we remain inside the "small experience", in the perspective of the interests of those who detain power and control over the global communication system can we believe that such a wealth of means does not deserve to be used for something better.

References

Bakhtin, Mikhail M. 1963. *Problemy poetiki Dostoevskogo*. 2nd revised and enlarged edition of Bakhtin 1929. Moscow: Sovetskijpisatel'. [*Problems of Dostoevsky's Poetics*, C. Emerson (ed. & trans.), W. C. Booth (Introduction). Manchester, Manchester University Press, 1984.].

Bakhtin, Mikhail M. 1965. *Tvorčestvo Fransua Rable i narodnaja kul'tura srednevekov'ja i Renessansa*. Moscow: Chudozevennaja literature. [*L'opera di Rabelais e la cultura popolare*, M. Romano (trans.). Torino: Einaudi, 1979.] [*Rabelais and His World*, H. Iswolsky (trans.). Bloomington, Indiana University, Press, 1984.].

Benjamin, Walter. 1986 [1931]. The Destructive Character. In Walter Benjamin. *Reflections. Essays, Aphorisms, Autobiographical Writings*, 301–303. Edited by and Intro. P. Demetz. New York: Schocken Books. [Original Publication: Benjamin, Walter. 1931. Der destruktive Charakter. In Walter Benjamin, Rolf Tiedemann & Hermann Schweppenhäuser (eds.), *Gesammelte Schriften, 1972*, IV, 1: 396–401. Frankfurt am Main: Suhrkamp.].

Cortelazzo, Michele, A., Leso, Erasmo, Ivano Paccagnella & Fabio Foresti. 1977. *La lingua italiana e il fascismo*. Bologna: Consorzio Provinciale Pubblica Lettura.

Derrida, Jacques. 1998. Formation professionelle, Revue Européenne 14, 1998/II, dedicated to the theme: *Peut-on mésurer les bénéfices de l'investiment dans le ressources humaines?*

Foresti, Fabio. 1977. *Proposte interpretative e di ricercasu lingua e fascismo: la "politica linguistica"*. In Michele A. Cortelazzo *et alii* (eds.), 111–148. Bologna: Consorzio Provinciale Pubblica Lettura.

Foucault, Michel. 1988a. *Tecnologie del Sé*. In AA. VV., *Un seminario con M. Foucault, 1992*, 11–47. Torino: Bollati Boringhieri.

Foucault, Michel. 1988b. *La tecnologia politica degli individui*. In AA. VV., *Un seminario con M. Foucault, 1992*, 135–153. Torino: Bollati Boringhieri.

Foucault, Michel. 1994. *Poteri e strategie. L'assoggettamento dei corpi e l'elemento sfuggente*, edited by Pierre Dalla Vigna. Milan: Mimesis.

Jachia, Paolo & Augusto Ponzio (eds.). 1993. *Bachtin &. . .* Bari-Roma: Laterza.

Levinas, Emmanuel. 1961. *Totalité et Infini*, The Hague: Martinus Nijhoff. [*Totality and Infinity*, eng. trans. by A. Lingis, Intro. by J. Wild, Pittsburgh, Duquesne University Press, 1969; Dordrecht, Kluwer, 1991.].

Levinas, Emmanuel. 1972. *Humanisme de l'autre homme*, Montepellier: Fata Morgana. *Humanism and the Other*, Eng. trans. N. Poller, Champaign, Illinois: University of Illinois Press, 2003.

Levinas, Emmanuel. 1985. *L'invisibilité des droits de l'homme*. Friburg: Editions Universitaires.

Levinas, Emmanuel. 1987. *Hors Sujet*. Montpellier: Fata Morgana. [Eng. trans. M. B. Smith, *Outside the Subject*, London: The Athlone Press, 1993.].

Marx, Karl. 1959 [1844]. *Manoscritti economico filosofici*, It. trans. by N. Bobbio, Turin, Einaudi, 1978.

Marx, Karl.1970 [1875]. *Critica del programma di Gotha*. In Marx e Engels, *Operescelte*, Rome, Editori Riuniti, 1966, 951–976.

Morris, Charles. 1948. *The Open Self* [*L'io aperto. Semiotica delsoggetto e delle sue metamorfosi*, It. trans. & Intro., Charles Morris e la scienza dell'uomo.Conoscenza, libertà, responsability, vii–xxvi, by S. Petrilli. Bari: Graphis, 2002; Lecce, Pensa MultiMedia, 2017]. New York: Prentice-Hall.

Ponzio, Augusto. 1996. *Sujet et altérité. Sur Emmanuel Levinas*, suivi de Deux dialogues avec Emmanuel Levinas. Paris: L'Harmattan.

Ponzio, Augusto. 1997a. La rivoluzione bachtiniana. Il pensiero di Bachtin e l'ideologia *contemporanea* [Spanish trans. *La revolución bajtiniana. El pensamiento de Bajtin y la ideologia contempoanea, 1998*. Madrid: Universitat de València]. Bari: levante Editori.

Ponzio, Augusto. 1997b. *Elogio dell'infunzionale. Critica dell'ideologia della produttività.* Rome: Castelvecchi. New ed. Bari, Graphis, 2006.

Ponzio, Augusto. 2004 [1999]. *La comunicazione*, 2nd edn. Bari: Graphis.

Ponzio, Augusto. 2006. *The Dialogic Nature of Sign*. Ottawa: Legas.

Ponzio, Augusto. 2006b. *The I Questioned: Emmanuel Levinas and the Critique of OccidentalReason. Subject Matters* (2006), Special Issue, 3(1). 1–42. [This issue includes contributions commenting A. Ponzio's essay, by A. Z Newton, M. B. Smith, R. Bernasconi, G. Ward, R. Burggraeve, B. Bergo, W. P. Simmons, A. Aronowicz, 43–127.].

Ponzio, Augusto. 2007. *Fuori luogo. L'esorbitante nella riproduzione dell'identico*. Rome: Mimesis.

Ponzio, Augusto. 2009a. *Da dove verso dove. La parola altra nella comunicazione globale.* Perugia: Guerra Edizioni.

Ponzio, Augusto. 2009b. *Emmanuel Levinas, Globalisation, and Preventive Peace*. Ottawa: Legas.

Ponzio, Augusto. 2009c. *L'écoute de l'autre*. Paris: L'Harmattan.

Ponzio, Augusto. 2010a. *Rencontres de paroles*. Paris: Alain Baudry & Cie.

Ponzio, Augusto. 2010b. *Procurando uma palavra outra*. San Carlos (Brasile): Pedro & João Editores.

Ponzio, Augusto. 2010c. *Enunciazione e testo letteraio nell'insegnamento dell'italiano come LS*. Perugia: Edizioni Guerra.

Ponzio, Augusto. 2012. *Dialogando sobre diálogo na perspectiva bakhtiniana*. São Carlos-SP., Brazil: Pedro & João Editores.

Ponzio, Augusto. 2013. *No circulo com Mikhail Bakhtin*. São Carlos: Pedro & João Editores.

Ponzio, Augusto. 2015a. Philology and Philosophy in Michail Bachtin, *Philology* 1. 121–150.

Ponzio, Augusto. 2015c. *Trasemiotica e letteratura. Introduzione a Michail Bachtin*, Milan, Bompiani, new revised end enlarged edition of Ponzio 1992.

Ponzio, Augusto. 2015d. *Illinguaggio e le lingue. Introduzione a una linguistica generale*, Milan, Mimesis [From the 1st 2002 edition, now reviewed and enlarged].

Ponzio, Augusto. 2016b. *La coda dell'occhio. Letture del linguaggio letterario senza confine nazionali*. Canterano Rome: Aracne.

Susan Petrilli, Augusto Ponzio
Difference and similarity in the I-Other relation: between two individuals or two singularities? A semioethic approach

Abstract: In *Forms of Meaning: Modelling System Theory and Semiotic Analysis*, Thomas Sebeok and Marcel Danesi (2000) distinguish between two forms of similarity: *cohesive form* and *connective form*. Similarity as *cohesive form* assigns given individuals to the same class, group, collectivity, assemblage, affiliation. Concepts, types, species, identities are all formed on the basis of similarity understood as cohesive form. *Cohesive form*, also called *assemblative form*, subtends similarity and difference in the I-Other relation. In this case the I-Other relation is determined by the identity of each of its parts, by one's belonging to this or that group, assemblage, organization, to this or that set, to this or that type or kind. In this type of relation the I and the other are determined by signs, depend upon signs that distinguish and differentiate between them.

But is this really what the I-other relation must be reduced to? We do not believe so. And this is the belief that orientates the particular bend in semiotics we have denominated as *semioethics*.

With reference to the distinction proposed by Thomas Sebeok and Marcel Danesi, the other type of similarity and difference is that indicated as *connective form*. Following Charles S. Peirce it can also be denominated as *agapastic form*. This type of similarity does not involve individuals assigned to this or this other class, assemblage, type, classified in terms of a fixed identity.

Connective form is not related to what presents itself as the *same* or as *different* in terms of cohesive form. In the case of connective form, similarity in the I-other relation involves relations *among singularities* and not among individuals considered as representatives of some form of identity. In the case of connective form the relation is among irreducible, unique single individuals, where uniqueness, singularity cannot be reduced or eliminated as occurs, instead, when grouped together in an assemblage, a class, a group identity of some sort.

When a question of *connective or agapastic form*, difference is characterized in terms of irreducible alterity, otherness, uniqueness, non-interchangeability among singularities. The condition of irreducible otherness, of irreducible singularity

Susan Petrilli, Augusto Ponzio, University of Bari "Aldo Moro", Italy

https://doi.org/10.1515/9783110662900-003

is inevitably accompanied by responsibility of each single individual for every other single individual in one's uniqueness, singularity. This is a form of responsibility that cannot be delegated.

Keywords: Depicture, dialogism, difference, iconicity, modelling, otherness, responsibility, similarity

> For Solomon Marcus

1 Modeling and figuration

According to the tradition of thought delineated by the research of John Locke and Charles S. Peirce, and more recently Charles Morris, Roman Jakobson and Thomas Sebeok, semiotics contributes to explaining the workings of associative-metaphorical thought processes in the formation of concepts. Giambattista Vico's role in twentieth century semiotics is evidenced by Sebeok (2000).

Metaphor is the motor behind human reasoning which does not merely consist in representing objects (indicational modeling) but in depicting, portraying, figurating them (modeling proper to language and modeling systems based on language, that is, "secondary" modeling proper to historical languages, and "tertiary" modeling proper to human cultural systems and capable of highly abstract symbolically structured processes (Sebeok and Danesi 2000).

Adequate explanations of a theoretical order cannot come from linguistics given that the most advanced currents in this science, that is, with claims to the status of "philosophy of language," Chomskyian generative-transformational theory, is deaf to the question of metaphor, which it considers as an abherration.[1] In this light, Marcel Danesi's (2000) appeal to examine Vico and his "new science" is worthy of consideration given Vico's description of metaphor as the main mechanism in the formation of concepts. Danesi associates Vico's reflections on signs to *linguistics*, with a special focus on modern day *cognitive linguistics*. However, this is not simply a question of evidencing a similarity or a precedent. Vico's reflections can effectively contribute to present-day research in linguistics and to an explanation in theoretical terms of associative-metaphorical processes

---

[1] From this perspective things do not change when passing from Chomsky's earlier essays to the more recent ones collected in *Knowledge of Language*. For a discussion of Chomskyian theory, see A. Ponzio's book of 1992, *Production linguistique et idéologie sociale*.

characteristic of thought and language, that is, of modeling specific to human beings.

The Vichian notion of "poetic logic" maintains that the human mind is predisposed to intuit and to express synthetically and holistically. This position provides a valid alternative to the Chomskyan model, to "Cartesian linguistics," and is on the same line of thought as recent trends in cognitive linguistics and neuropsychology, as well as with modeling theory in semiotics.

The human mind moves among meanings and concepts (indeed with Danesi we may speak of imaginative mental navigation) within a network of interpretive routes made of associative connections which, in turn, are part of that complex system or "macro-web" commonly called "culture".

At this point we should be able to glimpse the inadequacy, or insufficiency, of the notion of "linguistic competence" (Chomsky) and of "communicative competence" (by contrast with or as a completion of Chomskyan theory) to explain thinking or speaking behavior, that is, the capacity to verbalize and to reason: both competencies are part of an organic *conceptual competence*. This consists in the ability to appropriately metaphorize a concept, to select the structures and linguistic categories which reflect appropriate conceptual dominions, and to move through appropriate fields of discourse and conceptual dominions.

"Linguistic creativity" is the capacity to form new metaphorical associations, to propose new cognitive combinations, and invent new figurations. This is not a prerogative exclusive to poets, scientists, and writers, but is a capacity which each one of us possesses thanks to imagination, ingenuity, and memory, as Vico would say, insofar as we are all capable of metaphorical associations. This is part of "primary modeling," what Sebeok calls "language," the preliminary basis of human symbolic behavior, that is, a structurally constitutive element of the primary, secondary and tertiary systems with which human beings are endowed.

Thanks to the associative character of verbal language and thought, with Peirce, and differently from the Cartesian model of the thinking subject, we may claim that guessing is a characteristic of reasoning, and that reasoning is ever more capable of inventiveness and innovation the more it attempts associations among terms that are distant from each other and belong to different and distant fields in the macro-web of culture.

2 Vichian linguistics versus Cartesian linguistics?

An issue we wish to at least hint at here is whether or not on the basis of Vico's contribution to the theoretical framework of current research in cognitive linguistics we may speak of a "Vichian linguistics", as Danesi suggests. This contrasts with "Cartesian linguistics," to which Chomsky associates his own theory of generative-transformation grammar given its innastic assumptions.

However, in spite of being a pioneer in research into the metaphorical character of thought and speech (apart from founding the historical sciences), we do not believe that Vico can be indicated as the guiding light of Crocean historicism (nor do we think we should risk unfortunate associations between cognitive linguistics and the oversimplifying aesthetic and linguistic ideas expressed by Benedetto Croce in *Estetica come scienza dell'espressione e linguistica generale*, 1902). When a question of working scientifically, it is best not to commit to guiding lights.

As testified by Sebeok's paper "Some Reflections of Vico in Semiotics," research on the relations between Vico and semiotics has flourished, many ideas held by exponents of semiotics, philosophy of language and other sign sciences are influenced by Vico, whether directly or indirectly, or at least present analogies with his thought system. Studies comparing Peirce and Vico are not lacking as in the case, for example, of studies on the concept of "common sense," or the critique of Descartes, or the relation between Peirce's pragmatism and Vico's formula "verum factum convertuntur". Concerning such associations, in the present context we can only limit ourselves to declaring that we have a few doubts.

Vico's critique of Descartes is based on motivations and argumentations and above all is formulated in a context entirely different from Peirce's. And restriction of the cognitive sphere to human circles is a far cry from the extensiveness of Peircean semiotics and its current developments beyond the boundaries of anthroposemiosis and of the "semiosphere" as understood by Lotman, that is, limited to the world of human culture. Sebeok develops the Peircean idea that the whole universe is perfused with signs. And with all those scholars who work in the sphere of biosemiotics, has made an important contribution to extending the boundaries of Lotman's semiosphere to the point of establishing that it coincides with the biosphere (Vernadsky).

Alongside important ideas to develop, and not only in the field of cognitive linguistics, Vico's *New Science* presents apologetic and rhetorical expedients used to reject and restrain the new emerging vision of the world and of man as perspected by progress in the physical and mathematical sciences. Vico was intent upon searching for firm points and unviolable boundaries in the realms of religious tradition and common sense (see, e.g., *Scienza Nuova*, I, Degnità XII and XIII). In Vico's view: "common sense is a judgment without reflection, commonly felt by a whole order, by a whole people, by a whole nation, by the whole of the human race, and is taught to nations by divine providence" (Vico 1999: 142).

As observed by Giuseppe Semerari (see "Sulla metafisica di Vico and Intorno all'anticartesianesimo di Vico," in Semerari 1969: 252, 271, 239–240), to the power of critique Vico opposes and juxtaposes faith in common sense considered as a system of judgments of nonhuman and divine origin, enhanced by the rhetorical expedient of quantity on the basis of which the validity of anything is established according to the criterion of universal consensus, or consensus from the highest number of people. Vichian anti-Cartesianism is also the expression of "an attitude of resistance and defense against the philosophical development of the new mathematical and experimental science [. . .], a cultural tactics invented, more or less consciously, to the end of *quieta non movere*, leaving things as they are, as far as possible limiting the field of action of the new methodology which seems dangerous for the natural course of ideas and for common sense" (Semerari 1969: 239–240).

3 Iconicity, syntactics and metaphor

According to Charles S. Peirce's typology of signs, metaphor is a type of icon. The iconic relation between that which is interpreted and which as an interpreted is a sign, and that which interprets it and which as an interpretant is also a sign, may well be a relation between that which is not originally or naturally related. "An icon is a sign", says Peirce, "which would possess the character which renders it significant, even though its object had no existence; such as a lead pencil streak representing a geometrical line" (*CP* 2.304).

Peirce divides the icon into three subclasses: 1) *images*, 2) *diagrams* and 3) *metaphors* (*CP* 2.277). In the image likeness is overall and direct; in the diagram it concerns the relation between the parts represented through analogical relations; in metaphor is consists of parallelism, a comparison.

Verbal iconicity concerns relations of likeness that depend on modeling by language, that is to say that are part of the world modelled by a historical-natural language and of worlds modelled by its special languages. Likeness is internal to modeling of historical-natural language, as such it is not determined by a relation of analogy or isomorphism with objects external to such modeling processes. The relation between signs and the real world is the relation between signs and the reality they model.

An aspect which strongly evidences the iconic character of verbal language is metaphor. In this case iconic similarity consists in a comparison and only concerns a few characteristics of what is being compared, which are sometimes superficial and other times more profound (simple and superficial *analogy*, or a structural and/or genetic *homology*), leaving aside all the rest.

In *Principi di scienza scienza nuova*, Vico (1999: 444) observes that philologists in good faith believe that natural languages signify "*a placito*", that is, by convention. On the contrary, he observes that most words are formed through metaphors and are generated by the senses. Vico cites Aristotle: 'Nihil est in intellectu quin prius fuerit in sensu.' In other words, the human mind understands nothing that has not been first perceived by the senses (Vico 1999: 363). Languages form words through metaphors; and metaphors generally carry out a central role in all languages (Vico 1999: 444). Vico then goes on to say that in the face of words that produce confused and indistinct ideas, and whose origins are unknown, the grammarians universally established the rule that articulate human words signify *a placito*. This solution served to remedy their ignorance, and was also attributed to Aristotle, Galen, as well as to other philosophers:

> Ma i grammatici, abbattutisi in gran numero di vocaboli che dànno idee confuse e indistinte di cose, non sappiendone le origini, che le dovettero dapprima formare luminose e distinte, per dar pace alla loro ignoranza, stabilirono universalmente la massima che le voci umane articolate significano a placito, e vi trassero Aristotele con Galeno ed altri filosofi [. . .]. (Vico 1999: 444)

As anticipated metaphor is a type of icon; it is an expressive modality that cuts across all verbal language and connects it to the nonverbal, activating interpretive routes that relate sections in the sign network that are even very distant from each other, similarly to inference of the abductive type. With Vico, it becomes clear once and for all that metaphor cannot be reduced to a mere rhetorical device, decorative covering with respect to a given "nucleus of meaning," to presumed "simple and literal" meaning (Vailati, Welby). On the contrary, in Vico metaphor emerges as the place where sense is generated. From this point of view his work is particularly important. Like abductive inference, the cognitive capacity of metaphor depends on the type of similarity (simple and superficial *analogy* or structural and/or genetic *homology*) established among things that are different from each other. Meaning is developed through metaphor, through relations of '*interinanimation*' among words (Richards 1936). The processes of metaphorization are present in discourse even when we are not aware of them. In fact, we may distinguish between metaphorical trajectories that are practiced automatically by speakers and would seem to present simple, 'literal' meaning, on the one hand, and metaphorical trajectories, on the other, that are immediately recognizable as such, with a strong charge of inventiveness, creativity and innovation thanks to the association (similarly to abduction) of interpretants that are distant from each other in relations that are altogether new and unexpected (see Ponzio 2004: 63–68, 83–88, 182–183).

As Vico knew verbal language is particularly important for our understanding of the workings of the human mind. And as we know today, one of the main objectives of present-day "cognitive linguistics" is to understand the workings of human thought, particularly how concepts form in the human mind. From this perspective and similarly to Vico, cognitive linguistics focuses on the *metaphor* understood as a particular type of icon (with reference to Peirce's tripartition of the sign into *icon*, *index* and *symbol*). But cognitive linguistics does not stretch beyond the empirical finding that the basis of verbal communication as much as of symbolic expression in general, is given by *metaphorical interconnections* that characterize human thought. On the contrary, with his "poetic logic" Vico provides a good starting point for an explanation of the functioning of associative-metaphorical processes in the formation of concepts in theoretical terms beyond the level of empirical phenomena.

In the second chapter of his *New Science,* Vico writes that tropes are a part of "poetic logic," metaphor being the most luminous, most necessary and most used (see Vico 1999: 404).

The issue we are addressing is our full understanding of the decisive role played by tropes and in particular metaphor in thought, verbal communication and symbolic expression in general. Metaphor is figure of speech, figurative discourse, an associative modality of signification which has wrongly and at length been considered as a rhetorical device, poetic embellishment, while in fact it is central to the development and functioning of human language-thought.

In *Traité des Tropes*, César Chesneau Du Marsais ([1730] 1977: 11) acknowledges that figures that are tropes are already metaphors (see [1730] 1977: 8), and as speech modalities they are not distant from natural and ordinary figures; on the contrary, there is nothing more natural, ordinary and common in human language than figures (see ([1730] 1977: 11). And similarly to Vico he refers to the immediately perceptible character of tropes and observes that like the figures of the body they have particular forms which render them immediately identifiable. As says Dumarsais, it is not figures of discourse that take their distances from human ordinary language, but rather discourse without figures conceding that discourse is possible without figurative espression (*Ibid.*: 8; Petrilli and Ponzio 2019b).

Such a position is extraordinarily close to Jean-Jacques Rousseau's own reflections on tropes in his essay *Essai sur l'origine des langues* ([1781] 1989) as well as Vico's. And this association was in fact signaled by Cassirer in his *Philosophy of symbolic forms* (1923), discussed by J. Derrida in *De la grammatologie* (1967). According to Rousseau man was originally pushed to speak by the passions so that his first expressions were tropes. Figurative language appeared first, while proper sense came last (see Rousseau 1989: 18).

As regards reflection on the role of metaphor in thought and language, studies by Victoria Welby and Giovanni Vailati on the figures of speech, published toward the end of the nineteenth century and beginning of the twentieth should also be remembered. Welby developed an important correspondence with Peirce (some of his most innovative writings are part of this correspondence) and worked at a theory of meaning she denominated "significs." In *What Is Meaning?* (1903), *Significs and Language* (1911) and her essays "Meaning and Metaphor," and "Sense, Meaning, and Interpretation" (1896), she describes metaphor as a vital part of thought and verbal language whose essential characteristic is what she calls plasticity. Rather than consider figurativity as something that must be repressed or literalized, or rather than consider images and analogies as indistinct abstractions or mere rhetorical devices, Welby maintains that we must get free of the "plain meaning" fallacy, from the idea of being able to refer directly to hard dry fact. She promotes scientific research on the necessary use of metaphor in thought and discourse to the end of enhancing its instrumental value in reasoning, knowledge and communication.

Vailati was also aware of the need to reflect on the functioning of metaphor. He worked with Mario Calderoni and was in contact with Welby. He used Welby's and Peirce's research (he was among the first in Italy to appreciate the importance of the latter's writings) for his own reflections on logic and meaning in the spheres of ordinary and scientific discourse. He also underlined the need to reflect on the functioning of metaphor.

In "I tropi della logica" (1905), occasioned by Welby's *What Is Meaning?* Vailati examines metaphors used to discuss reasoning, that is, logical operations. In fact, even when we speak of discourse and thought, our discourse (or metadiscourse) is tied to metaphors which condition the way we understand linguistic and logical operations. In relation to metaphors Vailati distinguishes between three types of images: images that 1) *support* (as when we speak of conclusions that are 'founded,' 'based,' 'depend on,' 'connect up with'); 2) *contain or include* (conclusions 'contained' in the premises); 3) *come from* or *go to* (conclusions 'coming from' given principles). Vailati interrogates such images used to describe reasoning and underlines their connection with a hierarchical view of things (to base, to be founded on), or with the mere distribution of certainties included (in premises) and which must simply be explicited. In order to describe the relation among concepts in terms of associative-metaphorical relations, Vailati speaks of attraction and mutual support. The spread of certainty is bidirectional, not unidirectional (see Vailati 2000: 80).

Vailati does not use the Peircean term 'abduction,' but speaks of a "particular type of deduction" used in thought processes which has enabled the development of modern science. In this particular type of deduction, says Vailati in "Il metodo deduttivo come strumento di ricerca," propositions taken as the starting point need proof more than those which are reached, consequently final propositions, or conclusions, must communicate what certainty is reached through experimental verification to initial hypotheses. This is a special type of deduction based on supposition, conjecture, guessing, hypothesis, "deduction as a means of anticipating experience," and which, differently from proper deduction, "leads to unsuspected conclusions" (Vailati 1971: 80). In this new type of deduction, or abduction, observes Vailati, relations of similarity are established among things which are not immediately given, identifying analogies among things which to immediate experience would not seem to be related. This enables cognition to progress beyond the processes of induction, says Vailati, so that, as an effect of hypothetical deduction, or abduction, "we are able to discover intimate analogies among facts that would seem to be different, and that immediate observation is incapable of revealing" (Vailati 2000: 80).

4 Assemblative likeness, poetic logic and mathesis singularis

In a very interesting part of his lessons on the *Neutre* (1977–78, in Barthes 2002: 199–201), Roland Barthes (who keeps account of Vico, as emerges from the references section of his lessons) opposes the *concept* to the *metaphor*. Both concept and metaphor are constructed by association on the basis of likeness. To explain the difference we shall leave Barthes and refer to William Shakespeare's *The Merchant of Venice*.

Choice of the right casket in this artwork is based on recognition of the relation between the lead casket and love, on recognition of the likeness between choosing this casket and the beloved. This type of likeness is not the same as that which allows for assigning certain individuals to the same class, to a group, at the basis of classifying an individual according to a given genus; this is not a question of what we may call the *assemblative* type of likeness. The type of likeness subtending the strategy used to solve the enigma of the caskets is rather the same as that subtending the metaphor. It uses *elective* likeness, *attraction*, likeness on the basis of *affinity*. Likeness in this sense does not concern that which presents itself as the same, as belonging to the same category, as identical, but rather that which is different, refractory to the assemblative form, that which is other. In *The Merchant of Venice* procedure according to this type of likeness is not only the reasoning behind the solution to the puzzle of the casket, that is, according to which the choice of the casket is similar to the choice of the beloved: to choose one is to find the other. It also characterizes some of the decisive arguments developed by Portia. Portia appeals to this type of reasoning, what we may call by "elective affinity" and which leaves the terms associated to each other in their mutual otherness, in her effort to induce Shylock to clemency: given that *mercy* is an attribute of God himself, earthly power seems more similar to divine power when mercy softens justice, and therefore if Antonio recognizes *the bond*, Shylock must be *merciful*. But Portia's decision to help Antonio is also based on this type of likeness: the fact that Antonio is tied to Bassanio by profound friendship, "makes me think that this Antonio [. . .] must needs be like my lord". Indeed if Antonio, whom Bassanio loves, is like Bassanio, then he is like Porzia, who loves Bassanio: in Antonio Portia says she recognizes "the semblance of my soul".

This type of likeness involves a movement towards the other as other, analogous to giving, forgiving, giving at a loss. The logic of this movement by attraction is the same as that of unconditioned risk for the other, a one-way movement, without return, without gain, towards the other, of being one for the other, of "substitution" (Levinas 1974). This is a question of likeness where the terms are not indifferent to each other, where differences are not cancelled, as in the case of the indifference of assemblative likeness, which identifies, homologates, equals. On the contrary, likeness by attraction is characterized by a relation of undifferent difference with the other: likeness by otherness contrasts with likeness by identity characteristic of greedy exchange, of justice (on the basis of justice we cannot any of us sell our salvation). This is the only type of salvation Shylock knows, when he affirms his ethnic, religious identity, when he complains of being betrayed by his daughter, who as such belongs to him, is his own "flesh and blood", when he claims what was established by contract. Shylock refers to cohesive logic, identity logic, the logic of just exchange, greedy logic, the logic of indifference, justice and revenge:

> I am a Jew. Hath not a Jew eyes? Hath not a Jew hands, organs, dimensions, senses, affections passions? – fed with the same food, hurt with the same weapons, subject to the same diseases, healed by the same means, warmed and cooled by the same winter and summer as a Christian is? If you prick us do we not bleed? If you tickle us, do we not laugh? If you poison us do we not die? And if you wrong us shall we not revenge? If we are like you in the rest, we will resemble you in that. (Act III, Sc. I, ll. 56–65)

In his analysis of the relation between concept and metaphor Barthes identifies two logics, assemblative logic of the concept, which proceeds on the basis of genera e species, paradigms, only recognizing individuals that belong to genera and not singularities, assimilating that which cannot be assimilated, and the logic of association by attraction and affinity (Vico's "poetic logics"– Vico is listed in Barthes bibliography under the title *Intertexte* his first course, 18 February 1978 – which characterizes metaphor (as the expression of "iconicity" and "firstness", and which we may claim following Peirce is based on the "agapastic" relation, see Peirce 1923). According to the logic of elective affinity, *agapasm* likeness leaves the terms of the relation in their otherness, irreducible singularity. With reference to what Barthes says in *La chambre claire*, we may observe that this logic renders plausible what he qualifies as a "bizarre" question: why cannot there be, in a certain sense, a new science for every subject, an affectively "new science", in the Vichian sense? A *Mathesis singularis* (and not *universalis)?* (Barthes 1980, It. trans.: 10).

"All concepts", says Barthes (2002: 201) arise from "identification of the non identical". The concept: "reducing force of that which is different". Therefore "if we want to refuse reduction, we must say no to the concept". So how do we speak then? Barthes' reply is: "Through metaphors. Replace the concept with the metaphor: write" (Barthes 2002: 201).

Writing, literary writing, is precisely that practice that redimensions, avoids, cringes from the arrogance of discourse (Barthes 2002: 201, see also Petrilli & Ponzio 2006), whether individual or collective, including Vico's "boria delle nazioni" (arrogance of nations). The only dialectic action against arrogance with the same assertive nature as discourse is the transition from discourse to writing, the practice of writing: the Neutral of writing, the desire of writing.

5 Mute language and speaking

The idea that man was created in the likeness of God involves a species-specific trait essential to the human being, that is, language. In this context 'language' does not mean verbal language. The species-specific trait of the human being is a modeling device capable of inventing many worlds, differently from other animals. Human beings was endowed with this particular modeling device when they first appeared in evolutionary development; and this specific device conditions evolution determining the capacity for articulate speech in *Homo sapiens* and *Sapiens sapiens*. Sebeok calls this species-specific human modeling capacity (a capacity for the "play of musement") "language" and distinguishes this from "speech": orignally language was mute. As a modeling device capable of modeling different worlds, it allowed for communication. Therefore, language thus understood is at the basis of the different historical natural languages. This is the "*lingua mutola*" (mute language) discussed by Vico: originally all nations were mute and expressed themselves through acts or body language (1999: 434). What Vico calls "language of the Gods" was almost completely mute, and gradually became articulate language (Vico 1999: 446). We would do injustice to deaf-mutes if we said that man is the animal that speaks. However, phonocentrism is hard to die, as most forms of discrimination.

We could make the claim that the human being is endowed with language, "poetic logic" understood as a species-specific modeling procedure. Language described as a faculty is more precisely a modeling device. Language as a species-specific modeling capacity is at the basis of verbal language which only exists concretely in the different historical languages and in the different special languages that compose it. Verbal language also consists of nonverbal languages (gestural, pictorial, photographic, musical language, etc.) which in accord with the distinction made in certain languages (e.g. Italian and French) between historical natural language and language in general (It. *lingua* and *linguaggio*, Fr. *langue* and *langage*) are called 'languages' though they are nonverbal. Historical natural languages and nonverbal languages depend on the faculty of language proper to human beings. At the same time, language understood as pre-verbal modeling also subtends manipulative activity of verbal and nonverbal languages (see Rossi-landi 1985: 217–269). The production of artifacts and transformation of material objects into signs proceed together (on a phylogenetic level as well, that is, in the process of homination). And if they presuppose language as primary modeling, the central element in such transformation, as Vico underlines, is the human body.

For whoever tries to explain the origin of language with Chomskyian concepts, as does Liebermann (1975), the lack of distinction between 'language' and 'verbal language' gives rise to forms of 'psychological reductionism.' "Complex anthropogenetic processes are limited to the linear development of certain cognitive capacities, and described in the language of traditional syntactics" (Rossi-Landi 1985: 229).

According to Sebeok's modeling theory, language (the primary modeling system of the species *Homo*) appeared and developed through adaptation much earlier than speech in the course of evolution of the human species through to *Homo sapiens*. Language was not originally a communicative device. Chomsky also maintained that language was not essentially comunicative, but by 'language' he understood 'verbal language,' or 'speech,' as Sebeok would say. Instead, according to Sebeok, verbal language has had a specific comunicative function through adaptation from the very moment of its appearance. Chomsky's theory of verbal language does not keep account of the difference between language and verbal language, and without this difference it is not possible to adequately explain the origin, nor the functioning of verbal language.

Other traits characterizing the human being derive from language. These include, 'creativity,' the capacity for inventiveness and innovation, for deconstruction and reconstruction, the *semiotic* character of semiosis, that is, the capacity for *metasemiosis*, for using signs to reflect on signs, and therefore the capacity for being aware, and consequently of being 'condemned' to responsibility. The divine vocation of the human, that is proper to the human, consists in such characteristics and belong to human beings insofar as they are endowed with language.

Language thus understood is a species-specific capacity of the human being: on a phylogenetic level, *homo habilis* (mute) was endowed with language before *homo sapiens* used speech to communicate; on a ontological level, the infant is already endowed with language, though it is *in-fans*, that is, it does not speak; from the perspective of certain pathologies, deaf-mutes have language as much as they cannot use speech as a means of communication. We can claim that the dog most able to make itself understood does not only lack the word, as people generally say, but first of all it lacks language. Instead, the deaf-mute only lacks the word, but is not at all devoid of the capacity of language because of this. If, as occurs in the fun scene imagined by De Mauro (1994: 24), two of our ancestors could complain about the appearance of speech and its 'drawbacks' (De Mauro does not ask himself how they managed), this is because they were endowed with language, and on this basis, with other communicative means with respect to speech.

The question of the origin of verbal language has generally been undervalued by the scientific community and considered as not worthy of discussion because of the unfounded solutions it has given rise to (an exception is a book by Giorgio Fano entitled *Origini e natura del linguaggio*, 1972). Sebeok has reproposed this problem developing the concept of 'modeling' as introduced by the Moscow-Tartu school. And Vico had already offered a correct approach to this problem. Not only: he also focuses on what we may call the 'enigma of Babel,' that is, the question of the multiplicity of languages. Why many languages and not just one, as instead occurs in the case of the thesis of origin through convention, *ad placitum*; and above all when the thesis is that languages derive from the same universal grammar, as Chomsky would have us believe? Vico maintains that an enormous problem remains to be explained, the fact that there are as many different natural languages as there are peoples (Vico 1999: 445).

The multiplicity of languages (as well as 'internal plurilingualism' in all languages) derives from the human modeling capacity and ability to invent many worlds, that is, from the human disposition for the 'play of musement' or, as Vico would say, for 'poetic logic' proper to the human being. Instead, we have seen that Chomsky's linguistics which resorts to the (Cartesian and biologistic) assumption of an innate Universal Grammar is unable to explain all this in spite of insistence on the 'creative character of language' (that is, verbal language).

The human modeling procedure, language, differs totally from that of all other animals, while the *type* of sign it uses does not (signals, icons, indexes, symbols, names, as Sebeok in particular has demonstrated). Its specific characteristic is *articulation*, or as says Sebeok, *syntax*, that is, the possibility of producing different significations that avail themselves of the same objects with an interpretant-interpreted function. 'Articulation' makes us think of deconstruction into elements. 'Syntax' best expresses the spatio-temporal organization of these objects. It would be best to speak of *'syntactics'* which is a term taken from the typology of the dimensions of semiosis and of semiotics, as proposed by Charles Morris (1938), in order to avoid confusion with syntax in the linguistic-verbal sense and in the sense of neopositivist logic. The syntactics of language determines the possibility of combining a finite number of elements in an infinity of different ways, producing different meanings each time. Aristotle maintained a similar position when he indicated as a specific characteristic of the 'properly human signifying voice,' the fact that it is *suntheté* (*Poetica* 1457 a 1415) or *katà sunthéken* (*De interpretazione* 16a 26–29; see Lo Piparo 2003) that is, reached through composition and combination, despite interpretations that reduce the Aristotelian conception of language to convention.

We prefer to speak of 'writing' to indicate the *syntactics of language*. Writing is the combination procedure that enables us to produce an unlimited number of senses and meanings through a finite number of elements (see Petrilli and Ponzio 2016). In this sense writing is antecedent to and the condition for speech. In fact, the phonetic sign itself is writing because it functions uniquely on the basis of combination.

Language is already writing, which therefore subsists *avant la lettre* (before the letter), even before the invention of writing as a system for the transcription of vocal semiosis, indeed even before the connection of language to phonation and the formation of historical natural languages. Writing is part of language 'before the stilet or quill imprints it as letters on tablets, parchment or paper' (Levinas 1982, Eng. trans.: xi).

Language as it now presents itself has been influenced by its development resulting from the use of phonetic material, and all the same it has not lost its characteristics of writing antecedent to transcription. These include its articulation into verbal writing, its iconic character (signification through position, extension, as when adjectives become longer in the superlative, or verbs in the plural, etc., as shown by Jakobson1968). When writing returned subsequently as a secondary cover to fix vocalism, it used space, as says Kristeva (1982), to preserve the oral word through time by giving it, that is, a spatial configuration.

The articulation of verbal language (Martinet's double articulation, see Martinet (1965, 1967, 1985) is an aspect of the modeling procedure of language which articulates the world through difference and deferral – *différence/différance* (Derrida 1967). Articulation is first of all distancing, *espacement*, operated by language understood as a modeling procedure insofar as it is writing. To signify with the same elements through different positions is already writing, and articulation *of* verbal language and *through* verbal language (secondary modeling) takes place on the basis of this type of signification through position.

As syntax, or – to avoid a term used by the linguists and neopositivists (Carnap's 'logical syntax') and the misunderstandings it may give rise to –, as *syntactics*, or, more precisely, *writing* antecedent to phonation and independently from the comunicative function of transcription, modeling through language uses pieces that may be combined in an infinite number of different ways. This means to say that an indeterminate number of models can be deconstructed in order to construct different models with the same pieces. Therefore, as says Sebeok (1986), thanks to language human beings are not only able to produce their worlds, like other animals, but also an infinite number of other possible worlds: this is 'the play of musement,' which carries out an important role in scientific research and in all forms of investigation, as in simulation, from lying to fiction, and in all forms of artistic creation. Creativity understood by Chomsky as a specific characteristic of verbal language in reality is derivative, while instead it is proper to language understood as writing, as a primary modeling procedure.

The formation itself of speech and relative verbal systems, the historical natural languages, presuppose writing. Without the writing capacity, the human being would not be able to articulate sounds or identify a limited number of distinctive features, *phonemes*, to reproduce phonetically. Without the writing capacity the human being would not know how to compose phonemes in different ways to form a multiplicity of different words (*monemes*), nor to compose the latter syntactically in different ways to produce *utterances* that are always new, or *texts*, complex signs whose unitary meaning is qualitatively superior and irreducible to the sum of the parts that compose it.

Writing is inherent to language as a primary modeling procedure; its specific characteristic is to confer different meanings upon the same elements depending on their chronotopic position. In other words, writing is inherent to language as a signifying procedure insofar as it is characterized as *syntax*. The phonetic sign itself is writing. Language in itself is already writing, even before writing is invented as a system for transcribing vocal semiosis, indeed even before language is connected with phonation and the formation of (historical natural) languages.

The a priori is not speech. The a priori is language and its writing mechanism. Musical writing, for example, derives from the language modeling capacity similarly to verbal language, and therefore participates in the conditions for viewing, articulating, and relating without which a human world would not be possible.

Walter Benjamin also seems to insist on the idea of a connection between language and writing as understood above when he focuses on allegory and describes it in terms of writing, or when he reflects on hieroglyphics, on the ideogram and on the relation between thought and original writing, on verbal language that does not limit itself to simply serving communication, on the letter that withdraws from the conventional combination of writing atoms and assumes a sense in itself, as an 'image,' because it assumes an iconic *character*: in the 'baroque,' that which is written tends towards the image, and from a linguistic perspective this constitutes the unity of the linguistic and the figurative baroque (see Benjamin 1980: 162–229).

Another characteristic of verbal and nonverbal languages thanks to writing proper to language is the possibility of functioning as signs without an end, or rather, as an end in themselves, producing a sort of excess with respect to their cognitive, communicative, and manipulative function. This capacity is also present in animal behavior, but only in terms of repetition. The dialogism of interpretants, therefore their capacity to overcome signality in the direction of signness, and *signification* in the direction of *significance* (what R. Barthes in his book of 1982 calls *third sense*, with respect to sense in communication, in signification, in the message) is connected to the character of writing inherent in language. The idea of man being created in the likeness of God can no doubt also be interpreted in terms of the nonfunctionality of linguistic creation, of being an end in itself. However, this is not undertood in the sense of esthetics, that is, art for art's sake, nor in the productive sense of communication for production (profit), nor in the anthropocentric sense according to which all means used by man to affirm himself is justified by his ends, e.g., anthropization of the planet (that is, the process of adapting the environment to human needs). On the contrary, the nonfunctionality of creativity, of being an end in oneself is understood in a humanistic sense, which means to say that man's greatest wealth is to view himself from the perspective of otherness, to view otherness as an end and not as means.

6 Open identity

Identity matters are now very much at the centre of attention in the world today. To claim identity, to assert and to make a show of identity, to the point of obscenity nowadays would seem to be an important goal for human behaviour – if not *the* most important. This is true of the single individual as much as of the community and applies to the various spheres of life – sexual, ethnic, religious, racial, ethical, linguistic, cultural, national, etc. Identity aims to assert itself and achieve its goals at all costs, even at the cost of sacrificing otherness, whether one's own or the other's.

Contrary to all this, the single individual may be associated, instead, with otherness logic and considered in one's unrepeatability, uniqueness. Uniqueness, singularity, the *eachness* of each one of us is associated with otherness. Moreover, otherness as we are describing it involves the condition of responsibility without alibis, and not just technical responsibility, it is associated with unlimited, absolute responsibility. Absolute responsibility implies a relation of unindifference to the other.

Identity as it is commonly understood, that is, closed identity, tends to deny uniqueness, singularity: it asserts itself in its generality. Indifference is a characteristic of the whole, of the totality, of the group or community to which a given identity belongs, of which it is a part. Understood in these terms identity is short-sighted identity, indifferent identity, identity-difference indifferent to other identity-differences.

This type of identity, closed identity, implies well-defined boundaries that serve to exclude the other, that narrow and delimit responsibility towards the other. Dominant ideology and official discourse are grounded in the logic of closed identity. As such they assign roles and responsibilities, special or technical responsibilities, that is, responsibilities limited by alibis.

As part of a given totality or assemblage, identity-difference, that is, difference, the single individual considered in terms of closed identity, reduced to identity logic is achieved on the basis of the elimination of singularity, uniqueness. Externally, difference is achieved and maintained on the basis of the logic of opposition, general binary opposition, which means to say on the basis of relations that tend to be conflictual. Such relations are described by Charles Peirce as "obsistent" relations. In such relations opposition is a necessity, given that to eliminate opposition means to eliminate identity.

Difference based on closed identity will at the most succeed at tolerating the other, the opposite other, thus defined on the basis of such factors as sex, role, ethnic group, culture, language, religion, nation, colour of the skin, social status, etc. But to the extent that identity-difference implies a relation between oppositional identities, tolerance is always on the verge of degenerating into open conflict.

However, "interpretation semiotics" developed as global semiotics and as "semioethics" evidences the possibility of developing difference that is not oppositional, that is not conflictual, that is to say, identities and differences that do not necessarily involve relations of opposition, exclusion and mutual indifference – non-conflictual identity and difference.

Instead, the history of the conquest of identity is the history of the extromission of the other, the history of imposing monologism, univocal unity on the great plurality of different linguistic-cultural practices across the globe, conditioning goals, structures and orientations. Sacrifice of the other, the outsider, the foreigner, the stranger is structural to *identity*. Never before as in social reproduction today has the demand made by identity for sacrificing the other, for elimination of the other been so extreme, so obsessively insistent, so paroxysmal. Self-centred identity is manifest in the present day and age in all its unequivocal "obscenity".

Closed identity pervades the social in all its aspects, the problem of the relation between power, freedom, responsibility and difference in a global world where identity is specified as national, cultural, ethnic, and racial identity. With reference to race and ethnicity, the whiteness question today continues to be relevant. At the heart of identity today is the market, or better market logic, precisely so-called *equal exchange* market logic. Identity constructs its rhetoric on the logic and ideology of exchange, which means to say that identity excludes any movement towards the other that does not return to the identical and that does not produce profit.

Human society and its signs are constructed through the mediation of "linguistic work," social reproduction and ideology where the role of identity is an a priori. Language, that is, verbal language, acts as an essential means, as the material in which ideologies are constructed and in which the signs of identity are determined. Even more, beyond means and material, language models and conditions ideologies and identities. We are alluding here to "language" as language for communication, the language of different discourse genres, ordinary language, special language, the language of mass media and social networks, technological language. But "language" is also language as modelling (see Sebeok, Petrilli and Ponzio 2001).

"Signification" or better "significance" beyond and before the *hic et nuc* implies the *otherness dimension of semiosis*, Welby's "mother-sense", Peirce's "firstness", "iconicity", "orience" or "originality" and in terms of interpretation his "final interpretant", Levinas's *"jouissance"*, his "absolute otherness," Bakhtin's extralocality, unlimited responsibility. Thus conceived significance is not reducible to the empirical moment, to the *hic et nuc* of semiosis, but tells of our signifying potential before and beyond it, of the human capacity for transcendence and overcoming with respect to restrictive limitations and boundaries aiming to repress and exclude.

Semioethics predicates the need to resist the reduction of otherness – the potential for dialogue and opening to the other – to the world-as-it-is. Ultimately semioethics is a semio-philosophical method for resistance and dissent, for disobedience in the face of repression, power and control: challenging silence as imposed by the order of discourse, semioethics reproposes the power of critique and resistance through listening, dialogue, co-operative participation and responsibility.

As asserted and approved unanimously by participating states at *The Final Act of the Conference on Security and Co-operation in Europe*, 1975 (more simply known as the *Helsinki Final Act*), renewed in 1990 by the European Co-ordination Centre for Research and Documentation in Social Sciences) there is no justification for conflict. This document interdicted war as a solution to international conflict. War was never to be justified. Recourse to force in any form, or even just to a threatening attitude among states was not admitted, apart from whether those states had undersigned the accord or not.

Nonetheless, this document proved to be ineffectual, incapable of influencing international relations. Within a few years it was clear that the Helinski Conference with its idea of cooperation among states in Europe and across the world for the sake of peace, safety and security had failed. A major flaw in the Helsinki document is that despite good intentions, its discourse was grounded in identity logic (Petrilli and Ponzio 2005: 491–499; 2017: 159–176). By 2002 the *Helsinki Final Act* was supplanted by another document, *The National Security Strategy of the United States of America*, issued by the White House for the sake of national defence. This document introduces the concept of war as "just and necessary". Thereafter recourse to war was legitimated at a global level and considered instrumental for the resolution of conflict, whether national or international, and justified as "preventive war", as "humanitarian war", as "war on terror", ultimately as the means to maintaining "World Order".

Identity in today's world closes to the other, hiding beyond walls of fear and violence, and as such is inevitably orchestrated to generate conflict. But identity can also be conceived as *open identity* (Morris 1949), as non-indifference towards the other, before and after identity and its reduction to the empirical world, the world "at hand", to borrow the expression from Ronald Arnett (2017a).

A slogan we could launch for liberation from the obscenity of identity is the following: *not signs of difference as signs that make difference, but signs that make a difference*. Otherness-difference, that is, difference regulated by the logic of otherness, is unindifferent difference, difference unindifferent to the other. Difference regulated by the logic of otherness involves dialogical deferral among signs, continuous responsive *renvoi* from one sign to the next, beyond the limits of identity, understood as closed identity. Thus interpreted identity implies difference that is unindifferent to the other, unindifferent difference, *difference* or *différance* to evoke the language of our contemporary French philosopher, Jacques Derrida.

The sign is in becoming in the movement of *renvoi* and deferral among signs, according to the condition of difference thus described; the sign lives on signs and among signs. The sign is in becoming as it shifts across signs, is structured and constructed in the relation among signs, in the space among signs. In this sense the sign is *in translation*, in other words, the sign as such inevitably involves deferral to another sign.

Open identity: in other words, identity outside closed, short-sighted identity and outside the exclusiveness of belonging. With respect to the ontology of representation, being and immanence, open identity is out of place, out of order. Not only is such a shift necessary, but also urgent. Closed identity demands indifference towards the other, homologation of singularity, of the uniqueness of each one, thus it demands elimination of the other, negation of the other. Yet singularity, uniqueness can only be achieved in the relation with the other and for the other. And given that the other is witness to the actions of the identical, the subject, the self, only in the relation to the other can we aspire to the *properly human*.

7 Singularity as the dialogical space of responsible action

Extralocalization, exotopy are terms that translate the Russian expression *vnenakhodimost'*, to find oneself on the outside, to place oneself outside in a way that is unique, absolutely other, uncomparable, singular. "*Vnenakhodimost'*" is a basic concept in the thought of the Russian philosopher Mikhail M. Bakhtin (Orël 1875–Mosca 1975), in his aesthetic vision, present throughout all his works, from his very first paper, an essaylet of 1919, "Art and answerability", to that on the human sciences of 1974, "Toward a Methodology for the Human Sciences".

The concept of *vnenakhodimost'* enters Bakhtin's grand project for a "moral philosophy" as "first philosophy," which he had already elaborated during the 1920s. However his papers *à propos*, which at the time were never completed, were only published posthumously, as late as 1986 (more than ten years after his death) under the title "K filosofii postupka". "*Postupok*" means "act" and contains the root "*stup*" which means "step," therefore "act as a step," as an initiative, a move. As signifying "to take a step," "*postupok*" recalls another of Bakhtin's expressions from the 1920s, beginning from "Author and hero in aesthetic activity" (now in Bakhtin 1990: 4–256), "transgredient", which too implies "to take a step", a step beyond, outside reciprocity, outside equal exchange, outside *do ut des* logic, outside symmetry, synchrony, sameness, identification. In the architectonics of Bakhtinian thought, the word "transgredience" like "*postupok*" is central in outlining the concept of *vnenakhodimost'*.

Bakhtin estabilishes a relation of "complicity" between *moral philosophy as first philosophy,* understood as *philosophy of responsible* action, and *philosophy of verbal art,* that is to say philosophy of literary language. He refers this problematic specifically to artistic activity already in his first writing of 1919, subsequently elaborated in the "K filosofii postupka" in the 1920s (v. A. Ponzio's introduction to Bachtin e il suo circolo, *Opere 1919–1930*, 2014). Philosophy of responsible action and philosophy of the artistic text and specifically the literary text are connected in a relation of reciprocal implication. This connection is at the basis of all the fundamental concepts developed by Bakhtin during the course of his research, from these early writings through to the second half of the 1970s: in addition to "exotopy" or "extralocalization" (*vnenakodimost'*), "excess" (*izbytok*), "picturing" (*izobraženie*), "responsibility," "witness-judge," "dialogue".

The moral orientation and the artistic encounter each other in their common opening to alterity, singularity, to the uniqueness of each one. In fact, the expression "*Edinstvennyi*", which means singular, unique, unrepeatable, exceptional, uncomparable, *sui generis*, corresponding to the German *einzig*, is central in Bakhtin's philosophical discourse. But singularity interests Bakhtin not in the sense of individualism, of egocentric identity and righteous self-esteem. Instead, he develops the concept of singularity in the sense of opening to the other, the other of self and the other from self, listening to the other, responding to the other in the context of a semiosic network built on relations of dialogic intercorporeity interconnectedly with life throughout the whole universe.

Bakhtin, as he recounts himself, read Søren Kierkegaard (in German) (see Bakhtin 2008), who at the time was unknown in Russia. Consequently, Bakhtin was familiar with Kierkegaard's conception of the "single", of "singularity," which Bakhtin relates to Dostoevskij, a writer he loved as did Emmanuel Levinas. Bakhtin and Levinas shared a love of literature, literary writing, and both recognized in Dostoevskij the literary writer – whose artistic vision portrays the essential unity between responsible action and singularity – a master in "moral art". Developing Dostoevsky, Bakhtin too envisages a relation of reciprocal implication in the relation between art and answerability, art and umlimited responsibility at high degrees of dialogism (see Petrilli 2020a).

For Levinas as well responsibility is unalienable, and unlimited responsibility is inalienable in keeping with the uniqueness of the single individual, beyond the limits of the individual the representative a group or assemblage of some sort. I am responsible for the other, my neighbour without that other being responsible for me. In this sense responsibility is election beyond identification with the assemblage, the genre, a principle for individuation. Levinas supports the idea of individuation of man by means of responsibility for others. For Levinas I am responsible for the other more than anybody else, and like Bakhtin Levinas too recalls Dostoevskij author of the novel *Brothers Karamazov* in which one of the characters says: "We are guilty for everything and for everybody, and me more than anybody else". From the very beginning encounter with the face of the other is in the asymmetry of intersubjectivity, of responsibility for him (see Levinas 1991: 139–143).

From his early writings Bakhtin's problem was to evidence the profound connection between two worlds, the world of life and the world of culture. In effect, the cultural world – to know, dialogue, contemplate, create, take responsibility, a standpoint – is achieved in the general context of life, and in this sense the cultural world continues to be a vital world. We are in the life-world when we construct the world in which we objectify our life and identify it with a *given* sector of culture (see Ponzio, "Introduzione," in Bachtin e il suo Circolo 2014: 4). The question Bakhtin asks is what is it that unites these worlds?, and he responds that the connection is given by the act ("*postupok*"), the act as a unique event in which are decided the choices of each one, my choice as a singularity. In other words, the connection between culture and life is given in the responsible act.

Already in "K filosofii postupka" Bakhtin distinguishes between "special responsibility" and "moral responsibility". In the responsible act that unites the world of life with the world of culture, choices and standpoints are oriented according to a double responsibility: "special responsibility" or "technical responsibility" and "moral responsibility" or "absolute responsibility". *Special or technical responsibility* is relative to the objective unity of a sector of culture and therefore to a given role, a certain function. As such it is relative to the identity of the individual, that is to repeatable, replaceable, interchangeable identity. Instead, *moral responsibility, absolute responsibility* knows no limits, nor guarantees, nor alibis. Special responsibility responds to the limits of identity connected with social roles, with the abstract individual member of an assemblage, this type of responsibility can be deferred to the other, but is altogether indifferent to the other outside social role.

Indifference to the other, exclusion of the other, conflictual relations with the other are justified in the name of special or technical responsibility. Special responsibility justifies identity with its short-sighted egocentrisms. Instead, moral responsibility, absolute responsibility is what makes the action of each one unique, unrepeatable, *non deferrable*. It is not the responsibility of the individual and of alterity relative to the latter, but responsibility of uniqueness, singularity, absolute alterity that overcomes the boundaries of the identical.

Bakhtin establishes a relation between these two types of responsibility and two types of meaning: "meaning" differentiated as *identical, repeatable, objective meaning*, which corresponds to the sector of culture in which the act is objectified; and "actual sense," "theme" or "significance" which is the *unrepeatable meaning* of the act, the act as a unitary, unique and unclassifiable event. *What unites culture and life, cultural consciousness and singular consciousness is the non-indifference of the responsible act*. Detached from responsibility, cultural, cognitive, scientific, aesthetic, political vales rise to values in themselves and lose all possibility of verification, sense, transformation.

Evoking Levinas when he distinguishes between "relative alterity" and "absolute alterity" and Morris when he discusses the "open self" as opposed to the "closed self", we will distinguish between "relative identity" and "absolute identity", between "closed identity" and "open identity". Identity, difference recognized on an official, public level is associated with assemblative logic, identity of a set, genre, class, concept. Thus conceived identity is indifferent to singularity, uniqueness, to the unrepeatability of each one. This type of identity, of difference generally functions on the basis of a code, by opposition between one identity and another. It takes shape relatively to another identity, another difference representative of another group, as the condition for its very own identification. This is difference indifferent to the difference of absolute otherness and singularity (Ponzio 1994), identity-difference by contrast to otherness-difference.

Official socio-cultural relations, those recognized with a juridical status, are relations between identities representing a group of some sort, between differences distributed in binary relations regulated by secondness. Assemblative identity of the genre, class, concept emerges in the oppositional relation with another identity, another difference, another relative alterity. To subsist as identity thus conceived, on the one hand presupposes singularity, on the other imposes indifference. All this creates the conditions for the constitution of binary, oppostive relations of identity, where the problem is not binarism in itself, binarism as opposed to triadism, but the lack of dialogue in relations dominated by confrontation and conflict.

Even when we profess tolerance towards the other (Pasolini), this is the tolerance of genre, of the abstract other, whose alterity is relative to a group of some sort, whose identity is shaped in terms of belonging or not belonging. The other who tolerates calls for the tolerated other. Thus the relation between "tolerator" and "tolerated" is one of opposition, where the "tolerated" is easily perceived as undesirable and cumbersome. All communities, identities, differences are constituted on the basis of indifference towards the other. Consequently, they each generate their own "extracommunitarian" whom in the best of cases is tolerated. By contrast to the community of citizens, the comunitarians, on the one hand, migrants, illegals, asylum seekers, foreigners, the extra-comnunitarians, on the other; by contrast to the community of workers, on the one hand, the unemployed, the parasite, exploiter, on the other; by contrast to white skin, black skin; by contrast to the Christian the Muslim; by contrast to the healthy, "normal," "able" members of a community, the disabled, the abnormal, the unable. Without dialogue, substantial dialogism among elements, the step to open conflict is short.

8 Semioethics of love and listening

Aesthetic value, in turn, increases the sign's autonomy with respect to pre-established and immediately productive ends, privileging, instead, reflection, imagination, Peirce's "play of musement," process, projectuality, inventiveness, and together the capacity for critique, for response, *responsive understanding*, and of responsibility for the other, the other in me as much as from me. And as excess and escape from the constrictions of identity – closed identity, egocentric identity –, from the obligations of social role, the abductive capacity involves what evoking Peirce we may call the agapastic capacity (*CP* 6.302).

Agape is the capacity to recognize in relations of reciprocal implication, in the inexorable condition of intercorporeal interconnection, of intercorporeity, *the singularity* of each one of us. Consequently, agape involves elevation and transcendence (not in the metaphysical sense, but earthly transcendence, semio-material transcendence, the transcendence of humanism oriented by otherness) with respect to the limits of the identical, of the same; and it involves liberation from the pretentious claim to "reciprocity" and "symmetry" between parts, to equal exchange among signs, meanings and linguistic values, as posited by code semiotics (see Saussure 1916; for a critique in this sense, see Ponzio 1981: 95ff.).

Agape, love, is resistance and dissent in front of the attempt at reducing singularity to identity, the each of each and every one of us to the code, to convention, to social roles, social programs and their distorted, mystifying ideologies. In fact, singularity is associated with non-functionality and what has been specified as the "right to non-functionality" (see Ponzio 1997), described as a fundamental human right. Non-fuctionality is the right to life which is not reduced to mechanical schemes, statistics and calculations, to a life which is not reduced to mere subsistence levels, to relations of self-interest, to relations governed by the rules of official social programming, today in terms of efficiency, productivity, competitiveness.

Love is responsibility towards the otherness of the other (*autrui*), outside the schemes of functionality, unindifference towards the other as other, outside the schemes of self-interest, hospitality, and listening to the other, encounter with the other – even the "unalike," the "odd" other, the "different" other, the "distant" other, this other whom in a globalized world is getting closer and closer, is ever more "my neighbour". As such love is the capacity to respond to the other as other, to account to the other and for the other, and not only for one's own self (see Petrilli 2020b).

Singularity calls for *unlimited responsibility* as thematized by Levinas, but also by Bakhtin, responsibility without alibis or justifications in support of identity, that is, short-sighted and egocentric identity. With Levinas love, love as unindifference to the other, as implication with the other, as hospitality towards the other is the original relation. In other words, the originating condition of the human, the original consciousness of the human is love (see *Philosophy, Justice Love*, 1982, in Levinas 1991, 137–156).

Contrary to prejudices and limits implied by perspectives that are ultimately anthropocentric, ethnocentric, glottocentric, our humanity (humane humanity) is determined in responsibility towards the other, by a capacity for responsibility/responsivity that is unlimited, unconditional. Such an approach, which is founded semiotically, scientifically, evidences how the life of each one of us, how life in its singularity is effectively implicated not only in the life – in the sense of the quality of life as well – of other human beings, but in all life-forms over the planet. We are all, each as a singularity, implicated, inexorably interconnected in the great global semiosic network. So that dialogicality, more than a kind concession towards the other, is inevitable exposition to the other, involvement with the other, implication in the relation of intercorporeity with the other, impossibility of closing to the other, of withdrawing from one's implication, of each and every one of use, in the destiny of the other, and all this notwithstanding any efforts to the contrary, impossibility of barricading oneself behind walls of indifference in relation to the other, precisely.

The human individual is conditioned semiotically, that is to say beyond organic and biological terms, in historical and social terms as well. As a "semiotic animal" capable not only of communicating with signs, but of distancing ourselves from signs and of reflecting upon them, we are called to respond to and to account not only for our own self, but for the other. And the more we are conscious of the dialogical nature of our own self, of the condition of inevitable implication with the other, of the ineludibility of involvement with the other, as foreseen and prescribed by the sign, by semiosis, by life that converges with semiosis, the greater is our responsibility for the other, a responsibility that is "unlimited", "absolute", in relationships without alibis, without parapets, without justifications.

Dialogue and encounter is no less than prescribed by the sign nature of cultures and languages. Dante in the XIII century, as a poet and writer, knew this and at the time, through the extralocalized gaze of the literary work, had already perspected the solution to the problems that afflict the world today, the only one possible for the health of humanity: encounter, dialogic encounter. There is no other solution. War generates war, violence generates violence, conflict generates conflict. And on creating one of the most extraordinary artistic chronotopes of all times – as testified by the ongoing multiplication of interpretants that *The Divine Comedy* as an artwork has continued to generate across the centuries – Dante's cosmic vision depicts exactly this, salvation for life in the journey of love for love, among unique human beings, among singularities: Dante and Virgil, Dante and Beatrice, towards God, a multiethnic, multicultural and plurilinguistic God without bounds.

The Divine Comedy is the expression of man's desire to cross over boundaries in the search for the other. It's magnificence metonimized in the splendour of Beatrice's eyes and in her smile, inspire and elevate the poet carrying him well beyond the petty boundaries of identity in the boundlessness of the chronotope of absolute, unlimited love. The unique, singular act, Bakhtin's "step" (*postupok*, act as a step), his *responsible action* is here depicted in the "*transumanar*" (transhumanizing) of a journey guided, first in the dialogue with Virgil and after in the flight with Beatrice towards absolute light.

In the system of values of the Christian world light is associated with knowledge, therefore with God as the highest knowledge and as love, hence light is associated with charity as the ultimate end of knowledge itself which reaches its apex, its maximum expression and signifying potential in divine love, a divine love of this world, with its gaze turned to the human.

To evoke the words of Bakhtin in which the aesthetic, philosophical and theological visions intersect, united in dialogue, unindifference, responsibility and love:

> . . . the center of value in the event-architectonic of aesthetic seeing is man as a lovingly affirmed concrete actuality, and not as a something with self-identical content. Moreover, aesthetic seeing does not abstract in any way from the possible standpoints of various values; it does not erase the boundary between good and evil, beauty and ugliness, truth and falsehood. Aesthetic seeing knows all these distinctions and finds them in the world contemplated [. . .] they remain within that world as constituent moments of its

architectonic and are all equally encompassed by an all-accepting loving affirmation of the human being. [. . .]

In this sense one could speak of objective aesthetic love as constituting the principle of aesthetic seeing [. . .] The valued manifoldness of existence as human [. . .] can present itself only to a loving contemplation. Only love is capable of holding and making fast all this multiformity and diversity [. . .] Only un-self-interested love on the principle of "I love him not because he is good, but he is good because I love him," only lovingly interested attention, is capable of generating a sufficiently intent power to encompass and retain the concrete manifoldness of existence, without impoverishing and schematizing it. [. . .]

Lovelessness, indifference, will never be able to generate sufficient power to slow down and linger intently over an object, to hold and sculpt every detail and particular in it, however minute. Only love is capable of being aesthetically productive; only in correlation with the loved is fullness of the manifold possible. (Bakhtin 1993: 63–64)

References

Arnett, Ronald C. 2013. *Communication Ethics in Dark Times*. Carbondale: Southern Illinois University Press.
Arnett, Ronald C. 2017a. *Levinas's Rethorical Demand. The Unending Obligation of Communication Ethics*. Carbondale: Southern Illinois University Press.
Arnett, Ronald C. 2017b. Communication Ethics. The phenomenological sense of semioethics, *Language and Dialogue* 7(1). 80–99.
Bachtin e il suo Circolo. 2014. *Opere 1919–1930,* ed., commented and intro., vii–xlvii, by Augusto Ponzio, bilingual Russian/ Italian edition, trans. (in collab. Luciano Ponzio). Milan: Bompiani.
Bakhtin, Mikhail M. 1919. Iskusstvo i otvetstvennost. *Den" iskusstva*, Nevel," 13 September 3–4. In M. M. Bakhtin, 1979, It. trans. & Russian original now in Bachtin e il suo circolo 2014, 27–31; Eng. trans. Art and Answerability. In M. M. Bakhtin 1990, 1–3.
Bakhtin, Mikhail M. 1974. Basi filosofiche delle scienze umane, N. Marcialis (trans.). *Scienze umane* 4, 1980. 8–16. [Original Russian in *Kontext*, Vadim V. Kozinov (ed.). Moscow: Nauka, 1974, 375–377.].
Bakhtin, Mikhail M. 1979. *Estetica slovesnogo tvorčestva* (Aesthetics of verbal art). Moscow: Iskusstovo; *L'autore e l'eroe. Teoria letteraria e scienze umane*, *1988*, Clara Strada Janovič (trans.). Turin: Einaudi.
Bakhtin, Mikhail M. 1990. *Art and Answerability. Early Philosophical Essays by M. M. Bakhtin*, edited by Michael Holquist & Vadim Liapunov, Eng. trans. & notes by Vadim Liapunov, supplementary translation by Kennet Brostrom. Austin: Austin University of Texas Press.
Bakhtin, Mikhail M. 1993 [1920–1924]. *Toward a Philosophy of the Act*, English translation by V. Liapunov, edited by M. Holquist. Austin: Austin University of Texas Press.
Bakhtin, Mikhail M. 2008. *In dialogo. Conversazioni del 1973 con V. Duvakin*, R. S. Cassotti (trans.), A. Ponzio (ed.). Naples: Edizioni Scientifiche Italiane. [Russian original, *Besedy V. D. Duvakina s M.M. Bachtinym 1973*. Moscow: Soglasie 1st. ed. 1996, new ed. 2002].

Barthes, Roland. 1980. *La chambre claire. Note sur la photographie*, Paris, Seuil; [It. trans. *La camera chiara, 1982*. Turin: Einaudi.].
Barthes, Roland. 1982. *L'obvie et l'obtus. Essais critiques III*. Paris: Éditions du Seuil.
Barthes, Roland. 1984. *Le bruissement de la langue* [*The Rustle of Language, 1989*, by Richard Howard. Berkeley, Los Angeles: The University of California Press]. Paris: Editions du Seuil.
Barthes, Roland. 2002. *Le Neutre, Cours et séminaires au Collège de France* (1977–78), ed. by T. Clerc. Paris: Seuil.
Benjamin, Walter. 1980. *Il dramma barocco Tedesco*. Turin: Einaudi.
Benveniste, Émile, et *alii*. 1968. *I problemi attuali della linguistica*. Milan: Bompiani.
Cassirer, Ernst. 1965 [1923]. *Filosofia delle forme simboliche*. Florence: La Nuova Italia.
Chomsky, Noam. 1985. *Knowledge of Language*, It. trans. G. Longobardi & M. Piattelli-Palmarini, *La conoscenza del linguaggio*, Milan, Il Saggiatore.
Danesi, Marcel. 2000. *Lingua, metafora, concetto. Vico e la linguistica cognitiva*. Bari: Edizioni dal Sud.
DeMauro, Tullio. 1994. *Capire le parole*. Rome-Bari: Laterza.
Derrida, Jacques. 1967. *De la grammatologie*. Milan: Jaca Book, 1969.
Derrida, Jacques. 1967. *L'écriture et la différence*. Paris: Seuil; Eng. trans. A. Bass, *Writing and Difference*, Chicago: The University of Chicago Press, 1978.
Dumarsais, César Chesneau. 1977 [1730]. *Traité des Tropes*. Paris: Le Nouveau Commerce.
Fano Giorgio. 1972. *Origini e natura del linguaggio*. Turin: Einaudi; Eng. trans., ed. and intro. Susan Petrilli, *Origins and Nature of Language*. Bloomington: Indiana University Press.
Jakobson, Roman. 1966. *Saggi di linguistica generale*, ed. L. Heilmann and L. Grassi, Milan: Feltrinelli.
Jakobson, Roman. 1968. Alla ricerca dell'essenza del linguaggio. In Émile Benveniste et alii, *I problemi attuali della linguistica*, 1968, 27–46. Milan: Bompiani.
Kristeva, Julia. 1982. *Le langage, cet inconnu*, Paris, Seuil: It. trans. A. Ponzio, *Il linguaggio, questo sconosciuto, 1992*, and an interview by A. Ponzio with J. Kristeva. Bari: Adriatica.
Levinas, Emmanuel. 1974. *Autrement qu'être ou au-dela de l'essence*. The Hague: Nijhoff; Eng. trans. A. Lingis, *Otherwise than Being or Beyond Essence*. Pittsburgh: Duquesne University Press, 2000; It. trans. by S. Petrosino and M. T. Aiello, intro. by S. Petrosino, *Altrimenti che essere o al di là dell'essenza, 1990*. Milan: Jaca Book.
Levinas, Emmanuel. 1982. *L'Au-delà du Verset*. Paris: Les Edition de Minuit; Eng. trans. Gary D. Mole, *Beyond the verset, 1994*. Bloomington: Indiana University Press.
Levinas, Emmanuel. 1991. *Entre nous. Essais sur le penser à l'autre*. Paris: Grasset; *Entre nous. On Thinking-of-the-Other, 1998*, Eng. trans. by M. B. Smith & B. Harshav. London: The Athlone Press.
Liberman, Ph. 1975. *On the Origins of Language*. New York: Macmillan.
Lo Piparo, Franco. 2003. *Aristotele e il linguaggio*. Rome-Bari: Laterza.
Lotman, Jurij. 1975. *La semiosfera*. ed. S. Salvestroni. Venice: Marsilio.
Martinet, André. 1965. *La considerazione funzionale del linguaggio*. Bologna: Il Mulino.
Martinet, André. 1967. *Elementi di linguistica generale*. Rome-Bari: Laterza.
Martinet, André. 1985. *Sintassi generale*. Rome-Bari: Laterza.
Morris, Charles. 1938. *Foundations of the Theory of Signs*. In *International Encyclopedia of Unified Science* I (2). Chicago: University of Chicago Press; It. trans. by F. Rossi-Landi. *Lineamenti di una teoria dei segni*. Turin Paravia, 1954. New ed. by S. Petrilli, Lecce: Manni, 1999; and Lecce: Pensa Multimedia, 2009.

Morris, Charles. 1948a. *The Open Self*. New York: Prentice-Hall; *L'io aperto. Semiotica del soggetto e delle sue metamorfosi*, It. trans. & Intro., "Charles Morris e la scienza dell'uomo.Conoscenza, libertà, responsabilità," vii–xxvi, by S. Petrilli. Bari: Graphis, 2002; new revised Italian edition Lecce: Pensa Multimedia, 2017.

Peirce, Charles S. 1923. *Chance, Love and Logic*, ed. and intro. by Morris R. Cohen. New York: Harcourt.

Peirce, Charles S. 1931–1958 *Collected Papers of Charles Sanders Peirce* (i 1866–1913), Vol. I–VI, ed. by C. Hartshorne & P. Weiss, 1931–1935, Vol. VII–VIII, ed. by A. W. Burks, 1958. Cambridge, MA: The Belknap Press of Harvard University Press. [In the text referred to as CP followed by volume and paragraph number.] All eight vols. in electronic form ed. John Deely. Charlottesville, VA: Intelex Corporation, 1994. Dating within the *CP* is based on the Burks Bibliography at the end of *CP* 8. The abbreviation followed by volume and paragraph numbers with a period between follows the standard *CP* reference form.

Petrilli, Susan. 2020a. Beyond Signs of Identity as Justification for Conflict. A Semioethic Approach. *Listening: Journal of Communication Ethics, Religion, and Culture*, Vol. 54, issue 3, 92–143, ed. by Susan Mancino, Spring, 2020. Department of Communication and Rhetorical Studies, Dusquene University, Pittsburgh, USA.

Petrilli, Susan. 2020b. At the Margins of Speaking of Love with Roland Barthes and Mikhail Bakhtin. In *MARGINALIA, Acta Translatologica Helsingiensia* (ATH), vol. 4, 21–57, 2020. Ritva Hartama-Heinonen & Pirjo Kukkonen (eds.). Nordica/Swedish Translation Studies, Department of Finnish, Finno-Ugrian and Scandinavian Studies. Helsinki: University of Helsinki.

Petrilli, Susan; Ponzio Augusto. 2005. *Semiotics Unbounded. Interpretive Routes through the Open Network of Signs*. Toronto: University of Toronto Press.

Petrilli, Susan & Ponzio, Augusto. 2006. *La raffigurazione letteraria*. Milan: Mimesis.

Petrilli, Susan & Ponzio, Augusto. 2016. *Lineamenti di semiotica e di filosofia del linguaggio. Un contributo all'interpretazione del segno e all'ascolto della parola*. Perugia: Guerra Edizioni.

Petrilli, Susan; Ponzio, Augusto. 2017. Building Nations and International Relations. The Peace of War. In Susan Petrilli (ed.), *Challenges to Living Together. Transculturalism, Migration, Exploitation. For a Semioethics of Human Relations*, 131–177. Milan: Mimesis.

Petrilli, Susan; Ponzio, Augusto. 2019a. *Identità e alterità. Per una semioetica della comunicazione globale*. Milan: Mimesis.

Petrilli, Susan; Ponzio, Augusto. 2019b. *Dizionario, Enciclopedia, Traduzione. Fr César Chesneau Dumarsais e Umberto Eco*. Fasano: AGA; Paris: L'Harmattan.

Ponzio, Augusto. 1981. *Segni e contraddizioni. Tra Marx e Bachtin*. Verona: Bertani.

Ponzio, Augusto. 1992. *Production linguistique et idéologie sociale*. Québec, Canada: Les Editions Balzac. [French Translation of the Italian original, *Produzione linguistica e ideologiasociale*. Bari: De Donato, 1973].

Ponzio, Augusto. 1994. *La differenza non indifferente. Comunicazione, migrazione, guerra*. Milan: Mimesis; new enlarged edition 2002.

Ponzio, Augusto. 1997 *Elogio dell'infunzionale. Critica dell'ideologia della produttività*. Rome: Castelvecchi; new enlarged ed. Milan: Mimesis, 2004.

Ponzio, Augusto. 2004. *Semiotica e dialettica*. Bari: Edizioni dal Sud.

Richards, Ivor A. 1936. *The Philosophy of Rhetoric*. London-Oxford-New York: Oxford University Press.

Rossi-Landi, Ferruccio. 1985. *Metodica filosofica e scienza dei segni*. Milan: Bompiani.

Rousseau, Jean Jacques. 1989. *Saggio sull'origine delle lingue* [*Essai sur l'origine des langues*, 1781], It. trans. P. Bora. Turin: Einaudi.

Saussure, Ferdinand de. 1916. *Cours de linguistique générale*, ed. by C. Bally and A. Secheaye. Paris: Payot, 1922; Critical edition by R. Engler. Wiesbaden: Otto Harrassowitz, 4 Vol., 1968–1974; *Course in General Linguistics*, Eng. trans. by W. Baskin, Intro. by J. Culler. London: Peter Owen, 1959, 1974; & by R. Harris. London: Duckworth, 1983; *Corso di linguistica generale*, Intro., It. trans. & comment by Tullio De Mauro. Rome: Latera, 1967, 24th ed. 2011.

Sebeok, Thomas A. 1986a. *I Think I Am a Verb: More Contributions to the Doctrine of Signs*. New York and London: Plenum Press; *Penso di essere un verbo*, It. trans., ed. & introd., 11–18, by S. Petrilli. Palermo: Sellerio, 1990.

Sebeok, Thomas A. 2000. Some Reflections of Vico in Semiotics. In D. G. Lockwood, P.H. Fries, J.E. Copeland (eds.), *Fuctional Approaches to Language, Culture and Cognition*, 555–568. Amsterdam: John Benjamins.

Sebeok, A. Thomas & Marcel Danesi. 2000. *The Forms of Meanings. Modeling Systems Theory and Semiotic Analysis*. Berlin: Mouton de Gruyer.

Sebeok, Thomas A., Petrilli, Susan; Ponzio, Augusto. 2001. *Semiotica dell'io*. Rome: Meltemi.

Semerari, Giuseppe. 1969. *Esperienze del pensiero moderno*. Urbino: Argalia.

Vailati, Giovanni. 1971. *Epistolario 1891–1909*, ed. by G. Lanaro, intro. M. Dal Pra. Turin: Einaudi.

Vailati, Giovanni. 2000. *Il metodo della filosofia. Saggi di critica del linguaggio*, ed. F. Rossi-Landi, new ed. A.Ponzio. Bari: Graphis.

Vico, Giambattista. 1999. *Scienza nuova*, in *Opere*, vol. 1, 2. ed. A. Battistini. Milan: Mondadori. [published as *Principi di scienza nuova*, ed. F. Nicolini, Classici Ricciardi, 1953; ried. Turin: Einaudi, 1976.

Shakespeare, William. 2003. *Il mercante di Venezia*, bilingual text, It. trans. by A. Serpieri, Milan, Garzanti.

Vernadsky, Victor I. 1926. *Biosfera*. Leningrad: Nauka. [French version, Paris, 1929.].

Welby, Victoria. 1893. Meaning and Metaphor. *The Monist* 3(4): 510–525. Now in V. Welby 1985a. It. trans. in V. Welby 1985b. 79–107.

Welby, Victoria. 1896. Sense, Meaning and Interpretation. *Mind*. 5 (17). 24–37; 5 (18).186–202. Now in V. Welby 1985a. It. trans. in V. Welby 1985b. 109–170.

Welby, Victoria. 1903. *What Is Meaning? Studies in the Development of Significance*, London: Macmillan [new ed. with addition of essays by others, see Welby 1983].

Welby, Victoria. 1911a. *Significs and Language. The Articulate Form of Our Expressive and Interpretative Resources*, London: Macmillan [new ed. now included in the volume Significs and Language, Welby 1985].

Welby, Victoria. 1983 [1903] *What Is Meaning? Studies in the Development of Significance*, ed. and Preface by A. Eschbach, ix–xxxii, Intro. by G. Mannoury, xxxiv–xlii [= Foundations of Semiotics 2.] Amsterdam/Philadelphia: John Benjamins.

Welby, Victoria. 1985a. *Significs and Language*, ed. and Intro. by H. W. Schmitz, ix–ccxxxvii. Foundations of Semiotics 5. Amsterdam-Philadelphia: John Benjamins. [Includes Welby's monograph of 1911, *Significs and Laguage* and a selection of other writings by her.].

Welby, Victoria. 1985b. *Significato, metafora, interpretazione*, It. trans., ed., and Intro. By S. Petrilli. Bari: Adriatica.

Welby, Victoria. 2007. *Senso, significato, significatività*, ed., It. trans., & intro. by S. Petrilli. Bari: Graphis.

Welby, Victoria. 2010. *Interpretare, comprendere, comunicare*, It. trans., ed. & Intro., 11–96, by S. Petrilli. Rome: Carocci.

Susan Petrilli
Global semiotics and its developments in the direction of semioethics

Abstract: Following an ideal tradition in sign studies that leads from Hippocrates and Galen through John Locke to Charles S. Peirce and Charles Morris, Thomas Sebeok with *global semiotics* reaches the broadest vision ever concerning what to regard as semiosis or communication: communication or semiosis coincides with life. Where there is life, there are signs, where there is life there is communication, semiosis. And human semiosis? Unlike Saussure's pioneering *sémiologie*, in global semiotics, human semiosis is "semiotics" in a special sense. "Semiotics" not only names the general science of signs, but also *properly human semiosis*. All living beings indifferently are semiosical animals. Instead, *anthropos* is a *semiotic animal*, which means to say that humans not only use signs, but they reflect on signs, are endowed with metasemiosis, which invests them with a capacity for conscious awareness and responsibility. But throughout history limitations have been placed on responsibility on the basis of distinctions and separations based on the logic of "identity" and "belonging". *Semioethics* evidences how separatism is unwarranted and inevitably leads to conflict. Reconnecting with ancient *medical semeiotics* and its vocation for health and listening (auscultation), semioethics proposes to interpret symptoms of social disease, through listening and tuning into otherness, and thereby contribute to the health of semiosis, for the sake of the quality of life globally.

Keywords: global semiotics, language, listening, medical semeiotics, metasemiosis, responsibility, semioethics, semiotic animal

---

Note: This text was originally drafted as a keynote lecture for delivery at the International Conference "Semiosis and Communication. Differences and Similarities", Southeast European Center for Semiotic Studies, Bucharest, 14–16 June 2018. It is divided into the following sections: 1. Semiotics, semeiotics, global semiotics; 2. Enter semioethics; 3. The semiotic animal, a semioethic animal; 4. Critical global semiotics. Caring and responsibility as the secret of sociality; 5. Global communication as global listening. An ethical demand.

Susan Petrilli, University of Bari "Aldo Moro", Italy

https://doi.org/10.1515/9783110662900-004

1 Semiotics, semeiotics, global semiotics

Biology and the social sciences, ethology and linguistics, psychology and the health sciences, their internal specializations – from genetics to medical semeiotics (symptomatology), psychoanalysis, gerontology and immunology – all find in "global semiotics," as conceived by Thomas A. Sebeok (2001), the place of encounter and reciprocal exchange, as well as systematization and unification. Important to note, however, is that "systematization" and "unification" are not understood neopositivistically, in the static terms of an "encyclopedia" (whether a question of juxtaposing knowledge and linguistic practices, or of reducing knowledge to a single scientific field and its relative languages, as in neopositivistic physicalism).

Global semiotics is a metascience concerned with all academic disciplines insofar as they are sign-related. It cannot be reduced to the status of "philosophy of (ideoscopic) science," though of course as a *cenoscopic* science[1] it is dialogically engaged with – is indeed intrinsic to – philosophy. Global semiotics unites what other fields of knowledge and human praxis generally keep apart either for justified needs of a specialized order, or because of a useless and even harmful tendency towards short-sighted sectorialization (which is not free of ideological implications, most often poorly masked by motivations alleged to be of a scientific order).

Instead, the continuous and creative shift in perspective that the global approach to semiosis makes possible favours the identification of new interdisciplinary relationships and new interpretive practices, as foreseen by Charles Morris among others. Sign relations are identified where it was thought there were none: that is, where no more than mere "facts" and relations among things had been identified, as if independently from communication and interpretive processes. Moreover, this continuous shift in perspective favours the discovery of new cognitive fields and languages which interact dialogically – in truth a question of dialogical relations among signs that already exist and call for recognition. It is not just a question of building bridges *ex novo*, but of recognizing in the interconnectedness, the dialogical intercorporeality that is already present and structural to the existent. Characterized by the capacity to explore the boundaries and margins

---

[1] Peirce 1908, see *CP* 8.342 and 8.343, from a letter to Lady Welby: "the cenoscopic studies (i.e., those studies which do not depend upon new special observations) of all signs remain one undivided science"; and "one of the first useful steps towards a science of semeiotic, or the cenoscopic science of signs, must be the accurate definition, or logical analysis, of the concepts of the science".

of an array of different sciences, by its capacity for opening to the other, semiotics has also been dubbed by Sebeok as the "doctrine of signs" (1976, 1986).

The expression "semiotics" refers to both the *specificity of human semiosis* and the *general science of signs*. Under the first meaning semiotics relates to the specific human capacity for *metasemiosis*. In the world of life that encompasses semiosis,[2] human semiosis is characterized as metasemiosis, that is, as the possiblity of reflecting on signs. We can approach signs as objects of interpretation indistinguishable from our responses to them. But we can also approach signs in such a way that we suspend our responses to them so that deliberation is possible.

Semiotics as metasemiosis is connected to responsibility: the human being, the only "semiotic animal" to exist, is the only animal capable of accounting for signs and sign behaviour. Consequently, the semiotic animal, the human animal is subject *to* responsibility and subject *of* responsibility. Under this aspect, the critical instance of the philosophy of language towards the science of signs consists concretely in not limiting its attention to the gnoseological-theoretistic aspect of semiosis, but of focusing on the pragmatic dimension as well, and on the well-being of semiosis, therefore of life, on *caring* for life, for the health of semiosis generally.

From this point of view general semiotics, which in its current configuration as "global semiotics"[3] posits that semiosis and life converge and consequently

---

2 Semiosis is the process, or relation, or situation, whereby something serves as a sign. The sign is inseparable from semiosis. For something to be a sign, something else must be present to it. This second thing is referred to as an interpretant. The interpretant itself is a sign and is thus connected to another interpretant, and so on in an open and infinite chain of interpretants (see Petrilli and Ponzio 2005: 1.3). All of this means that for every sign there is a semiosis. Every sign is a portion of its own semiosis and cannot be detached from it. This is similar to the relationship between a cell and the cell tissue it helps form. And every semiosis is in turn connected with other semioses. Signs are linked together in an infinite chain; for their part, semioses form something like a network. In the same way that the sign is a portion of semiosis, semiosis is a portion of the sign network.

3 Semiotics studies the different forms of semiosis, sign processes, sign activity forming the semiosphere. From the perspective of so-called "global semiotics" where *semiosis* is described as converging with *life* (in this sense global semiotics is "semiotics of life"), the *semiosphere* identifies with the *biosphere* (term coined in Russian by Vladimir Vernadskij in 1926) to emerge, therefore, as the *semiobiosphere*. The semiosphere thus extended is articulated into different subspheres which overlap and converge with the great kingdoms of life: the zoosphere whose material is zoosemiosis, the object of study of zoosemiotics; the phytosphere made of phytosemiosic processes studied by phytosemiotics; and the mycosphere whose mycosemiosic activities are the object of study of mycosemiotics. Semiosis in the human world, anthroposemiosis and its various articulations, enters the more general and inclusive sphere of zoosemiosis, it forms the anthroposphere and is studied by anthroposemiotics.

concerns all of life over the planet, recovers its relationship with ancient *medical semeiotics*. This is not only a question of the historical order, involving knowledge of the origins, but rather it concerns our approach to reality today in a globalized world where the implication of each one's destiny in the destiny of all of life, of all others over the planet is rendered manifest as never before. We have denominated this orientation, this special bend in the study of signs, "semioethics" (see Petrilli 2014; Petrilli and Ponzio 2003a, 2010).

2 Enter semioethics

Semioethics is a crucial part of the answer to the question regarding the future of semiotics, the destiny of semiosis, proposed by Thomas A. Sebeok in "Semiosis and Semiotics: What Lies in Their Future?" (in Sebeok 1991: 97–99). In fact, it evidences the responsibility of semiotics towards semiosis, thereby proposing that "global semiotics," which is founded in the general science of signs as conceived by Charles S. Peirce, now be further developed in this direction.

Semioethics is closely associated with the question of listening, where "listening" is also understood in the sense of *medical semeiotics*, auscultation. We must listen to the symptoms of today's globalized world and identify the various voices of social *malaise*, and thus counteract the race towards our own destruction.

Our future is the "future anterior of semiotics". We decide today for the future of semiotics, not only as a *science*, but also as a *human species-specific capacity for using signs to reflect on signs and making decisions as a consequence*. The problem is not only of a theoretical order, but inevitably involves semiotics as *semeiotics*, symptomatology, and as semioethics, thus the future of semiotics (indeed of life) is a problem of the cognitive, pragmatic and ethical orders.

But all this only concerns a part of global semiotics, that involving the world of "eukariots," leaving aside the enormous quantity of "prokariots" with which life arises on earth and continues to flourish and evolve to this very day. All this occurs thanks to an incredibly refined communication system which interconnects all life forms in a network that covers the entire planet. Prokariots are the object of study of a branch of general semiotics known as microsemiotics and specified as endosemiotic. As the expression itself explains endosemiotics focusses on semiosis and communication inside the organisms populating the great kingdoms. In addition to prokariots, endosemiotics studies intercellular communication in larger organisms including the genetic code, the immunitary system, the neuronal system, all communication systems which allow for the reproduction, maintainance and overall behaviour appropriate to a specific *Umwelt*.

Never has the present (as today, our own present) been so charged with responsibility towards the future, and so capable of putting the possibility itself of a future at risk. Today decides our tomorrow, today's decisions condition the life of signs and the signs of life, their future, the continuity of semiosis on the planet Earth. As a semiotic animal, the human being is the only animal responsible for semiosis, for life. And the person involved in the study of signs as a profession is even more responsible than any other.

Reformulating Terence's famous saying, "*homo sum: umani nihil a me alienum puto*", Roman Jakobson (1959) asserted that "*linguista sum: linguistici nihil a me alienum puto*". This commitment by the semiotician to all that is linguistic, indeed, to all that is sign material (not only relative to anthroposemiosis or more extensively to zoosemiosis, but to the whole semiobiosphere) is not only intended in a cognitive sense, but also in the ethical. Such a commitment involves concern for the other, not only in the sense of "to be concerned with . . . ." but also in the sense of "to be concerned for . . . ." "to care for . . . ". Viewed from such a perspective, concern for the other, care for the other imply a capacity for responsibility without limitations of belonging, proximity, or community. In truth, this capacity is not exclusive to the "linguist" or "semiotician". Developing Jakobson's intuition, we could claim that it is not as professional linguists or semioticians, but more significantly as human beings that no sign is "*a me alienum*". And leaving the first part of Terence's saying unmodified, "*homo sum*," we may now continue with the statement that insofar as we are human beings not only are we *semiosical* animals (like all other animals), but we are also *semiotic* animals. From this point of view humans are unique by comparison to the rest of the animal kingdom. The consequence is that nothing semiosical, including the biosphere and the evolutionary cosmos whence it sprang, "*a me alienum puto*". Paraphrasing Terence: "I deal with signs, so nothing in the life of signs is indifferent to me".

To summarize: the human being is a "semiotic animal" thus denominated to the extent that we are endowed with a capacity for semiotics understood as "metasemiosis," with the capacity for making decisions, taking a stand, intervening upon the course of semiosis, which implies that the human being is invested with a unique capacity for responsibility towards semiosis. From this point of view, the semiotic animal is also a "semioethic animal". The expression "semioethics", then, indicates a propensity in semiotics to recover its ancient vocation as "semeiotics" (or symptomatology) interested in symptoms and the quality of life. It is not intended as a discipline in its own right, but as an orientation in the study of signs developed in the framework of "global semiotics".

"Global semiotics developed in the direction of semioethics" describes a research itinerary that studies signs in relation to values. Though a constant focus in sign studies across the twentieth century (with such figure as Victoria Welby,

Charles Morris, Ferruccio Rossi-Landi and on the background others still like Mikhail Bakhtin and Charles Peirce), the relation between signs and values has not been a mainstream interest, but today, in a globalized world, this relation has become ever more urgent to address and foster.

A fundamental claim made by semioethics is that semiotics most not only describe and explain signs, but must also search for adequate methods of inquiry for the acquisition of knowledge as well as make proposals for human behaviour and social programming. As the general (cenoscopic) science of signs, semiotics should overcome parochial specialisms – that is, all forms of separatism among the sciences (see Perron, Sbrocchi, Colilli, Danesi, eds. 2000; Rossi-Landi 1968, 1972, 1992). The ethical aspect of semiotics is projectual and critiques human practice generally with reference to all aspects of life from the biological to the socio-cultural, paying attention to reconnect that which is generally considered to be separate. For an approach to semiotic studies intending to interrogate not only the sense of science, but the sense of life for humankind, the capacity for criticism, social awareness, and responsible behaviour are central issues. Developing Sebeok's standpoint and proceeding beyond him, semioethics evidences the ethical implications of global semiotics and their importance for communication and life overall (see Cobley 2010b; Petrilli 2014).

3 The semiotic animal, a semioethic animal

Thanks to the "human modelling capacity" and its "syntactics," also designated as "language," a species-specific characteristic, the human being can be described as a "semiotic animal". Because of this special modelling device, we now know that the human being is not only endowed with *semiosis*, but also with *metasemiosis* or *semiotics*, that is, a capacity for using signs to reflect on signs, for critical awareness (Deely, Petrilli, Ponzio 2005). In this proposition the expression "language"[4] is used to denominate the human capacity for *modelling* as

---

4 The term *language* is introduced by Thomas Sebeok for the primary modelling system specific to the genus *Homo*. The *primary modelling system* is not natural language (Fr. *langue* / It. *lingua*), as instead the Moscow-Tartu school maintains, but rather language in the sense of the French *langage* and Italian *linguaggio*. Instead, natural language (Fr. *langue* / It. *lingua*) appears quite late in human evolution and is a *secondary modelling system*. Consequently, cultural sign systems that presuppose natural languages are *tertiary modelling system*.

The concept of *modelling* comes from the so-called Tartu-Moscow school (A.A. Zaliznjak, V.V. Ivanov, V.N. Toporov, Ju.M. Lotman, cf. Lucid 1977; Rudy 1986) where it is applied to natural language (Fr. *langue* / It. *lingua*), which it describes as a "primary modelling system" (cf.

distinct from *communication*; whilst the expression "semiotics" in addition to the name for the general *science of signs* indicates the *specificity of human semiosis, metasemiosis*. "Semiotics" in this second sense qualifies human animals as "semiotic animals" and connects human behaviour with conscious awareness and the capacity for responsibility, where "responsibility" is understood both in the sense of the ability to respond, "response-ability", of responsiveness, and of answer-ability, account-ability.

The capacity for language understood as modelling and characterized by syntax (better, syntactics) endows human beings with the capacity to construct not only one world, like all other animal species, but numerous possible worlds. This species-specific modelling capacity appeared with hominids and determined their evolution during the whole course of development from *Homo habilis* to *Homo erectus* to *Homo sapiens* and now *Homo sapiens sapiens*. Syntax or writing (*ante litteram* writing, that is, writing before the letter, *avant la lettre*, to use an expression introduced by Emmanuel Levinas (1972), writing before verbal transcription) involves the capacity to (mutely) construct multiple meanings and senses, multiple registers, that is, multiple meanings relative to different registers, with a finite number of elements. From this point of view, oral verbal language can also be discussed in terms of "writing" (Petrilli and Ponzio 2003b: 7–10, 11–26; see also Petrilli 2012: 122–123). Parallel to activation of the modelling capacity (language) in the evolutionary development of *Homo*, nonverbal signs were also used for communication as in all other animals, but with the difference that in humans they were rooted in (mute) language (modelling). In this sense these nonverbal signs are linguistic nonverbal signs (Posner et al., 1997–2004, Art. 18, §5, §6).

As "semiosic animals" human beings interpret signs without distinguishing between the levels of immediate interpretation, what I propose we call *direct semiosis, primary semiosis*, and the understanding of interpretation; instead, as "semiotic animals" or "metasemiosic animals," human beings can suspend the

---

Deely 2007), and to the other human cultural systems described as "secondary modelling systems". On our part we implement the term *modelling* following Sebeok who extends the concept beyond the sphere of anthroposemiosis and connects it to the biologist Jakob von Uexküll and his concept of *Umwelt* ("surrounding world") (cf. Kull 2010a, b). In Sebeok's interpretation, *Umwelt* means "external world model". On the basis of research in biosemiotics, we know that the modelling capacity can be observed in all life-forms (cf. Sebeok 1979: 49–58, 68–82 and Sebeok 1991a: 117–127). "Modelling systems theory" has recently been reformulated by Sebeok in collaboration with Marcel Danesi (Danesi and Sebeok 2000). They study semiotic phenomena as modelling processes. In light of semiotics oriented in the sense of modelling systems theory, semiosis can be defined as a capacity in all life-forms which produces and understands signs according to specific models, organizing perceptive input as established by each species (Danesi and Sebeok 2000: 5).

immediate, direct interpretation of signs and set the conditions for reflection and deliberation, for what we can call *indirect or secondary semiosis, complex semiosis*. In fact, that in the life-world human semiosis should be characterized as metasemiosis means that not only can we approach signs as the object of interpretation undistinguished from our response to them, but that we can also suspend our responses and deliberate.

Metasemiosis is a biosemiosic and phylogenetic endowment that, thanks to syntactics, or language understood as modelling, favours a unique capacity for creative and critical intervention on semiosis. Thus equipped the human being is uniquely capable of assuming a responsible attitude to life, signs and sign behaviour; and attending to the quality of life. This is connected to the human capacity for listening and accountability, for caring for life in its joyous and dialogical multiplicity, where "caring for" implies an object, the other, to be concerned for that other, but without making claims to power and control through therapy and cure. Such a propensity arises in the context of global intercorporeality, dialogical interrelatedness, and creative awareness of others as actors in the same semiosic web that is life.

Remembering the axiom formulated by Sebeok with his "global semiotics" which recites that where there is life there is semiosis, that life and semiosis coincide (Sebeok 1986a, b, 1994, 2001a), the "semiotic animal," the rational animal that is *homo* is uniquely capable of reflection, deliberation, of making critical choices and taking a standpoint. Consequently, the semiotic animal is capable of taking responsibility for semiosis and life over the entire planet, for their health and good functioning. In this sense, we are both subject *to* and *of* responsibility.

As semiotic animals human beings are capable of a global view of life and communication:[5] hence the question "What is our responsibility to life and the universe in its globality?" (Petrilli and Ponzio 2010: 157). This question is central to the orientation in semiotics designated as "semioethics," an expression introduced

---

5 "Communication," "modelling" and "dialogism" are three fundamental concepts in semiotics where the first is generally privileged over the other two, but cannot be understood without them.

Communication presupposes modelling, given that communication occurs internally to a world produced by the modelling processes it presupposes. Modelling systems, in turn, also evolve from communication as it occurs in the species, and from the environment – being the context of modelling produced by adaptation. But communication always occurs on the basis of the type of modelling that characterizes a species.

By *dialogue* is understood the way in which an organism in its specific *Umwelt* relates to the intraspecific and extraspecific organic, and to the inorganic. Semiosis is generally *dialogic*. The notion of *dialogism* does not contradict, but rather supplements and confirms those notions that insist on the autonomy of the living organism, for example, Jakob von Uexküll's *functional cycle* and Humberto Maturana and Francisco Varela's *autopoiesis*. Furthermore,

to indicate what we consider to be an evitable turn in semiotics studies today in relation to the human world (more exactly, the multiple human worlds, real and possible, that characterize anthroposemiosis) (Petrilli 2010; Petrilli and Ponzio 2005: 562). The capacity for reflection that is at once creative and critical (as stated, a specific characteristic of human semiosis), for metasemiosis, effectively contributes to a better understanding of why, in what sense we are responsible for semiosis, for life throughout the "semio(bio)sphere".

The semiotic (i.e. metasemiosic) capacity implies a third human species-specific modality of being-in-the-world beyond the *biosemiosic* and the *semiotic*, what we have denominated the *semioethic*. Viewed together these different perspectives on sign activity in the global communication network afford a fuller understanding of the extent to which human beings are responsible for the health of semiosis generally in all its forms, for the *quality of life*, human and nonhuman, over the planet. The "semiotic animal" is also a "semioethic animal".

The idea that *homo* is not only a "semiotic animal," but also a "semioethic animal" has been elaborated keeping account of Charles S. Peirce when he thematizes the concept of "reasonableness" beyond "reason". "Reasonableness" is understood here as open-ended dialectic-dialogic semiosic activity, unfinished and unfinalizable, unbiased by prejudice and regulated by the logic of love, otherness and continuity or what he also calles "synechism" (*CP* 1.615, 2.195, 5.3). The concept of reasonableness is intended to supersede the limits of abstract gnoseologism and to orient semiotic research in a pragmatic-ethic or evaluative-operative sense. In his Preface to his 1903 *Lectures on Pragmatism* (*CP*, Vol. 5, Bk. I), Peirce makes the following statement (cited from his 1902 dictionary entry "Pragmatic and Pragmatism"):

> Almost everybody will now agree that the ultimate good lies in the evolutionary process in some way. If so, it is not in individual reactions in their segregation, but in something

---

dialogue must be distinguished from *communication*. Communication is only one aspect of semiosis. The other two are *modelling* and *dialogism*.

Dialogism, modelling and communication – which in the human being are characterized species-specifically – belong to semiosis in general and for this reason can be traced, in different forms, degrees and modalities, in all living beings. The dialogic character of verbal semiosis, its modelling and communicative functions, are specific characterizations of the human species of capacities that can be traced in semiosis generally in any living being. A more detailed study of the semiosis of language understood as "*langage / linguaggio*" (primary modelling), and as *langue / lingua* (secondary modelling), and of other cultural sign systems that presuppose language understood as "*langue / lingua*" (tertiary modelling) are available in the chapter "Language as primary modelling and natural languages: a biosemiotic perspective," co-authored with Augusto Ponzio, for the volume *Biosemiotic Perspectives on Language and Linguistic*, 2015.

> general or continuous. Synechism is founded on the notion that the coalescence, the becoming continuous, the becoming governed by laws, the becoming instinct with general ideas, are but phases of one and the same process of the growth of reasonableness. This is first shown to be true with mathematical exactitude in the field of logic, and is thence inferred to hold good metaphysically. It is not opposed to pragmatism in the manner in which C. S. Peirce applied it, but includes that procedure as a step. (*CP* 5.4)

According to Peirce the most advanced developments in reason and knowledge are achieved through the creative power of reasonableness, governed by the forces of agapasm (on the relation between logic and love, see Boole 1931b [1905]).[6] He maintains that love is directed to the concrete and not to abstractions; towards one's neighbour, not necessarily in a spatial sense, locally, but in the sense of affinity, a person "we live near [. . .] in life and feeling" (*CP* 6.288). Love is a driving force in logical procedure characterized in terms of abduction, iconicity and creativity. The development of mind occurs largely through the power of love thus understood. The type of evolution foreseen by synechism, the principle of continuity, is evolution through the agency of love. On such issues Peirce refers us directly to his essay of 1893, "The Law of Mind" (*CP* 6.289). Furthermore, Peirce polemically contrasts progress as achieved through a relation of sympathy among neighbours, the "Gospel of Christ," with what he calls the "Gospel of Greed" which reflects the dominant ideology of his day and encourages the individual to assert one's own rights and interests, its own individuality or egoistic identity over the other (*CP* 6.294).

Love, reasonableness and creativity are all grounded in the logic of otherness and dialogism and together move the evolutionary dynamics of semiosis in

---

6 Peirce identified the ultimate end of semiosis in the human world neither in individual pleasure (hedonism), nor in the good of society (Utilitarianism), but rather in a principle regulating the evolutionary development of the universe, what he calls "reasonableness" (*CP* 5.4). In Peirce's view, the ultimate value of the concept of the *summum bonum* is reason and the development of reason, that is, reason understood as an open, dialectic process, as unprejudiced research, or as Bakhtin would say, as an ongoing dialectical-dialogical process, a movement oriented by the logic of otherness. This process is never complete or finished, but rather is rooted in the principle of continuity or synechism (*CP* 1.172). Therefore Peirce himself transcended the limits of a merely gnoseological semiotics working in the direction of what can be described as an ethical-pragmatic or valuative-operative approach to the study of signs and human behaviour. In addition to his *Collected Papers*, here we shall simply recall the telling title of his posthumous collection of essays, which is indicative of his orientation: *Chance, Love and Logic* (1923). In the final phase of his production (which overall spans approximately from 1887 to 1914) – what Gérard Deledalle in his 1987 monograph on Peirce calls the Arisbe period (the name Peirce gave to his home in Milford, Pennsylvania, where he lived to the end of his days) – Peirce specifically turned his attention to the normative sciences: in addition to logic these include aesthetics and ethics and hence the question of ultimate ends or of the *summum bonum*.

the human world. And given their unique, species-specific capacity as semiotic animals, human beings, as anticipated, are also invested with a major role in terms of responsibility towards semiosis generally, which means to say towards life in all its forms over the entire planet.

From the point of view of human social semiotics, our own approach to sign studies, linguistic and nonlinguistic, verbal and nonverbal, is oriented "semioethically" to embrace questions traditionally pertaining to ethics, aesthetics and ideology (see Rossi-Landi 1978, 1992a). Indications in this sense can be traced in Peirce who, coherently with his pragmatism, developed a cognitive approach to semiotics in close relation to the study of the social behaviour of human beings and the totality of their interests. From a Peircean perspective, the problem of knowledge necessarily involves considerations of a valuational and pragmatical order. Semioethics in fact extends its gaze beyond the logico-cognitive and epistemological boundaries of semiotics to focus on the relation of signs to values and thus on the axiological dimension of sign activity, which includes the human disposition for evaluation, critique, creativity and responsibility, thereby overcoming any tendency towards dogmatism and unquestioning acceptance.

This orientation is also prefigured by Victoria Welby with her *significs*[7] which saw important developments across the first half of the twentieth century (see Petrilli 2009a, 2015). The term "significs" indicates the human disposition for evaluation, the import conferred upon something, the signifying potential and significance of human behaviour, participation in the life of signs not only on the cognitive and logical levels, but also in corporeal, emotional, pragmatic and ethical terms.[8]

Creative love and reasonableness associate knowledge and experience to the pragmatic-ethical dimension. If we do not persist in proceeding in a contrary

---

7 "Significs" is a neologism coined by Welby in the 1890s for her theory of meaning and special approach to the study of signs in all their forms and relations with a special focus on their relation to values. Significs transcends pure descriptivism and gnoseological or logico-epistemological boundaries in the direction of axiology and study of the conditions that make meaningful behaviour possible (cf. Welby 1983, 1985a, b, 2009). As Welby claimed in a letter of 18 November 1903 to Peirce (in which she mentions her intellectual solidarity with the Italian philosopher and mathematician Giovanni Vailati, 1863–1909), "significs" is a "practical extension" of semiotics: "Prof. G. Vailati, . . . shares your view of the importance of that – may I call it, practical extension? – of the office and field of Logic proper, which I have called Significs" (in Hardwick 1977: 5–8; see also Vailati 1971, 1987). Though this specification may seem superfluous given that the pragmatic dimension is inscribed in Peirce's approach to semiotics, that the ethical-valuational aspects of signifying processes are closely interrelated with the operative-pragmatic is important to underline.
8 Significs elects "significance" as its ultimate object of study with respect to "sense" and "meaning," the other two terms forming her meaning triad. "Sense" corresponds to the most

sense and separate, even juxtapose processes that, instead, should integrate and complete each other, we soon realize that to transcend the limits of a strictly

primitive level of pre-rational life, that of one's response to the environment, it concerns the use of signs and emerges as a necessary condition for all experience; "meaning" concerns rational life, the intentional, volitional aspects of signification; "significance" implies both sense and meaning and extends beyond these to concern the "import" and "value" that signs have for each one of us. As such, this notion can be associated with Morris's own interpretation of the concept of "significance" (Welby 1983 [1903: 5–6, in Petrilli 2009a: 264, see also 265–272]). According to Welby, "sense," "meaning" and "significance" indicate three simultaneous and interacting dimensions in the development of expressiveness, interpretive capacity and operative force (cf. Heijerman and Schmitz 1991; Schmitz 1985, 1990).

In the Preface to her monograph *Significs and Language* (1911), Welby describes significs as "the study of the nature of Significance in all its forms and relations, and thus of its workings in every possible sphere of human interest and purpose"; and the interpretive function as "that which naturally precedes and is the very condition of human intercourse, as of man's mastery of his world" (Welby 1985: vii). In *Significs and Language*, as in all her writings, the problem of analyzing signifying processes is also the problem of investigating the processes of the production of values as a structural part of the production of meaning in human sign activity. The epistemological, ethical and pragmatic dimension of signifying processes finds expression in unconsciously philosophical questions asked by the "man in the street," as Welby says, in everyday language: "What do you mean by . . . ?," "What does it signify?," "What is the meaning of . . . ," etc. In what may be described as her most complete published work on the problem of signs and meaning, *What is Meaning?* (1903), Welby observes that "Man questions and an answer is waiting for him. . . . He must discover, observe, analyse, appraise, first the sense of all that he senses through touch, hearing, sight, and realize its interest, what it practically signifies for him; then the meaning – the intention – of action, the motive of conduct, the cause of each effect. Thus at last he will see the Significance, the ultimate hearing, the central value, the vital implication – of what? of all experience, all knowledge, all fact, and all thought" (Welby 1983: 5–6).

Further on in the same volume she specifies that "significs in a special sense aims at the concentration of intellectual activities on that which we tacitly assume to be the main value of all study, and vaguely call 'meaning'"; (Welby 1983: 83). Therefore, in the face of accumulating knowledge and experience, the so-called "significian," whether scientist, philosopher, or everyday person, is urged to ask such questions as: "What is the sense of . . . ?," "What do we intend by . . . ?," "What is the meaning of . . . ?," "Why do we take an interest in such things as beauty, truth, goodness?," "Why do we give value to experience?," "What is the expression value of a certain experience?". In Welby's view, such questions and their responses concern the sense of science and philosophy, and are at the basis of all controversies concerning aesthetics, ethics, and religion. Consequently, significs is relevant to all spheres of life not because it claims semiotic omniscience, but because it turns its attention to interpretation and meaning value as the condition of experience and understanding.

As the study of significance, significs advocates an approach to everyday life and to science that is oriented by the capacity for critique and creativity, release from dogmatism, dialectic-dialogic answerability, by the capacity for listening and responsibility. Significs results from relating the study of signs and sense to ethics. Ethics not only constitutes the object of study,

gnoseological approach in the study of sign activity is not only appropriate, but necessary (see Petrilli 2014a: 67–83).

In sum, these considerations present general semiotics with a plan that is not related to any particular ideological orientation. The semiotic animal is a properly responsible actor, capable of *signs of signs*, mediation, reflection, awareness, of suspending action and of deliberation. As such, the semiotic animal is capable of critical, creative, and responsible awareness as regards semiosis over the entire planet, of taking a critical standpoint with respect to semiosis in its various aspects, and on this basis of acting rationally and reasonably. From this perspective we have seen that the semiotic animal can also be described as a semioethic animal.

---

but is also the perspective. The measure itself of the semantico-pragmatical validity of all human knowledge and experience is ethical insofar as they produce sense and value.

The term "significance" designates the *disposition towards valuation*. Reference is to the value we confer upon something, the *relevance, import, and value* of meaning itself, the *condition of being significant*. This is determined by the involvement of human beings in the life of signs at the theoretical, emotional, ethical and pragmatic levels together. Welby oriented a large part of her own research in the sense of the relation of signs to values, what we have indicates as "semioethics" as a development on "global semiotics" and preferred the term "significs" to underline the direction of her studies rather than "semiotics" and other similar expressions such as "semantics" (Bréal 1897), "semasiology" (Reisig 1839), or "sematology" (Smart 1831, 1837), etc.

In a letter to Welby dated the 14 March 1909 (in Hardwick 1977: 108–130), Peirce established a correspondence between Welby's triad, "sense," "meaning" and "significance" and his own that distinguishes between "immediate interpretant," "dynamical interpretant," and "final interpretant". Peirce's "immediate interpretant" concernsmeaning as it is normally used by the interpreter. As Welby says in relation to sense, it concerns the interpreter's immediate response to signs. The "dynamical interpretant" concerns the sign's signification in a specific context. So, as Welby claims in relation to meaning, it is used according to a specific intention. But even more interesting is the connection established by Peirce between his concept of "final interpretant" and Welby's "significance" (Petrilli 2009: 288–293). According to Peirce, the final interpretant concerns the sign at the extreme limits of its interpretive possibilities. In other words, it concerns all possible responses to a sign in a potentially unlimited sequence of interpretants. As attested by the correspondence to Welby's "significance," the "final interpretant" also alludes to signifying potential, to the capacity for creativity and critique and is fundamentally concerned with valuational attitudes.

4 Critical global semiotics. Caring and responsibility as the secret of sociality

Semiotics conceived as the general science of signs needs to refine its auditory and critical functions, the capacity for listening and critique, and "semioethics" can contribute to the task. From this perspective, "global semiotics,"as anticipated, is not limited to a gnoseological-theoretistic approach to semiosic processes, but is also sensitive to the pragmatic-ethical dimension of sign activity. Global semiotics is founded in cognitive semiotics, but as we have claimed must also be open to a third dimension of semiosis beyond the quantitative and the theoretical which is the ethical. This third dimension concerns the ends towards which we strive: in fact, other expressions previously introduced for this particular dimension of semiosis include in addition to "ethosemiotics," the expressions "teleosemiotics," and "telosemiotics," though for the relation between semiotics and ethics we now prefer the expression "semioethics" (Petrilli 1998: 180–186; Petrilli and Ponzio 2016: 223–259).

Semioethics is not intended as a discipline in its own right, but as an orientation, a perspective in the study of signs, which inherits the critical instance of philosophy of language, the quest for sense.[9] The expression "semioethics" indicates

---

[9] In response to a query from John Deely *à propos* the term "semioethics," Ponzio explains as follows in an e-mail exchange between 4 and 5 January 2010:

Semioethics was born in the early 1980s in connection with the introductions (written by Susan Petrilli) to the Italian translations of works by Thomas Sebeok, Charles Morris, Victoria Welby and my own introduction and interpretation of works by Mikhail Bakhtin, Ferruccio Rossi-Landi, Giovanni Vailati, and Peirce (see my Bibliography). The problem was to find, with Susan, a term which indicates the study of the relation between signs and values, ancient semeiotics and semiotics, meaning and significance, and which somehow translates Welby's "Signifies" into Italian: we coined terms and expressions such as "teleosemiotica," "etosemiotica," "semiotica etica" in contrast with "semiotica cognitiva" (see the Italian edition by Massimo Bonfantini of Peirce, *La semiotica cognitiva*, 1980, Einaudi, Turin).

The beginning of semioethics is in the introductions by myself and Susan to the Italian editions (translation by Susan) of Sebeok, *Il segno e i suoi maestri*, Bari, Adriatica, 1985, of Welby, *Significato, metafora e interpretazione*, Adriatica, 1985, in the essays by Susan and me published in H. Walter Schmitz (ed.), *Essays in Significs*, Amsterdam, John Benjamins, 1990, in Susan's books of the 1980s, such as *Significs, semiotica, significazione*, "Prefazione" by Thomas Sebeok, Adriatica, 1988, and Ponzio"s, such as *Filosofia del linguaggio*, Adriatica, 1985.

In a private note written in the context of the International Colloquium, "Refractions. Literary Criticism, Philosophy and the Human Sciences in Contemporary Italy in the 1970s and the 1980s," held at the Department of Comparative Literature, Carleton University, Ottawa, 27–29 September 1990 (in the discussion following delivery of my paper "Rossi-Landi tra *Ideologie e Scienze umane*"), I used the Italian term "Semioetica" playing on the displacement of "e" in

a propensity in semiotics to recover its ancient vocation as "semeiotics" or "symptomatology" which focuses on symptoms.

Metasemiosis is a condition for global responsibility and implies the capacity for listening, for listening to the other. Semiotics (therefore semioticians) can commit to the health of semiosis and cultivate the capacity for responsible and responsive listening towards the semiosic universe, which means to say towards life.

We have mentioned that a major issue for semioethics (like semeiotics) is care for life, and given that in a global semiotic framework semiosis and life converge, as postulated by Sebeok, our perspectives on life and our responsibilities towards life take on global dimensions.

The expression to "care for life" does not imply any form of therapeutic power, the power *to cure*, but far more essentially the capacity for involvement with the other, interest in the other, unindifference to the other. It is in this sense that general semiotics can be related to ancient medical semeiotics or symptomatology with Hippocrates and Galen and their vocation for the health of semiosis, for life (on sign conceptions in medicine in Ancient Greece, see Langhoff 1997; on the medical origin of semiotics, see Sebeok 1994: 50–54; on Galen in medical semeiotics, see Sebeok 2001a: 44–58). Given that semiosis coincides with life (at least), the focus on the health of life practiced by this ancient branch of the medical sciences can be recovered by "semiotics" understood as the general science of signs and reorganized in terms of "semioethics".

The semiotician concerned with the health of semiosis, the health of life (human and nonhuman), focuses on symptoms (of illness, malaise, and individual and social disorders), but not as a physician, a general practitioner, or some type of specialist. He does not prescribe drugs or administer therapeutic treatments of any sort. Indeed, the widespread condition of medicalization in present-day society must be challenged, as does uncritical recourse to such paradigms as normal/abnormal, healthy/sick (view how to ignore warnings in this sense from

---

the Italian word "semeiotica": indicating in Semiotics the ancient vocation of Semeiotics (as conceived by Hippocrates and Galenus) for improving life, bettering it.

But in the title of 3 lessons delivered with Susan at Curtin University of Technology, Perth in Australia, we still used the term "teleosemiotica": "Teleosemiotics and global semiotics" (July–September, 1999, Australian lecture tour: Adelaide University, Monash University, in Melbourne, Sydney University, Curtin University, in Perth, Northern Territory University, Darwin).

The book *Semioetica,* co-authored by Susan and me, was published in 2003 and is the landing achievement of this long crossing of texts, conceptions, and words, as results from our bibliographic references [. . .].

It is very difficult to say exactly when an idea is born with its name: "universal gravitation" was born when an apple fell from a tree on Newton's head: isn't that so?

Thomas Szasz in the United States, see, e.g., Szasz, 1961, 2001, 2007; Petrilli and Ponzio 2017a; Schaler, Lothane, Vatz, eds., 2017).

The semiotician's interest in symptoms bears a certain resemblance to Freudian analysis given the central role played by interpretation in both cases and the inclination to listen to the other which is decisive for interpretation. But listening here is not understood in the medical sense: to listen to the other is not to auscultate. And if semiotic or, better, semioethic analysis of symptoms is similar to Freudian analysis, it shares nothing with the practice of institutionalized and medicalized psychiatry, with medicalized and "psychiatrized" psychoanalysis, with psychiatric patients, psychiatric treatment, administration of drugs, and sundry concoctions, that is, it shares nothing with the medicalization and psychiatrization of life as practiced ever more in today's globalized world.

Another connection can be established here with Victoria Welby and her original approach to the study of signs, "significs," when she observes the following, thereby developing her theory of meaning in full consideration of the critical, pragmatic, and ethical dimensions of semiosis:

> It is unfortunate that custom decrees the limitation of the term diagnosis to the pathological field. It would be difficult to find a better one for that power of "knowing through," which a training in Significs would carry. We must be brought up to take for granted that we are diagnosts, that we are to cultivate to the utmost the power to see real distinctions and to read the signs, however faint, which reveal sense and meaning. Diagnostic may be called the typical process of Significs as Translation is its typical form. (Welby 1983 [1903]: 51)

Analogically "diagnostic" can be associated with the semioethic orientation in semiotics. In fact, semioethics derives its inspiration from Welby's significs and its focus on sense, meaning and significance, from Peirce's interest in ethics, and from Charles Morris's focus on the relation between signs and values, signification and significance, semiotics and axiology (Morris 1964[10]), as much as from the focus on otherness and dialogism thematized by Mikhail M. Bakhtin and

---

10 The title of Morris's 1964 book, *Signification and Significance. A Study of the Relations of Signs to Values* is significant in itself. In ithe draws attention to the relation between signs and values as anticipated by the subtitle. Morris dealt with values almost as much as he dealt with signs and opposed the idea that the mere fact of describing signs would give an insight into values (Rossi-Landi 1953, 1975^2; 1992: Chs. 2, 3; Petrilli 1992: 1–36). Morris devoted a large part of his research to the problem of ethical and aesthetic value: after his *Foundations of the Theory of Signs* (1938) and *Signs, Language and Behaviour* (1946), he concentrated specifically on value theory in his book *Varieties of Human Value* (1956).

He opens *Signification and Significance* describing two senses according to which the expression "to have meaning" can be understood: as having value, of being significant, on the one hand, and as having a given linguistic meaning, a given signification, on the other. Morris uses the term

Emmanuel Levinas.[11] By contrast with a strictly cognitive, descriptive, and ideologically neutral approach to signs, language, and behaviour as it has traditionally

---

"meaning" to indicate a global concept analyzable into "signification" and "significance". He aimed to recover the semiotic consistency of signifying processes in the human world as testified by the ambiguity of the term "meaning". Meaning understood as signification is the object of semiotics, while significance is the object of axiology. An important aspect of the relation of signs to values is that it calls for recognition of the inevitable relation of semiotics to axiology. Though working from different perspectives, these disciplines converge in their object of study, namely human behavioural processes. Morris was intent upon rediscovering the semiotic consistency of the signifying process to which the ambiguity of the term itself "meaning" testifies. As he explains in the Preface to the volume in question: "That there are close relations between the terms 'signification' and 'significance' is evident. In many languages there is a term like the English term 'meaning' which has two poles: that which something signifies and the value or significance of what is signified. Thus if we ask what is the meaning of life, we may be asking a question about the value or significance of living or both. The fact that such terms as "meaning" are so widespread in many languages (with the polarity mentioned) suggests that there is a basic relation between what we shall distinguish as *signification* and *significance*" (Morris 1964: vii).

11 Irrespective of the philosophical importance of dealing with the relation between signs and values, there are at least another two reasons – the first historical, the second theoretical – for treating the question of values in the context of sign theory: (1) research in this direction has already been inaugurated (especially by Peirceans); (2) an adequate critique of decodification semiotics calls for close study of the value theory that subtends it.

Sign theory as elaborated by Saussure in his *Cours de Linguistique générale* (1916), the "official Saussure," but actually written by a handful of students on the course, is based on the theory of equal exchange value formulated by the School of Lausanne with such representatives as Leon Walras and Vilfredo Pareto and marginalist economics (Ponzio 1986, 1990: 117–118). Saussure associates language with the market in an ideal state of equilibrium. Language is analyzed using the same categories developed by "pure economics" which studies the laws that regulate the market leaving aside the social relations of production, what Rossi-Landi (1968, 1975a, 1992) calls "social linguistic work" and its social structures. This approach orients the Saussurean sign model in the direction of equal exchange logic, establishing a relation of equivalence between *signifiant* and *signifié* and between communicative intention, on the one hand, and interpretation understood as decodification, on the other.

This particular sign model and the value theory it implies had already been critiqued by Rossi-Landi by the mid-1960s. In the light of historico-dialectical materialism he evidenced the limits of language theories that ground instead linguistic value in equal exchange logic. He applied theoretical instruments originally developed in the context of the Marxian critique of exchange value in relation to questions of a more strictly socio-economic order to the analysis of language (Rossi-Landi 1972, 1985). However, his critique can be traced back even further to his monograph, *Comunicazione, significato, e parlare comune*, 1961, where he discusses what he calls (with ironic overtones) the "postal package theory". This expression underlined the inadequacy of those approaches that describe signs, language and communication as messages that, like a postal package, are sent off from one post office and received by another. With this metaphor, Rossi-Landi critiqued communication analyzed in terms of univocal

characterized semiotic studies, an important task for semiotics today is to recover the ethical–axiological dimension of human semiosis.

---

intentionality, as though formed from pieces of communicative intention neatly assembled by the sender and just as neatly identified by the receiver.

Rossi-Land translated Morris onto the scene of semiotic studies in Italy. He inaugurated his commitment to semiotic inquiry with an early monograph on Morris, 1953, followed the year after with his translation of Morris's *Foundations of the Theory of Signs* (1938). *Signs, Language and Behaviour* (1946) had already appeared in Italy in 1949, translated by Silvio Ceccato. But despite such input, as Rossi-Landi recounts in "A fragment in the history of semiotics" (1984, in Rossi-Landi 1992: 7–16), in Italy the times were not ripe for Morris and his work was not as well received as he had hoped for. Since then Morris's research has proven to be nothing short of seminal for semiotic inquiry internationally. In 1975 Rossi-Landi's monograph on Morris appeared in a new enlarged edition with Feltrinelli (Milan), at last receiving the attention it deserved. Reflecting on the conditions that make for successful cultural communication, Rossi-Landi explains like this: "For cultural communication to obtain, the codes and subcodes must be sufficiently similar *already*; and noise and disturbance must be relatively low. Alternatively, an enormous redundancy is required. To make clearer what I mean: if one wants to be properly understood, one has to repeat the same things in a high number of different occasions, through a high number of different channels. Cultural communication must become a sort of propaganda. Each author is then compelled to choose between concentrating on the production of ideas and waging a sort of warfare for conquering an audience. Here, again, we can see how inextricably fortuitous the tangle of theoretical and practical factors can be. And, as Caesar put it, "*multum cum in omnibus rebus, tum in re militari potest fortuna*"" (Rossi-Landi 1992: 14–15).

Rossi-Landi's work can also be related to Mikhail M. Bakhtin's (1895–1975) research. Bakhtin's name is commonly associated with a monograph *Marxism and the Philosophy of Language*, published in 1929, by Valentin N. Voloshinov (1895–1936), his friend and collaborator. In this book, but even earlier, in 1927 with *Freudianism. A Critical Sketch* (Voloshinov 1927), Bakhtin and Voloshinov critique Saussure's *Cours*, illustrating how it does not account for real interpretation processes, for the specificity of human communicative interaction, that is, for phenomena that qualify human communication as such. The phenomena alluded to include, for example, the capacity for plurilingualism or heteroglossia, plurivocality, ambiguity, polysemy, dialogism, and otherness. Bakhtin-Voloshinov maintain that the complex life of language is not contained between two poles, the "unitary language system" and "individual speaking," that the signifier and the signified do not relate to each other on a one-to-one basis, that the sign is not at the service of meaning pre-established outside the signifying process (Voloshinov 1929: Part II, Chs. II, III).

In this perspective, "linguistic work" (Rossi-Landi 1968, 1992), which is "interpretive work" (Bakhtin, Voloshinov) is not limited to decodification, to the mechanical substitution of an interpreted sign with an interpretant sign; in other words, interpretation is not merely a question of recognizing the interpreted sign. In contrast, interpretive work develops through complex processes which may be described in terms of "infinite semiosis" (Peirce) and "unending deferral" (Derrida 1967) (on the difference between these two concepts, see Eco 1990), of "*renvoi*" (Jakobson 1963) from one sign to another, activated in the dialectic-dialogic relation among signs.

Bakhtin-Voloshinov place the sign in the context of dialogism, responsive understanding, and otherness, thereby describing interpretive work in terms of dialogic responsiveness among

To the question why each human being must be responsible for semiosis, for life over the whole planet, why and in what sense, which is pivotal in semioethics, our response distinguishes between ethics and semioethics. In fact, from the point of view of ethics, this question does not necessarily require an answer: to be responsible for life on the planet is a moral principle, a categorical imperative. Instead, from the point of view of semioethics this question does require an answer: unlike ethics, semioethics involves scientific research, argumentation, interpretation, a dialogic response regulated by the logic of otherness, and questioning. It formulates a definition of the human being as a "semiotic animal" which, as we have seen, also implies a "semioethic animal" (Petrilli 2019, 2021).

In our discussion of responsibility, the reference is not to *limited responsibility, responsibility with alibis*, but to *unlimited responsibility, responsibility without alibis, absolute responsibility*. Responsibility towards life (which converges with signs and communication) in the late capitalist communication-production phase of development is unbounded, also in the sense that responsibility is not limited to human life, but involves all life-forms in the planetary ecosystem with which human life is inextricably interconnected. As the study of signs, semiotics cannot evade this issue. The task of recovering the semioethical dimension of semiosis is now urgent, considering the nature of communication[12] between the historical-social sphere and the

---

the parts in communication. Thus analyzed, interpretive work is articulated through the action of deferral, in this sense translation, constitutive of sign activity or semiosis. In such a framework, the focus is on interpretation/translation viewed in terms of signifying excess with regard to communicative intention, that is, in terms of the generation of signifying surplus value in the dialectic-dialogic relation between the interpreted sign and the interpretant sign.

Bakhtin already saw in the 1920s what interpretation semiotics recognizes today: in real signifying processes the sign does not function in a state of equilibrium or on the basis of equal exchange between the signified and the signifier. Interpretation semiotics proposes a sign model that is far broader, more flexible, and inseparable from its pragmatic and valuative components; and that with its analyses of sense, signification, and significance is able to better account for the specificity of human signifying processes and communicative interaction.

[12] When considering the philosophical question of "communication" with reference to semiotics, presentday theorizers think less and less in terms of "sender," "message," "code," "channel," and "receiver," while practitioners of the popular version of the sign science still tend to cling to such concepts. This particular way of presenting the communication process mainly derives from the semiological approach to "sign studies," thus tagged given its prevalently Saussurean matrix. This approach is commonly identified with such expressions as "code semiotics," "decodification semiotics," "code and message semiotics" (Bonfantini 1981), or

biological, the cultural sphere and the natural, between the semiosphere and the biosphere, where interference is ever more destructive at a planetary level. Just to cite a relatively recent example, think of the devastating effects on the environment worldwide, natural and cultural, caused by the petrol platform explosion of 29 April 2010 in the Mexican Gulf; but think also of the anthropological derangement, anthropological r/evolution provoked by mass migration today over the globe, think of the causes and of its effects on socio-economic systems worldwide, of its implications for humanity(Petrilli 2017; Petrilli and Ponzio 2019).

According to Levinas, the sense of human life, the properly human, is founded on responsibility of the I for the other. Responsibility thus understood is more ancient than the *conatus essendi*, than beginnings, *in principium*; in other words, responsibility is *an-archical*, prior to being and to ontological categories. This type of responsibility is not stated in ontological categories. The shortcoming of modern antihumanism, as Levinas says in the conclusion to his 1968 essay, "Humanism and Anarchy," is in not finding in man, lost in history and in the totality, the traces of this prehistorical and an-archical responsibility. Responsibility for the other is

---

"equal exchange" (Ponzio 1973, 1977). It was amply criticized by Ferruccio Rossi-Landi (1921–1985) as early as the 1960s with his groundbreaking monograph, *Il linguaggio come lavoro e come mercato*, 1968 (Eng. trans. 1983).

This orientation is now counteracted by "interpretation semiotics," thanks in particular to the recovery of Charles S. Peirce (1931–1958) and his writings, therefore of such concepts as "infinite semiosis" and the dialogic relation between signs and interpretation. The interpretive approach describes interpretation as a phenomenon that results from the dialogic interrelation among "interpretants," or, more precisely, among "interpreted signs" and "interpretant signs" (Ponzio 1990a: 15–62). Meaning is not preestablished outside sign processes, but rather is identified in the "interpretant," that is, in another sign that takes the place of the preceding sign. The interpretant, as a sign, subsists uniquely by virtue of another interpretant, and so forth, in an open chain of deferrals. This movement represents semiosis as an open process dependent on the potential creativity of the interpretant in the dialectic-dialogic relation with the interpretive "habit," convention, or "encyclopedia" of a given social community (Eco 1990; Eco et al. 1992). Unlike decodification, or code and message, or equal exchange semiotics, in interpretation semiotics sign activity is not guaranteed by a code. The code only comes into play as a part of the interpretive process, as a result of interpretive practice, and is susceptible to revision and substitution.

However, in terms of commitment to a global understanding of humanity and its signs, to the totality of human relations to itself, to the world and to others, interpretation semiotics has its limits. Semiotics characteristically tends to concentrate on the gnoseological aspect of signs, and neglect the problem of the relation between signs and values which cannot be reduced to the cognitive problem of "truth" merely in a gnoseological sense. From this point of view, semiotics has often presented itself in terms of theoretism, adopting a unilaterally and abstractly gnoseological approach to the life of signs, which implies neglect of those aspects that concern values different from truth value.

the original relation with the other and is unlimited, absolute responsibility (in Levinas 1987a: 138–39). Responsibility thus described, as Levinas says in "Diachrony and Representation," is the "secret of sociality" (Levinas 1991, Eng. trans.: 169).

Encounter with the other from the very beginning, *in principium*, is responsibility for the other, for one's "neighbor," whomever this is, the other for whom one is responsible. As Levinas says in *Entre nous*, precisely in the section entitled "Philosophy, Justice, and Love," love as unindifference, as charity, is original and is original peace (Levinas 1991: 103–121). Absolute responsibility is responsibility for the other, responsibility understood as answering to the other and for the other. This type of responsibility allows for neither rest nor peace. Peace functional to war, peace intrinsic to war, a truce, is fully revealed in its misery and vanity in the light of absolute responsibility. The relation to the other is asymmetrical, unequal: the other is disproportionate with respect to the power and freedom of the I. Moral consciousness is this very lack of proportion. It interrogates the self's freedom (Ponzio 2006a).

General semiotics conceived in the framework of global semiotics presents itself as a metascience which overcomes artificial separations established between the human sciences and the natural sciences and, instead, favors a transversal and interdisciplinary approach which evidences the condition of interconnectedness among the sciences.[13] General semiotics in a global semiotic framework also continues its philosophical search for sense, as indicated above all by teachings in phenomenology, with special reference to the work of Edmund Husserl and Maurice Merleau-Ponty. In our own interpretation, the question of the sense for man of scientific research in general and of semiotics in particular is oriented by Husserl's distinction between the "exact sciences" and the "rigorous sciences" as thematized in his essay, "Philosophy as a Rigorous Science," and in his monograph, *The Crisis of the European Sciences* (1954). Husserl interrogates

---

[13] As stated earlier, from a diachronic perspective, the origins of general semiotics understood as global semiotics can be traced back at least to the rise of the medical sciences and specifically to symptomatology, see Petrilli 2014: 4.1 and 4.4). That the genesis of semiotics be identified, following Sebeok, in medical semeiotics or symptomatology, according to the tradition that leads from Hippocrates to Galen, is not only a question of agnition, that is, knowledge about origins. To relate semiotics to the medical sciences, therefore to the study of symptoms also means to recover the ethical instance of studies on signs. In other words, it means to recover the ancient vocation of "semeiotics" for the health of life which is an immediate concern for semiotics given that, as Sebeok posits, semiosis and life, that is, life globally over the entire planet, are coextensive. Semiotics is semioethics in this sense too. As anticipated, the ethical instance of semiotics as developed by semioethics also revolves around the work of Emmanuel Levinas and Mikhail Bakhtin.

the sense for man of scientific knowledge, avoiding all forms of scientism and technicalism, all forms of separation between means and conscious awareness of ends, by contrast to the alienated subject and false consciousness. From this point of view, semiotics is also "semioethics".

The trichotomy "global semiotics," "cognitive semiotics," and "semioethics" is decisive in our understanding of semiosis not only in theoretical terms, but also for ethical-pragmatic reasons. Semiotics must constantly refine its auditory and critical functions, its capacity for listening and critique in order to turn its attention to the semiosic universe in its globality and meet its commitment to the "health of semiosis," apart from understanding in cognitive and analytical terms. To accomplish this task, therefore, we believe that semiotics must be nothing less than 1) global semiotics, 2) cognitive semiotics, and 3) semioethics.

Global semiotics provides both a *phenomenological* and *ontological* context. However, as discussed earlier, reference to the *socioeconomic* context is also necessary for a proper understanding of communication today, especially when understood in terms of "communication-production". A semioethic approach must keep account of the fact that global communication-production converges with the socioeconomic context. These three contexts – the phenomenological, ontological, and socioeconomic – are all closely interconnected from the point of view of semioethics. And an important task today for general semiotics conceived as global semiotics and semioethics is to denounce any incongruities in the global sign system and, therefore, any threats to life over the planet produced by that system.

When developed in the direction of semioethics, global semiotics underlines the human capacity to care for life, which implies the *quality of life*. As anticipated, this approach does not orient semiotics in any particular ideological sense, but rather it focuses on human behaviour as sign behaviour interrelated with values. Semioethics is the result of two thrusts: one is biosemiotics (the complex of sciences that study living beings as signs), and the other is bioethics. Semioethics can offer a unified and critical point of view on ethical problems connected with progress in the biological and medical sciences – for example, in such areas as genetic engineering, microbiology, neurobiology, and pharmaceutical research. With bioethics, ethical problems become the object of study of a specific discipline. But prior to the introduction of this new discipline, ethical problems were already part of two totalities which together contribute to their characterization: the *semio(bio)sphere* and the *global socioeconomic communication-production system*. General semiotics developed in terms of global semiotics and semioethics must keep account of this dual context when addressing problems at the center of its attention. In this sense, it can also contribute to the philosophical vocation of semioethics and to the possibility of critical reformulation, therefore to an approach to the life of signs and method of research that is both foundational and critical.

The founder of biosemiotics, the Estonian born, German biologist Jakob von Uexküll (1864–1944) made an extraordinary contribution to research on signs and meaning, communication and understanding in the human world. He conducted his research in biology in dialogue with the sign sciences and evidenced the species-specific character of human modelling – which precedes and is the condition for human communication through verbal and nonverbal signs. According to Sebeok, Uexküll's work has carried out a crucial role in renewing the sign science itself, or "doctrine of signs" (Sebeok 1976, 1979), especially when it elects such issues as its object of research. "Biosemiotics," a relatively new branch of semiotics (which includes zoosemiotics and anthroposemiotics) and is also a foundational dimension of general semiotics (Favareau 2010; Petrilli 1998: 3–14, 29–37).

According to Uexküll, every organism enacts different inward and outward modelling processes for the construction of its *Umwelt*, its species-specific world. *Umwelt*, a characteristic endownment of each living organism of any species, concerns the species in general, whether human or nonhuman. But while in non-human living beings *Umwelt* is stable, in human beings it allows for change and involves each individual in its singularity. In other words, a species-specific feature of the human *Umwelt* and modelling is the capacity for creativity and innovation (see Kull 2001, 2010a, b; Merrell 1996, 1999).

This led Uexküll, the biologist specialized in zoology, physiology, ethology, to move beyond the field of biology and the life sciences strictly speaking to focus on problems of an ethical-political order in the human world. As he stated explicity – e.g. towards the conclusion in *Streifzüge durch die Umwelten von Tieren and Menschen* (1934) –, the human *Umwelt* is a prerogative that endows humans with an advantage by comparison with other living beings. However, it also exposes humanity, puts it at risk and in danger. In fact, not only is our species-specific *Umwelt* the condition for collaboration in its different forms, but also for competition and conflict, to the point even of programming war. As early as 1920, the biologist Uexküll published a book entitled *Staatsbiologie. Anatomie-Physiologie-Pathologie des Staates*.

In the light of a semiotic theory of modelling, semiotics referred to human behaviour and environments (human *Umwelten*) clearly cannot avoid taking a turn in the direction of ethics understood in a broad sense. "Ethics understood in a broad sense" means to include all that which concerns human social behaviour according to models, projects and programs, that is, according to social planning, in this sense according to ideologies (Rossi-Landi 1972, 5th edition 2011: 203–204), with reference to ethics, religion, politics, etc. And as claimed above, another interpretant for the word "ethics," or better "semioethics" is "responsibility". The open character of human modelling favours deferral from one individual to another and inevitably involves the question of choice, taking a standpoint, and of taking responsibility for that standpoint.

5 Global communication as global listening. An ethical demand

Global communication in today's world is dominated by the ideology of production and efficiency. The "interesting," "desire" are now determined in the capacity for homologation with respect to such values. In dominant ideology which in today's world converges with market logic, which is capitalist, or, if we prefer, post-capitalist exchange logic, the "interesting" is crudely substituted ever more by egotistical and vulgar "self-interest". This is in complete contrast with the "carnival" worldview, as thematized by the Mikhail Bakhtin.

But the world of global or better globalized communication celebrates individualism to an exasperated degree and with it the community understood as the "closed community," it too founded on individualism. Without the community understood as the "open community," the individual tends ever more to be alone, isolated and afraid. Without the open community, the individual is only an individual. Moreover, individualism is inevitably accompanied by the logic of competition, in the sense of shrewd and cunning competition. Infact, such values as productivity, efficiency, individualism, competitiveness and velocity represent dominant values in contemporary society over the "properly human," and consequently are inevitably accompanied by fear, that is, fear of the other.

However, despite such an orientation, that we are claiming is dominant in the globalized world today, the structural presence of the grotesque body, the condition of intercorporeality and involvement of self's body with the body of others, cannot be ignored. In this sense, the human being's vocation for the "carnivalesque," for excess with respect to the dominant order still resists, as testified, for example, by literary writing. In this sense literary writing, indeed artistic discourse in general, is and always will be carnivalized (Petrill and Ponzio 2003b, 2006).

Listening is decisive for global semiotics, for the capacity to tune into and synchronise with the semiosic universe. The capacity for listening is connected to music. In the first and second volumes of *Semiotik/Semiotics.A Handbook on the Sign-Theoretic Foundations of Nature and Culture* (Posner, Sebeok, Robering 1997–2004), music is treated as a topic in the study of signs and is analyzed in different cultures and successive eras in Western history: sign conceptions in music in Ancient Greece and Rome (Riethmüller 1997), in the Latin Middle Ages (Gallo 1997), from the Renaissance to the early 19th century (Baroni 1998), from the 19th century to the present (Tarasti 1998). As part of the discussion on the relationship between semiotics and the various individual disciplines, the third volume also includes an article on semiotics of music (see Mazzola 2003). As for other disciplines, reflections in musicology focus on the epistemologically relevant question of the extent to which the subject matter, methods, and forms of presentation in this discipline as well may be understood in terms of sign process.

Listening is necessary for a critical discussion of separatism and different trends that tend to exchange the part for the whole, whether by mistake or in bad faith, as in the case of exasperated individualism in social and cultural life, and the current "crisis of overspecialization" in scientific research. The capacity for listening is a condition for connecting semiotics to its early vocation as medical semeiotics and the interpretation of symptoms, as observed by Sebeok (1986; see also Petrilli and Ponzio 2001, 2002, 2008).

If semiotics is concerned with life over the whole planet given that life and semiosis coincide (for a critical discussion of the equation between the concepts of life process and sign process, see Kull 2002), and if the original reason for studying signs, symptoms precisely, is "health," the health of semiosis, *alias* the health of life, then a nonnegligible task for semiotics – especially today in the era of globalization – is to interpret the symptoms of social and linguistic alienation and call attention to the need to care for all of life and communication over the planet earth in its globality Social symptoms of *malaise* are on the rise globally and tell us as much.

Listening evokes auscultation, a medical attitude. In Ancient Greece music was invested with a therapeutic character, and still is today. And as we have already observed, semiotics possibly originated from semeiotics (or symptomatology), classified by Galen as one of the principal branches of medicine, whose task is to interpret symptoms of illness. In addition to auscultation and other ways of investigating symptoms, the activities of diagnosis and anamnesis, following Galen, include listening to the patient who is invited to discuss his ailments and to tell the story of his troubles.

But medicine today (as denounced by Michel Foucault) is functional to exercising what he calls "bio-power," to promoting techniques of subordination of the body to the knowledge-power of *biopolitics*. Medicine contributes to the controlled insertion of bodies into the production cycle. With its specialisms and manipulation of the body as a self-sufficient entity, medical discourse today strengthens the dominant conception of the individual as a separate sphere, efficient and self-contained, indifferent to the other, in pursuit of needs and aspirations that fail to keep account of the other or of the individual's own condition of dependency upon the other, ultimately of the inevitable semiosical, which in the anthroposemiosphere is also the semiotical, condition of intercorporeality, interconnectedness, in this sense interdepency with the other. And all this translates into failure to keep account of the individual's need for listening to the other, for responsivity towards the other, which in the human world is also responsibility towards that other, whether the other of self or the other from self, before and beyond self.

In such a context, listening itself becomes "direct, univocal listening," listening as imposed by the Law (Barthes and Havas 1977: 989), by the "order of discourse" (Foucault 1971), it becomes "applied listening," "wanting to hear," imposition to speak and, therefore, to say univocally. *Listening* is one thing, *wanting to hear* is another. Listening is responsive understanding (answering comprehension): "listening speaks," as Roland Barthes says (Barthes and Havas 1977: 900), similarly to Bakhtin; listening turns to signs in their constitutive dialogism.

On the contrary, to hear, that is, wanting to hear, "applied listening," excludes the capacity for responsive listening, which is dialogical listening. As such "wanting to hear" belongs to a "closeduniverse of discourse" (Marcuse 1964), which fixes interrogation and social roles and separates listening from responsive understanding. Unlike listening understood as dialogue and responsive understanding which continuously produces new signifiers and interpretants without ever fixing or freezing sense, "applied" listening freezes signifiers and interpretants in a rigid network of speech roles: it maintains the "ancient places of the believer, the disciple, the patient" (Barthes and Havas 1977: 990).

Rossi-Landi's philosophical methodics (1985) is a methodics of listening (Petrilli and Ponzio 2016: 11–37). Listening is an interpretant of responsive understanding, a disposition for hospitality, for welcoming the signs of the other, the signs of the other person, for welcoming signs that are other into the house of semiotics: signs that are other to such a high degree that generally we can only denominate them in the negative, that is, as "nonverbal signs". Listening is the condition for a general theory of signs.

Semiotics is a critical science, but not only in Kant's sense, that is, in the sense that it investigates its own conditions of possibility. Semiotics is a critical science in the sense that it interrogates the human world today on the assumption that it is not the only possible world, not the only world possible, it is not the definitive and finalized world, as established by some self-interested, individualistic, profit-oriented ideology. Critical semiotics looks at the world as a possible world, which means to say a world that is subject to confutation, therefore as one among many possible worlds.

As global semiotics, as metasemiotics, as critical semiotics, as semiotics subject to responsibility in a dual sense, that is, of "responding to" (rather than of indifference) and of "taking the blame for" (rather than of fleeing or cleansing, as in ethnic cleansing for example), semiotics must concern itself with life over the planet – not only in a cognitive sense, but also in the pragmatic and in the ethical. In other words, semiotics must care for life. From this point of view, semiotics must recover its relation with medical semeiotics. Nor is this is just a question of history, of remembering the origins. Far more radically, we are signaling a question of the ideologic-programmatic order.

Again, semiotics is listening, listening in the medical sense, and not just in the sense of general sign theory subtending semiotics; semiotics is listening in the sense of medical semeiotics or symptomatology. Semiotics must listen to the symptoms of today's globalized world and identify signs of unease and illness, as claimed earlier, in social relations, in international relations, in the life of single individuals, in the environment, in life generally over the planet. According to the orientation in semiotics baptized as "semioethics," we need to diagnose, prognose and indicate possible therapies for the future of globalization, for the health of semiosis globally, therefore of life, by contrast to a globalized world tending towards its very own destruction.

Semiotics shows how the other is inevitable and cannot be escaped. We can even go so far as to state that the vocation of the sign – the stuff of life and of the business of living, of communication (whether verbal or nonverbal), of human relationships – is the other. The other is the indistinct background from whence we, each one of us, are born into this world; the other testifies to my entry and to my exit from this world; the other is no less than the condition of possibility for life and communication to flourish. Extending the gaze beyond subsystems and microsystems, global semiotics evidences the condition of total interrelatedness and interdependency not only among the subsystems forming the anthroposphere and their porous boundaries, but between the latter and all other subsystems forming the great biosphere, ultimately between nature and culture as we know them, certainly as far as Gaia, and possibly beyond.

The quality of life and destiny of each and every one of us is determined by the relation with the other, irrevocably, and by our conscious awareness of this state of affairs. For as long as we are alive and connected to the sign network which accomodates us all, the other cannot be escaped and must be dealt with, in one way or another. The upshot is that in the bigger picture we do not choose the other, but if anything the other chooses us.

The world is nobody's if it is not everybody's. Indifference towards the other is not a reasonable option. In nature the tremors of the earth tell us as much, in culture the tremors of humanity, the symptoms of social disease also do, whether a question of wars, terrorism or alienation in its various forms, social and linguistic. And such a state of affairs implies the responsibility of each and everyone of us towards every other, whether a question of conquering lands and preexisting human and nonhuman societies, of creating new socio-political systems, of building nations and international relations, or simply caring for the most vulnerable, for the world's children, our own.

The contemporary world, the world-as-it-is, is overwhelmed by dominant ideology whose reach today is unbounded, global, thanks to a communication network that is just as unbounded, just as global, thereby acting as the perfect support for the overriding system. Semioethics underlines the need for conscious awareness of the role of values in our sign systems, our life systems. In the human world, signs and values come together, in the same packet: where there are values there a signs, the material of values are signs, values are construed and communicated through signs, whether verbal or nonverbal, and signs, properly human signs are perfused with values. This is another axiom we cannot escape. Our language, our behaviour, whether verbal or nonverbal, is intonated, accentuated, orientated in one direction or another, and is so before and beyond what are easily recognizable as the great ideological systems.

With reference to the citizens of the world, all this should not translate into a justification for passivity, a sense of fatality, of indifference towards the other simply because we enter an already given world, an already intonated world, a set social program. We have claimed that the vocation of the sign is otherness. The allusion here includes to the other that each one of us is, to the singularity, uniqueness of each one of us, therefore to absolute otherness and to the capacity it represents for creativity, critique and excess with respect to any one given system, for overflow, and for escape with respect to the order of discourse.

Involvement, participation in the life of the other, whether the other from us, or the other of us, is inevitable. How we process such inevitability will depend upon the values that drive our actions beyond immediate circumstance. Directly proportional to the global spread today is the need for critique, listening and love, not fear but love for one's neighbour, as close or as distant as that neighbour may be. And how we process that neighbour is a choice for each one of us to make, a responsibility for each one of us to take.

The "semioethic turn" proceeds from ongoing confrontation with different trends in semiotic inquiry, in dialogue with different figures as they have emerged on the semiotic scene. This orientation has a vocation for critique not only in relation to semiotics and its history, but towards itself as well. A whole philosophical tradition can be evoked here, beginning from Kant (1724–1804), where the expression "critique" resounds in a special sense, the "ethical" in the sense of the obligation to respond, to "answer to self" and to "answer for self," even before, or at least simultaneously to the request for reasons and justifications from others. Other key authors in this particular tradition of philosophical thought on the concept of "critique" include: Karl Marx with his "critique of political economy," an expression in the subtitle of most of his basic texts; Mikhail Bakhtin (1923) who recovers neokantism – critically – as developed by the Marburg School (headed by Hermann Cohen, and counting such prominent representatives as Ernst Cassirer, Paul Natorp); Victoria Welby and her Significs; Charles S. Peirce with his return to Kantism and critique of Cartesian dogmatism (see "On a New List of Categories," 1867, *CP* 1.545–567).

The approach we are outlining relates signs and values, semiotics and axiology, signification and significance, meaning and sense, semantics and pragmatics. It calls for a detailed study of the concepts of model and structure, and therefore of the relation between modelling systems theory and different positions that have gone under the name of "structuralism". This inevitably involves confrontation between so-called "global semiotics" as introduced by Sebeok and semiotics as practiced under the denomination of "semiology" at the beginning of the twentieth centrury. Semiology interrupted the connection not only with semiotics as conceived by John Locke, but also with much earlier roots, the origins as traced by Sebeok in ancient medical semeiotics (symptomatology) with the work of Hippocrates and Galen.

After various phases in the development of semiotics tagged "code semiotics" (or "decodification semiotics") and "interpretation semiotics" (see Bonfantini 1981), the boundaries of this science are now expanding to include studies that focus more closely upon the relation between signs and values. In truth, this relation is inscribed in the make-up of semiotics and in its very history. To concentrate on the relations of signs and values is important for a better understanding of expression, interpretation and communication.

Ferdinand de Saussure (1857–1913) – whose *Cours* is today in the process of being reread and reevaluated, in light of his unpublished writings as well, thanks above all to the work of a lifetime dedicated to Saussure's legacy by the Italian Tullio De Mauro (1966, 2005, 2011, 2014) – founded his sign theory on the theory of exchange value adapted from marginalist economics. Instead, Peirce breaks with the equilibrium of equal exchange logic thanks to a sign model based on the concept of infinite semiosis (or, if we prefer, infinite deferral from one sign to the next). This approach is oriented by the logic of otherness. It allows for opening to the other and for the concept of signifying surplus. Morris explicitly emphasized the need to address the relation between signs and values and oriented a large part of his research in this direction. However, official semiotics has largely emerged as a theoretistic or gnoseological science, as a descriptive science with claims to neutrality. With semioethics we propose to recover and develop that special slant in semiotics which is open to questions of an axiological order and is more focused on a global understanding of humanity and its signs.

Semioethics focuses on the relation between signs and sense and, therefore, on the question of significance as value. However, we have seen that Welby in the nineteenth century had already introduced the term "significs" for the same purpose, marking her distance from what was commonly understood at the time by both "semantics" and "semiotics". In addition to the renowned classics just mentioned – Saussure, Peirce and Morris –, Welby too deserves a place in the reconstruction of the history of semiotics for her invaluable contribution to furthering our understanding of signs and meaning not only from a historico-chronological perspective, but also in theoretical terms. And, in fact, she is now emerging as the mother-founder of modern semiotics alongside Peirce, recognized as the father-founder (Petrilli and Ponzio 2005: 35–79, 80–137).

Thinkers such as those mentioned so far can be considered as the representatives of a theoretical tendency which focuses on the relationship between social signs, values, and human behaviour in general, by contrast with philosophical analyses conducted exclusively in abstract epistemological terms divorced from social practice.

If, in agreement with Peirce we claim that man is a sign, a direct consequence is that with respect to signs, *humani nihil a me alienum puto (nothing human is alien to me)*. An important implication of this statement is that signs in the human world should not be studied separately from valuative orientations, nor should the focus be exclusively on truth value and its conditions. Instead, a general sign theory that is truly general should be capable of accounting for all aspects of human life and for all values, not just truth value. Signs are the material out of which the self is modeled and developed, just as they are the material of values. While signs can exist without values, values cannot exist without signs (Petrilli 2010a: 137–158). From the point of view of human social life, to evidence the *sign* nature of the human person has a counterpart (particularly on a practical level) in asserting the *human*, the *properly human* nature of signs.

To work in this direction leads to the possibility of identifying a new form of humanism which critiques the reification and hypostatization of signs and values and, instead, investigates the processes that produce them. The relation between signifying processes and values subtends the human capacity for establishing relations with the world, with the self and with others, and as such requires the critical work of demystification. In this framework, signs and values emerge as the live expression of historically specified human operations. With respect to social signs, this means to recover their sense and value for mankind, rather than accept them as naturally given. Ultimately, such an approach recovers a project originally conceived by Edmund Husserl with his transcendental constitutive phenomenology.

However, all this is possible on a condition: that any claim to pure descriptiveness, to neutrality, be left aside. Practiced in these terms, the general science of signs can contribute significantly to philosophical investigation for a better understanding of our relations to the world, to others, to the self. This means to recover our search as proposed by Husserl and his phenomenology for the sense of knowledge, experience, and practical action, and of the sciences that study them. It is well worth noting that Husserl authored an important essay entitled "Semiotik" and dealt extensively with signs and their typology in his *Logische Untersuchungen* (Husserl 1900–1901). Such a philosophical framework for the science of signs favours a more adequate understanding of the problem of communication, meaning, value and interpretation. And by working in this direction, the general science of signs or semiotics may operate more fully as a *human science*, where the "properly human" is a pivotal value (Petrilli 2010: 205–209).

Semioethics arises as a response and continuation of the critical approach to sign studies outlined in this book. It is intended to describe an approach to the study of signs that contrasts with approaches that tend towards abstract theoreticism characteristic of so-called "official semiotics". It is inevitably associated with the proposal of a new form of humanism identifiable as the "humanism of otherness" inscribed in the analysis, understanding and production of values relatedly to signs in signifying processes. As much as, strictly speaking, the term "semiotics" (understood as the global science of signs, hence as covering the domains of both signification and significance in Morris's sense relative to semiosis in the human world) should suffice, we believe that "semioethics" (which, as stated, indicates an approach to sign studies that is not purely descriptive, that does not make claims to neutrality but rather extends beyond abstract logico-epistemological boundaries to concentrate on problems of an axiological order, pertaining to values, therefore to ethics, aesthetics and ideology theory) signals more decisively the direction semiotics is called upon to follow today.

References

Bakhtin, Mikhail M. 1920–24a. K filosofii postupka. In S. G. Bočarov (ed.), *Filosofia i sociologia nauki i techniki Esegodnik 1984–85*, Moscow: Nauka, 1986; *Per una filosofia dell'azione responsabile*, Margherita De Michiel (trans.), A. Ponzio (ed.), Cosimo Caputo (Premise), with essays by A. Ponzio & Iris M. Zaala. Lecce: Manni, 1998; *Per una filosofia dell'atto responsabile*, Luciano Ponzio (new It. trans.), A. Ponzio (ed.). Lecce: Pensa Multimedia, 2009; now in *Bachtin e il suo circolo* 2014, 33–167; *Towards a Philosophy of the Act*,

Vladimir Liapunov (Eng. trans.), V. Liapunov and M. Holquist (eds.). Austin: University of Texas Press, 1993; *Pour une philosophie de l'acte* (Ghislaine Caèpogna Bardet, Fr. Trans.). Losanna: L'Age d'Homme, 2003.

Barthes, Roland. 1982. L'obvie et l'obtus. Essais critiques III. Paris: Éditions du Seuil.

Barthes, Roland; Havas, Roland. 1977. Ascolto. In Enciclopedia, Vol. 1, 982–991. Turin: Einaudi; now in Barthes 1982, 345–364.

Bonfantini, Massimo A. 1981. Le tre tendenze semiotiche del novecento. *Versus* 30. 273–294. Now in M. A. Bonfantini 1984.

Bonfantini, Massimo A. 1984. *Semiotica ai media*. Bari: Adriatica; 2nd ed. Bari: Graphis, 2004.

Boole, Mary Everest. 1931a. *Collected Works*, ed. E. M. Cobham, Pref. Ethel S. Dummer. London: C.W. Daniel.

Boole, Mary Everest. 1931b [1905] *Logic Taught by Love: Rhythm in Nature and in Education* (1890). In M. E. Boole, 1931a, Vol. II, 399–515.

Bréal, Michel. 1897. *Essai de sémantique. Science des significations*. Paris: Hachette.

Cobley, Paul. 2010a. *The Routledge Companion to Semiotics*, Intro., 3–12. London: Routledge.

Cobley, Paul. 2010b. Communication. In Paul Cobley, *The Routledge Companion to Semiotics*, 192–194. London: Routledge.

Danesi, Marcel & Thomas A. Sebeok. 2000. *The Forms of Meaning. Modelling Systems Theory and Semiotics*. Berlin, New York: Mouton De Gruyter.

Deely, John. 2007a. The Primary Modelling System in Animals. In Susan Petrilli (ed.), *La filosofia del linguaggio come arte dell'ascolto / Philosophy of Language as the Art of Listening*, 161–179. Bari: Edizioni dal Sud.

Deely, John, Susan Petrilli & Augusto Ponzio. 2005. *The Semiotic Animal*. Ottawa/Toronto/New York: Legas.

De Mauro, Tullio. 1966. *Introduzione alla semantica*. Rome, Bari: Laterza.

De Mauro, Tullio. 1991. Ancora Saussure e la semantica. In *Cahiers Ferdinand de Saussure* 45, 101–109.

De Mauro, Tullio. 2005. Introduzione, to Saussure 2016, It. trans. 2005, vii–xxvi.

De Mauro, Tullio. 2011. Introduzione, to Saussure 2016, It. trans. 2011, vii–xxxix.

Derrida, Jacques. 1967. *L'écriture et la différence*, Paris: Seuil; English translation A. Bass, Writing and Difference. Chicago: The University of Chicago Press, 1978.

Eco, Umberto. 1990. *I limiti dell'interpretazione* [The Limits of Intepretation, Bloomington: Indiana University Press, 1990]. Milan: Bompiani.

Eco, Umberto, Rorty, Richard, Jonathan Culler & Christine Brook-Rose. 1992. *Interpretation and Overinterpretation*, ed. by Stefan Collini. Cambridge: Cambridge University Press.

Favareau, Donald (ed.). 2010. *Essential Readings in Biosemiotics. Anthology and Commentary*, Preface, A Stroll Through the Worlds of Science and Signs, v–xii, and Introduction, An Evolutionary History of Biosemiotics, 1–80 (Biosemiotics 3, XII, 12). Dordrecht, Heidelberg, London, New York: Springer.

Foucault, Michel. 1971. *L'ordre du discours*. Leçon inaugurale au Collège de France prononcèe le 2 dècembre 1970. Paris: Gallimard.76

Hardwick, Charles (ed. with the assistance of J. Cook). 1977. *Semiotic and Significs. The Correspondence between Charles S. Peirce and Victoria Lady Welby*, Pref., ix–xiv, Intro., xv–xxxiv, by C. S. Hardwick. Bloomington-London: Indiana University Press.

Heijerman, Eric & Walter H. Schmitz (eds.). 1991. *Significs, Mathematics and Semiotics. The Signific Movement in the Netherlands*. Proceedings of the International Conference, Bonn, 19–21 November 1986. Münster: Nodus Publikationen.

Husserl, Edmund. 1900–1901. *Logische Untersuchungen*, Erster Teil, *Prolegomena zur reinen Logik*, Zweiter Teil, *Untersuchungen zur Ph.nomenologie und Theorie der Erkenntnis*. Halle a.d.S.: Max Niemeyer; *Logical Investigations*, Vols. 1, *Prolegomena to Pure Logic*, Vol. 2, *Studies in Phenomenology and the Theory of Knowledge*, Eng. trans. by J.N. Finlay. New York: Humanities Press, 1970; London: Routledge, 1973.

Husserl, Edmund. 1954. *The Crisis of the European Sciences and Transcendental Phenomenology*, Eng. trans. by D. Carr. Evanston: Northwestern University Press, 1970.

Jakobson, Roman. 1959. On Linguistics Aspects of Translation. In Reuben A. Brower (ed.), *On Translation*. 232–239. Cambridge, Mass.: Harvard University Press; Oxford, New York: Oxford University Press; also in R. Jakobson, *Selected Writings*, Vol. II, 260–266. The Hague: Mouton, 1971.

Jakobson, Roman. 1963. *Essais de linguistique générale*. Paris: Minuit.

Kull, Kalevi. 2001. Jakob von Uexküll: An Introduction. *Semiotica* 134(1/4). 1–59.

Kull, Kalevi. 2010a. Umwelt and Modelling. In Paul Cobley (ed.), *The Routledge Companion to Semiotics*, 43–56. London: Routledge.

Kull, Kalevi. 2010b. Umwelt. In Paul Cobley (ed.), *The Routledge Companion to Semiotics*, 347–348. London: Routledge.

Langhoff, Volker. 1997. Sign conceptions in medecine in Ancient Greece and Rome. In Roland Posner, Klaus Robering & Thomas A. Sebeok (eds.), 1997–2004, *Semiotik/Semiotics: A Handbook on the Sign-Theoretic Foundations of Nature and Culture*, Vol. 1, Art. 45, 912–922. Berlin: Walter de Gruyter.

Levinas, Emmanuel. 1972. *Humanisme de l'autre homme*. Montpellier: Fata Morgana; *Humanism of the Other*, Eng. trans. by Nidra Poller. Urbana: University of Illinois Press, 2003.

Levinas, Emmanuel. 1987. *Collected Philosophical Papers*, Eng. trans. & ed. by A. Lingis. Dordrecht: Martinus Nijhoff Publishers.

Levinas, Emmanuel. 1991. *Entre nous. Essais sur le penser à l'autre*. Paris: Grasset; *Entre nous. On Thinking-of-the-Other*, Eng. trans. by M. B. Smith & B. Harshav. London: The Athlone Press, 1998.

Lucid, Daniel P. (ed.). 1977. *Soviet Semiotics*. Baltimore: John Hopkins University Press.

Marcuse, Herbert. 1964. *One-dimensional Man*. Boston, MA: Beacon.

Merrell, Floyd. 1996. *Signs Grow: Semiosis and Life Process*. Toronto: University of Toronto Press.

Merrell, Floyd. 1999. Living signs. *Semiotica* 127(1/4). 453–479.

Morris, Charles. 1946. *Signs, Language, and Behaviour*. New York: Prentice Hall.

Morris, Charles. 1956. *Varieties of Human Value*. Chicago: The University of Chicago Press.

Morris, Charles. 1964. *Signification and Significance. A Study of the Relations of Signs and Values*. Cambridge (Mass.): MIT Press; It. trans. by S. Petrilli in Morris 1988; as an independent volume, see Morris 2000.

Peirce, Charles Sanders. 1923. *Chance, Love and Logic*, ed. and intro. by Morris R. Cohen. New York: Harcourt.

Peirce, Charles Sanders. 1931–1958. *Collected Papers of Charles Sanders Peirce* (i 1866–1913), Vols. I–VI, ed. by C. Hartshorne & P. Weiss, 1931–1935, Vols. VII–VIII, ed. by A. W. Burks, 1958. Cambridge, MA: The Belknap Press of Harvard University Press. [In the text referred to as CP followed by volume and paragraph number.] All eight vols. in electronic form ed. John Deely. Charlottesville, VA: Intelex Corporation, 1994. Dating within the *CP* is based on the Burks Bibliography at the end of *CP* 8. The abbreviation followed by volume and paragraph numbers with a period between follows the standard *CP* reference form.

Peirce, Charles Sanders. 1980. *Semiotica. I fondamenti della semiotica cognitiva*. Turin: Einaudi.

Perron, Paul, Leonard G. Sbrocchi, Paul Colilli & Marcel Danesi (eds). 2000. *Semiotics as a Bridge between the Humanities and the Sciences*. Ottawa: Legas Press.

Petrilli, Susan. 1988. *Significs, semiotica, significazione*, Intro. by Thomas A. Sebeok. Bari: Adriatica.

Petrilli, Susan (ed.). 1992. *Social Practice, Semiotics and the Sciences of Man: The Correspondence Between Morris and Rossi-Landi. Semiotica*. Special Issue 88 (1/2), Intro. by Susan Petrilli. 1–36.

Petrilli, Susan. 1998. *Teoria dei segni e del linguaggio*. Bari: Graphis, 2nd ed. 2001.

Petrilli, Susan (ed.). 2007. *La filosofia del linguaggio come arte dell'ascolto / Philosophy of Language as the Art of Listening*. Bari: Edizioni dal Sud.

Petrilli, Susan. 2009. *Signifying and Understanding. Reading the Works of Victoria Welby and the Signific Movement*. Foreword by Paul Cobley, xvii–x [=Semiotics, Communication and Cognition 2, Editor: Paul Cobley]. Berlin: Mouton.

Petrilli, Susan. 2010. *Sign Crossroads in Global Perspective. Semioethics and Responsibility*, John Deely (ed.), Preface, In Her Own Voice, vii–ix, & The Seventh Sebeok Fellow, xi–xiii. New Brunswick, London: Transaction Publishers.

Petrilli, Susan. 2012. *Expression and Interpretation in Language*, Pref. by Vincent Colapietro, xv–xviii. New Brunswick, London: Transaction Publishers.

Petrilli, Susan. 2014. *Sign Studies and Semioethics. Communication, Translation and Values*. In Paul Cobley & Kalevi Kull (eds.), Semiotics, Communication, Cognition series, Vol. 13. Berlin, Boston: Mouton De Gruyter.

Petrilli, Susan. 2015. *Victoria Welby and the Science of Signs. Significs, Semiotics, Philosophy of Language*, Foreword by Frank Nuessel, xi-xviii. New Brunswick, London: Transaction Publishers.

Petrilli, Susan. 2017. *Challenges to Living Together. Transculturalism, Migration, Exploitation. For a Semioethics of Human Relations*, ed., intro., 15–31. Milan: Mimesis.

Petrilli, Susan. 2019. *Signs, Language and Listening. Semioethic Perspectives*. Ottawa: Legas.

Petrilli, Susan. 2021. *Senza ripari. Segni, differenze, estraneità*, Presentazione. Milan Mimesis.

Petrilli, Susan; Ponzio, Augusto. 2002. *I segni e la vita. La semiotica globale di Thomas A. Sebeok*. Milan: Spirali.

Petrilli, Susan & Augusto Ponzio. 2003a. *Semioetica*. Rome: Meltemi.

Petrilli, Susan & Augusto Ponzio. 2003b. *Views in Literary Semiotics*, Eng. trans. & ed. by S. Petrilli. New York/Ottawa/Toronto: Legas.

Petrilli, Susan & Augusto Ponzio. 2005. *Semiotics Unbounded. Interpretive Routes in the Open Network of Signs*. Toronto: Toronto University Press.

Petrilli, Susan & Augusto Ponzio. 2010. "Semioethics". In Paul Cobley In *The Routledge Companion to Semiotics*, by Paul Cobley, 150–162. London: Routledge.

Petrilli, Susan; Ponzio, Augusto. 2011. A Tribute to Thomas A. Sebeok. *Biosemiotics*, Vol. 1, n. 1, April 2008, Editor-in-chief Marcello Barbieri, 25–40.

Petrilli, Susan; Ponzio, Augusto. 2015. "Language as primary modeling and natural languages: a biosemiotic perspective". In Ekaterina Velmezova, Kalevi Kull & Stephen J. Cowley (eds.), *Biosemiotic Perspectives on Language and Linguistics*, Book series Biosemiotics

13, 47–76. Springer Cham Heidelberg New York Dordrecht London: Springer International Publishing Switzerland.

Petrilli, Susan, Augusto Ponzio. 2016. *Lineamenti di semiotica e di filosofia del linguaggio. Un contributo all'interpretazione del segno e all'ascolto della parola*, Perugia: Guerra Edizioni.

Petrilli, Susan, Augusto Ponzio. 2017a. In Dialogue with Thomas Szasz. In J. A. Schaler, H. Z. Lothane, R. E. Vatz 2017, 25–47.

Petrilli, Susan; Ponzio, Augusto. 2019. *Identità e alterità. Per una semioetica della comunicazione globale*. Milan: Mimesis.

Ponzio, Augusto. 1973. *Produzione linguistica e ideologia sociale*. Bari: De Donato. New edition 2006. New expanded French edition, *Production linguistique et idéologie sociale*. Candiac (Québec): Les Editions Balzac, 1992.

Ponzio, Augusto. 1977. *Marxismo, scienza e problema dell'uomo*. Verona: Bertani.

Ponzio, Augusto. 1985. *Filosofia del linguaggio*. Bari: Adriatica.

Ponzio, Augusto. 1986. Economics. In Thomas A. Sebeok (ed.), *Encyclopedic Dictionary of Semiotics*, 215–217. Berlin: Mouton de Gruyter.

Ponzio, Augusto. 1990. *Man as a Sign. Essays on the Philosophy of Language*, Eng. trans., ed. & Intro. 1–13, Appendix I & II by Susan Petrilli, 313–392. Berlin: Mouton de Gruyter.

Ponzio, Augusto. 2006. *The Dialogic Nature of Sign*. Ottawa, Toronto, New York: Legas.

Posner, Roland, Klaus Robering & Thomas A. Sebeok (eds). 1997–2004. *Semiotik/Semiotics. A Handbook on the Sign-Theoretic Foundations of Nature and Culture*, 4 Vols. Berlin, New York: Walter de Gruyter.

Reisig, Christian Karl. 1839. *Vorlesungen über lateinische Sprachwissenschaft* (E. Lectures on Latin Linguistics) (1825), ed. by Friedrich Haase. Leipzig: Lehnhold.

Rossi-Landi, Ferruccio. 1953. *Charles Morris*. Milan: Bocca; Milan: Feltrinelli. [New revised and enlarged edition, F. Rossi-Landi 1975b.].

Rossi-Landi, Ferruccio. 1961. *Significato, comunicazione e parlare comune*. Padua: Marsilio, 1980. [New editions by Augusto Ponzio, 1998, 2006.].

Rossi-Landi, Ferruccio. 1968. *Il linguaggio come lavoro e come mercato*. Milan: Bompiani, new ed. 2007; *Language as Work and Trade*, Eng. trans. by M. Adams et al. South Hadley (Mass.): Bergin and Garvey, 1983.

Rossi-Landi, Ferruccio. 1972. *Semiotica e ideologia*. Milan: Bompiani. [5th ed. by Augusto Ponzio, 1994, 2011.].

Rossi-Landi, Ferruccio. 1975a. *Linguistics and Economics*. The Hague, Paris: Mouton. [2nd edition, 1977. For the first Italian edition, see Rossi-Landi 2016].

Rossi-Landi, Ferruccio. 1975b. *Charles Morris e la semiotica novecentesca*. Milan: Feltrinelli Bocca. [Revised and enlarged edition of Charles Morris, 1953.].

Rossi-Landi, Ferruccio. 1978. *Ideologia*. Milan: ISEDI; new expanded edition, Milan: Mondadori, 1982; new edition ed. by Augusto Ponzio, Rome: Meltemi, 2005. *Marxism and Ideology*. Eng. trans. from the 1982 edition by R. Griffin. Oxford: Clarendon Press.

Rossi-Landi, Ferruccio. 1985. *Metodica filosofica e scienza dei segni. Nuovi saggi sul linguaggio e l'ideologia*. Milan: Bompiani.

Rossi-Landi, Ferruccio. 1992. *Between Signs and Non-signs*, ed. and Intro. by Susan Petrilli, ix–xxix. Amsterdam: John Benjamins.

Rudy, Stephen. 1986. Semiotics in the USSR. In Thomas A. Sebeok & Jean Umiker-Sebeok, *The Semiotic Sphere*, 34–67. New York: Plenum.

Saussure, Ferdinand de. 1916. *Cours de linguistique générale*, ed. by C. Bally and A. Secheaye. Paris: Payot, 1922; Critical edition by R. Engler. Wiesbaden: Otto Harrassowitz, 4 Vols., 1968–1974; *Course in General Linguistics*, Eng. trans. by W. Baskin, Intro. by J. Culler. London: Peter Owen, 1959, 1974; & by R. Harris. London: Duckworth, 1983; *Corso di linguistica generale*, Intro., It. trans. & comment by Tullio De Mauro. Rome: Latera, 1967, 24th ed. 2011.

Schaler, Jeffrey A., Lothane, Henry Zvi & Richard E. Vatz (eds.) 2017. *Thomas S. Szasz. The Man and His Ideas*, Introduction by Henry Zvi Lothane, Brunswick, New Jersey, New York: Transaction Publishers; London: Routledge.

Schmitz, Walter H. 1985. Victoria Lady Welby's Significs: The origin of the signific movement. In V. Welby 1985a, ix–ccxxxv.

Schmitz, Walter H. (ed.). 1990. *Essays on Significs*. Papers Presented on the Occasion of the 150th Anniversary of the Birth of Victoria Lady Welby, Pref. by H. W. Schmitz, i–ix. Amsterdam: John Benjamins.

Sebeok, Thomas A. 1976. *Contributions to the Doctrine of Signs*. (Joint publication of) Lisse, Netherlands: The Peter de Ridder Press, and Bloomington, IN: Research Center for Language and Semiotic Studies of Indiana University. Indiana University Press / 2nd edition. Lanham, MD: reprinted as Vol. IV in the Sources in Semiotics, Lanham, MD: University Press of America, with a Preface by Brooke Williams "Challenging Signs at the Crossroads," xv–xlii, 1985. Thomas A. Sebeok, Thomas A. 1986a. *I Think I Am a Verb: More Contributions to the Doctrine of Signs*. New York and London: Plenum Press; *Penso di essere un verbo*, It. trans., ed. & introd., 11–18, by Susan Petrilli. Palermo: Sellerio, 1990.

Sebeok, Thomas A. 1979. *The Sign & Its Masters*. Texas: The University of Texas Press; 2nd ed. Intro. by Brooke Williams Deely, Lanham, MD: University Press of America, 1989; It. trans. of the 1979 ed. by Susan Petrilli, Pres. by Augusto Ponzio, 7–12 and Intro. by Susan Petrilli, 15–21, *Il segno e i suoi maestri*. Bari: Adriatica.

Sebeok, Thomas A. 1986a. *I Think I Am a Verb: More Contributions to the Doctrine of Signs*. New York and London: Plenum Press; *Penso di essere un verbo*, It. trans., ed. & introd., 11–18, by Susan Petrilli. Palermo: Sellerio, 1990.

Sebeok, Thomas A. ed. 1986b. *Encyclopedic Dictionary of Semiotics*. 3 Vols. Berlin, New York: Mouton de Gruyter.

Sebeok, Thomas A. 1994. *Signs. An Introduction to Semiotics*. Toronto: Toronto University Press. New edition, 2001; It. trans. & Intro., 11–44, by Susan Petrilli, *Segni. Introduzione alla semiotica*. Rome: Carrocci, 2004.

Sebeok, Thomas A. 1991a. *A Sign Is Just a Sign*. Bloomington-Indianapolis: Indiana University Press; *A Sign is just a sign. La semiotica globale*, It. trans. & Intro., 7–19, by Susan Petrilli. Milan: Spirali, 1998.

Sebeok, Thomas A. 2000. "Semiotics as a Bridge between Humanities and Sciences". In *Semiotics as a Bridge between the Humanities and the Sciences*, edited by Paul Perron, Leonard G. Sbrocchi, Paul Colilli and Marcel Danesi, 76–102. Ottawa: Legas Press.

Sebeok, Thomas A. 2001a. *Global Semiotics*. Bloomington: Indiana University Press.

Smart, Benjamin Humphrey. 1831. *An Outline of Sematology, or an Essay towards Establishing a New Theory of Grammar, Logic, and Rhetoric*. London: John Richardson, Royal Exchange.

Smart, Benjamin Humphrey. 1837. *A Sequel to Sematology. Being an Attempt to Clear the Way for the Regeneration of Metaphysics*. Unpubl. Copy. London: G. Woodfall, Angel Court, Skinner Street.

Szasz, Thomas S. 1961. *The Myth of Mental Illness. Foundations of a Theory of Personal Conduct*, New York: Harper Perennial, 2010.

Szasz, Thomas S. 2001. *Pharmacracy: Medicine and Politics in America*, Westport: Praeger.

Szasz, Thomas S. 2006. *My Madness Saved me. The Madness and Marriage of Virginia Woolf*, New York: Transaction; *La mia follia mi ha salvato. La follia e il matrimonio di Virginia Woolf*, It. trans. intro., 7–59, and ed., Susan Petrilli, Milan: Spirali, 2009.

Szasz, Thomas S. 2007. *The Medicalization of Everyday Life. Selected Essays*, Syracuse: Syracuse University Press.

Uexküll, Jakob von. 1934/1992. *Streifzüge durch Umwelten von Tieren und Menschen*. Reimbeck: Rowohlt, 1967; Eng. trans., *A Stroll through the Worlds of Animals and Men: A Picture Book of Invisible Worlds*. Semiotica 1992, 89(4). 319–391.

Vailati, Giovanni. 1971. *Epistolario 1891–1909*, ed. by G. Lanaro, Intro. by M. Dal Pra. Turin: Einaudi.

Vailati, Giovanni. 1987. *Scritti*, 3 Vols, ed. by M. Quaranta. Bologna: Forni.

Vernadsky, Victor I. 1926. *Biosfera*. Leningrad: Nauka. [French version, Paris, 1929.].

Voloshinov, Valentin N. 1927. *Frejdizm: Kriticeskij očerk*. Moscow, Leningrad: Gosizdat; *Freud e il freudismo*, A. Ponzio (ed.), L. Ponzio (trans.). Milan: Mimesis; now in Bachtin e il suo circolo 2014, 355–597; *Freudianism. A Critical Sketch*, Eng. trans. by I. R. Titunik, ed. by I. R. Titunik with N. H. Bruss. Bloomington: Indiana University Press, 1987. [*Freudianism: A Marxian Critique*, Eng. trans. New York, London: Seminar Press, 1973.].

Voloshinov, Valentin N. 1929. *Marksizm i filosofija jazyca. Osnovnye problemy sociologiceskogo metoda v nauke o jazyke*. Moscow-Leningrad: Priboj. 2nd ed. 1930; *Marxismo e filosofia del linguaggio*, M. De Michiel (trans.), A. Ponzio (ed.). Lecce: Manni, 2010; now in Russian & Italian in Bachtin e il suo circolo 2014, 1461–1839; *Marxism and the Philosophy of Language*, Eng. trans. L. Matejka and I. R. Titunik. New York, London: Seminar Press, 1973. New edition, 1986; *Marxisme et philosophie du langage* [1930], P. Sériot & I. Tylkowski-Ageeva (eds.), bilingual Russian/French edition. Limoges: Lambert-Lucas, 2010.

Welby, Victoria. 1983 [1903] *What Is Meaning? (Studies in the Development of Significance)*. Eschbach Achim (ed.), ix–xxxii, Intro. by G. Mannoury, xxxiv–xlii [= Foundations of Semiotics 2.] Amsterdam/Philadelphia: John Benjamins.

Welby, Victoria. 1985a. *Significs and Language*, ed. and Intro. by H. W. Schmitz, pp. ix–ccxxxvii. Foundations of Semiotics 5. Amsterdam-Philadelphia: John Benjamins. [Includes Welby's monograph of 1911, *Significs and Laguage* and a selection of other writings by her.].

Welby, Victoria. 1985b. *Significato, metafora, interpretazione*, It. trans., ed., and Intro. By S. Petrilli. Bari: Adriatica.

Welby, Victoria. 2007. *Senso, significato, significatività*, ed., It. trans., & intro. by S. Petrilli. Bari: Graphis.

Welby, Victoria. 2009. See Susan Petrilli. 2009.

Welby, Victoria. 2010. *Interpretare, comprendere, comunicare*, It. trans., ed. & Intro., pp. 11–96, by Susan Petrilli. Rome: Carocci.

Göran Sonesson
The relevance of the encyclopaedia. From semiosis to sedimentation and back again

Abstract: Unlike what is usually taken for granted in semiotics, communication, in the sense of semiosis, does not essentially depend on transport and/or recoding. Instead, it consists in the creation of an artefact, with the additional setting of a task of interpretation. This task may be set by the addressee, just as likely as by the addresser. In making this suggestion, we are inspired by Husserlean phenomenology, in particular as it was situated into a social framework by the Prague School. We cannot conceive of communication without sedimentation, the passive mnemonic remnants of earlier semiotic acts, which form the background to the interpretation of any current act. Although this notion was rediscovered by Lotman in terms of accumulation, phenomenology has been more thorough in its study of sedimentation. Giving a new twist to this idea, we are able to resolve some of the paradoxes of the current idea of extended mind. This paper also takes up the idea of semiotic acts being oriented either to the addresser or the addressee, as suggested, but in terms of cultures, by the Tartu School, then broadens the notion to stand for the shifting focus of attention, also as applied to the content of the semiotic act. Having recourse to Schütz's idea of a system of relevancies, combined with Gurwitsch's notion of the field of consciousness, we finally propose a resolution to the conundrum of the situated encyclopaedia, as it was characterized by Eco.

Keywords: Communication, sedimentation, interpretation, relevance, extended mind

As Michael Tomasello (2008) observes, communication is a kind of collaboration. Unlike other species, Tomasello (2009: 1f) notes, human beings are "born and bred to help". Other primates, it turns out, are able to co-operate when this is to their mutual benefit, and even, to some extent, to share food. But what they cannot do is to share information. This suggests that there is something particular about information, or, as I shall say, the pool of knowledge, but Tomasello has nothing to tell us about this specificity. Semiotics, on the whole, has not been of much help here either. Although trains and cars move, change of

---

Göran Sonesson, Division of Cognitive Semiotics, Lund University, Sweden

https://doi.org/10.1515/9783110662900-005

position in space is not a requisite of communication, in the sense that a meaning is communicated from one person to another, contrary to what is suggested by the mathematical theory of communication still current in semiotics, and promoted, notably, by Roman Jakobson (1960) and Umberto Eco (1976). Communication in the sense of presenting meanings has to be liberated from the sense in which it involves cars, trains, and the like, which change their position in space. Like the sharing of food resources, which is found in apes, the latter involves a movement from one position to another, and, if anything is shared, it is certainly not information. Two traditions from within semiotics may nevertheless lend us a hand here: There are things to be learnt about communication from the Prague model: that the receiver is equally active as the sender. And there are things to be learnt from the Tartu model. The latter is really concerned with relationships between cultures, but these can be reformulated in terms of the act of communicating. What the Tartu School says about sender and receiver cultures can be rephrased as two different positions in the act of communicating (Cf. Sonesson 1999).

1 Reconstruing communication

Let's start by distinguishing Communication^1 and Communication^2. Since this is the kind of communication in which we are interested at present, we will understand Communication^1 as the process by means of which some or other piece of knowledge (where knowledge is taken in such a broad manner as to include the fact of somebody just sneezing) only known to one (group of) person(s) is being shared with another (group of) person(s). Communication^2 involves the translocation of some item (person or thing) from one place to another. The tendency of understanding Communication^1 in terms of Communication^2 clearly antedates the formulation, by Claude Shannon & Warren Weaver (1949) of the mathematical model of communication. Indeed, well before the invention of communication theory, Valentin Vološinov (1929 [1973]) pointed out the fallacious implication of this comparison. Whether this "conduit metaphor", diagnosed by Michael Reddy (1979) really owes its popularity to being one of the "metaphors we live by", in the sense of George Lakoff and Mark Johnson (1980), is a moot question. In any case, it is not a metaphor in the sense of Aristotle, the vehicle of which makes you discover hitherto unknown and important properties of the tenor, as spelled out by Max Black (1962) comparing the metaphor to the (scientific) model. On the contrary, it is a metaphor which has the effect of obscuring the nature of Communication¹, the kind of communication which interests us here.[1]

1.1 The social phenomenology of the Prague School – and beyond

Neither Vološinov nor Reddy suggested any model apt to take the place of this beguiling metaphor. Reddy certainly proposed the "toolmaker's paradigm" as a better choice, but it is unclear how this conception could do the same job as the conduit metaphor. This is not the end of the confusion, however. One would think that the second metaphor, which has contributed to the muddle about the nature of Communication¹ must have come right out of the Shannon & Weaver model, according to which a message had to be transformed into another code, such as, in the case of telegraphy, the Morse code, in order to by communicated (in the sense, as it happens in the case of Communication¹ and most of the time also of Communication²). In the following, we will call this idea of communication Communication³. However, well before Shannon & Weaver (but after the invention of the telegraph), Peirce propounded an idea of the sign, which has been taken since then to be equivalent to Communication¹, according to which its meaning (interpretant, etc.) consists in its being exchanged for another sign, and so on (though not necessarily for ever, as Derrida understood it).[2] Unlike the translocation metaphor (Communication²), this comparison (Communication³) has something important to say about semiosis. Nevertheless, both metaphors serve to avoid the real issue about how meaning is produced and shared between different subjects.

It was in the social semiotics (or, as they said, semiology) of the Prague School in the 1930ies that a more relevant approach to the act of communication was sketched out. Jan Mukařovský (1974), one of the main figures of the Prague School, started out from the phenomenology of Edmund Husserl – or, more exactly, from that of his follower Roman Ingarden (1931 [1965]), – in order

---

[1] Here, and in the following, it is only for linguistic reasons that we speak about "kinds of communication". Communication¹ and Communication² (and, as we will see shortly, Communication³) are quite different in nature, although they may happen to occur together in some temporally and spatially situated acts.

[2] On the other hand, if we take Peirce's notion of sign to be a communication model, and if we downplay the idea of exchanging signs for signs, the triad of representamen-object-interpretant, may be taken to be equivalent to the phenomenological model which we are going to propose below.

to characterize communication, in particular as instantiated in a work of art, but then added to this a social dimension. The most important idea to retain from the Prague School, in my view, is that communication (in the sense of conveying information) is not necessarily about transportation or encoding, but it does involve the presentation of an artefact by somebody to somebody else, giving rise to the task of making sense of this artefact. This process by means of which an artefact is interpreted is called concretization. Mukařovský, like Ingarden, formulated this notion of concretization with reference to the work of art, but, in my view, this conception can be generalized to all kinds of communication processes in which information is shared or perhaps, better, jointly created. Since, to Mukařovský, this is a social act, the process of creating the artefact, as well as that of perceiving it, is determined by a set of norms, which may be aesthetic (and in works of art they would be predominantly so), but they can also be social, psychological, and so on. The work of art is that which transgresses these rules. Mukařovský points out, however, that these norms may be of any kind, going from simple regularities to written laws. We could conclude that there is a continuum from normalcy to normativity, without qualitative divisions being left out.

Already because this model builds on the phenomenological conception of perception, it can easily be generalized to the everyday case of communication. All kinds of communication consist in presenting an artefact to another subject and assigning him or her the task of transforming it by means of concretization into a percept. Simply put, what happens in communication, in the relevant sense, is that some subject creates an artefact, and another subject is faced with the task of furnishing an interpretation for this artefact (See Table 4.1; cf. Sonesson 1999; 2014). Displacement (Communication^2) may then be required, as when a letter is sent by train from one place to another, but it may also be the case that the addresser has to change location in order to initiate a semiotic act, such as sending a telegram (perhaps only relevant for such outmoded kinds of semiosis), or the addressee may have to take up some specific position to accommodate the act (as when going to a prehistoric cave, or even a museum, to see a picture, or to a theatre or a cinema to watch a play or a movie). As for recoding (Communication^3), it is sometimes needed, but most of the time, the same (or at least overlapping repertories of) signs may be used at both ends of the communication chain. The essential thing, in any case, remains the artefact and the instructions for bringing about the realization of its meaning.

In several earlier publications (Sonesson 1999; 2014; 2018), I have sketched this model of communication, but, so far, I have never pondered whether the task set by the addresser of the communicative act has something to do with the Gricean model of "non-natural meaning" (as opposed to the "natural meaning"

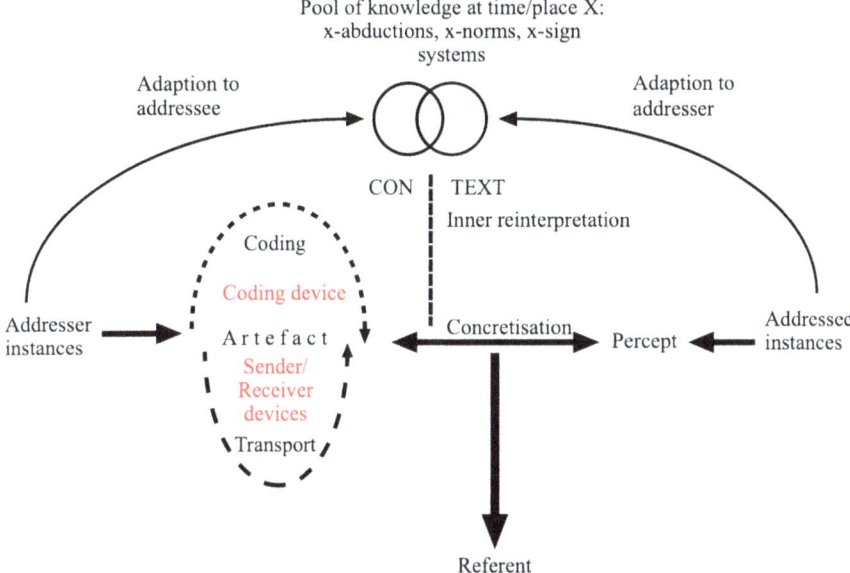

Figure 4.1: Model of communication integrating the Prague and the Tartu model, as proposed by Sonesson 1999.

of clouds, measles, animal tracks, etc.), according to which such meaning only comes into being if there is somebody around having an intention (in the sense of purpose) to convey the meaning; or, more precisely, if there is an intention to intend, and perhaps even an intention to intend the intending. Elsewhere (in Sonesson 2012; 2018), I have argued against this model, to my mind conclusively, suggesting that it stands the facts of our experience on its head. Thus, for instance, even if you are in the middle of a desert, and you recognize something as being an instance of writing, you will take for granted that it must have been caused by some person or other (including angels, djinns, and whatever) having had the purpose to produce it, whether directly, by writing it, or by means of what I have elsewhere called a *remote purpose*, which supposes there to be some kind of device that produces the writing, which, by increasing degrees of remoteness from the subject having the purpose, may be a set of seals, a printing press, a typewriter, a telegraph, or the printer of a computer (Sonesson 2002). On the other hand, if, on the same spot, you recognize some shapes in the sand as traces left by a camel or a horse, and of whatever else you may be able to identify from the shapes imprinted on the ground, if you are William of Baskerville, Zadig, or one of the Serendippus brothers, you take for granted (until proven otherwise, as in the Sherlock Holmes story "Silver Blaze") that these traces have not been produced on purpose, and you interpret them accordingly (See Figure 4.2). This may seem to be a kind of interpretation which is only relevant (or at least only arrives at the level of awareness) to hunter-gatherers, but, in fact, it is very relevant to scholars presently investigating possible indications of there being life elsewhere in the universe, i.e. biosignatures (See Dunér 2018).

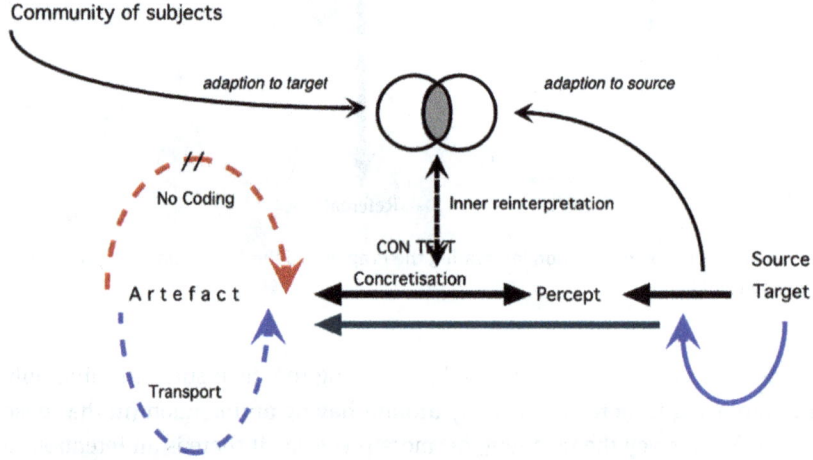

Figure 4.2: Model of (inadvertent) communication or, perhaps better, meaning conveyance, when there is no addressee, in the sense of a subject having a purpose in creating the meaning vehicle.

This suggests that at the beginning there is pattern recognition. In other words, there is a token which is mapped to its type, which does not necessarily mean, in this context, a specific type, but rather a type pertaining to a particular category of phenomena of the Lifeworld, such as, human beings or something kindred, in the first case, and things of nature, in the second case. But, in this second case, who sets the task of interpretation? Certainly not nature, which is always mute. The task is set by the collective knowledge of the community of which we are a part. In other words, William of Baskerville may be better than any other member of his community at interpreting the traces left on the ground by animals passing by, but his capacity in this respect can only be a refinement of the community knowledge existing of his sociocultural Lifeworld. If we generalize the counter-

description of the Gricean model which I have proposed elsewhere (Sonesson 2012), we may arrive at something like the following characterization:
1. X is perceived to be an instance I of the cultural object type O.
2. Our cultural experience tells us that object O a) is normally produced with a (more or less clearly articulated) purpose – or b) is the result of an act which did not include I as one of its purposes;
3. If 2a, there must be a conceivable subject S having such a purpose. If 2b, the subject to which the instance is assigned in not, at least in this particular case, supposed to harbour any purposes.
4. In the case of 2a, the purpose of producing O is normally to convey a message M from the subject S to some other (specified or non-specified) subject(s) S2. In the case of 2b, O is produced as a marginal result of some activity having a different purpose.
5. If there are no indications to the contrary, in case 2a, we have reasons to suppose that X has been produced with the purpose of conveying a message M from the subject S to the subject(s) $S2$. Similarly, in case 2b, if there are no indications to the contrary, we have reasons to suppose that X has been produced without any purpose of conveying a message of any kind, but as a result of an activity on the part of a subject which may be interesting in itself to $S2$.

Given this generalized model of the conveyance of meaning, it may be better not to say that the addresser sets a task of interpretation to the addressee, but rather that, along with the perceiving of the artefact, the addressee has the experience of a task of interpretation having been set for him, whatever the source of this task. This is coherent with the general thrust of the Prague school model, but it may be more specifically in line with the suggestions of Luis Prieto (1966; 1975a, b), although neither the Prague School, nor Prieto had to take into account the complications which we have just envisaged.

1.2 Meaning as sedimentation

According to Yuri Lotman (1976), the accumulation of information as well as of merchandise (for which read: material objects) precede their interchange and is a more elementary and more fundamental characteristic of a culture than communication. Material objects and information are similar to each other, in Lotman's view, and differ from other phenomena in two ways: they can be accumulated, whereas for example, sleep and breathing cannot, and they are not absorbed completely into the organism, because, unlike food, they remain separate objects after reception. At the time, Lotman may well have wanted to play on the ambiguity of the term information in the colloquial sense, and in the sense of the mathematical theory of communication. Here we will take it exclusively in the first sense, thus identifying it with meaning, knowledge, and even, in its aspect of being accumulated, with memory – and with what Husserl calls sedimentation (See further Sonesson 1999; 2010b).

Any present act of experiencing an object or state of affairs is embedded in patterns of understanding which modify these experiences, resulting from a process that Husserl (1954) calls sedimentation, a term made more famous by Maurice Merleau-Ponty (1945). This is the process in which previous experiences come to shape and condition more recent ones. In this context, sedimentation is, of course, a metaphor, like the conduit metaphor discussed above (in 1.1), but it is hopefully less misleading. To grasp the nature of sedimentation, and thus of the different kinds of memory, we will have to expand on the following judicious observation by Merlin Donald (1998: 11): "In humans there is a collective component to cognition that cannot be contained entirely within the individual brain. It is the accumulated product of individually acquired knowledge that has initially been expressed in a form comprehensible to other members of a society, tested in the public domain, filtered, and transmitted across generations." Husserl's idea (not referred to by Donald) is that such an accumulated product of experience can be reanimated in the phenomenological process, thus illuminating its validity, in the sense of its foundation.

A phenomenology geared to sedimentations should "inquire after how historical and intersubjective structures themselves become meaningful at all, how these structures are and can be generated" (Anthony Steinbock 2003: 300). In posthumous texts, Husserl distinguished between the *genetic* and *generative* dimensions of experience (Husserl 1973; Welton 2000; Steinbock 1995). Genetic phenomenology attempts to explore the origin and history of the sedimentation process in any given set of experiences. Every object in our experience has a genetic dimension: it results from the layering, or sedimentation, of the different acts that connect it with its origin in our personal experience, which gives it its validity. Thus, genetic phenomenology studies the genesis of meanings of things within one's own stream of consciousness. The genetic method enables us to plunge into layers of human existence that are pre-reflective, passive and anonymous, though nonetheless active. The term genetic is meant to evoke the idea of the life of an individual from the cradle to the grave.

There is also the further dimension of generativity, which pertains to all objects, and which results from the layering, or sedimentation, of the different acts in which they have become known, which may be acts of perception, memory, anticipation, imagination, and so on. Generative phenomenology studies how meaning, as found in our experience, is generated in historical processes of collective experience over time. The term generativity is meant to evoke the idea of generations following each other: "In distinction to genetic analysis, which is restricted to the becoming of individual subjectivity, a synchronic field of contemporary individuals, and intersubjectivity founded in an egology, generative phenomenology treats phenomena that are geo-historical, cultural, intersubjective, and normative." (Steinbock 2003: 292).

In his seminal paper on "The Origin of Geometry", Husserl (1954: 378 ff.) elucidates the way in which geometry derives from the praxis of land surveying. Although, in this paper, Husserl did not make this distinction, such an origin would only be genetic for people living at the time, but it must be considered the result of generative sedimentation for all subsequent generations. Taking all this into account, the return to the origin cannot amount to a reduction of geometry to land surveying, in which case non-Euclidean geometry would not only be impossible, but so would all of the "discoveries" of mathematics after the formalization of the practice of land surveying. As Husserl goes on to mention, though he fails to bring it into focus, geometry, as well as any other system of ideal structures, clearly has an existence beyond all the practice which is sedimented into them, because they are already present outside of time and space – or rather, in all times and spaces (after the foundational moment, or more precisely, the sequence of foundational moments; Husserl 1954: 371; see Sonesson 2015b).

It is important to note that the approach in terms of geneticity and generativity, unlike that preconized by Lotman, supposes accumulation/sedimentation to be as much a result of communication as vice versa. This does not only apply to semiotic acts, but to all acts accomplished by situated subjects. In other terms, each act of communication (and of meaning generally) adds to the sedimentation resulting in the pool of knowledge, and each act is also a realization of such a pool of knowledge (see Figure 4.3).

1.3 Sedimentation and extended mind

According to Dan Sperber (1996), sedimented meanings ("public representations") do not have any real existence, because, first, they are only material objects, until they are experienced by psychological subjects, that is, as "mental representations"; and, second, they subsist, and are distributed (and transformed) because they are reproduced as "mental representations": "Public representations are artefacts the function of which is to ensure a similarity between one of their mental causes in the communicator and one of their mental effects in the audience." The

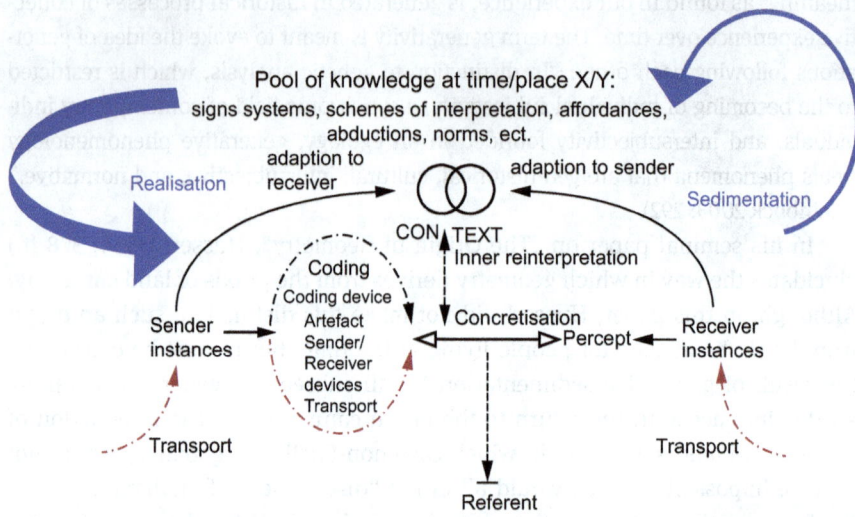

Figure 4.3: The act of communication, as construed in Sonesson 1999, with the addition of the process of sedimentation, which is the accumulated memory of historicized acts, and the process of realization, which recovers the structure of the act from the pool of knowledge which is sedimented.

first point is true in a way, but the second is not. In the case of systems (like "langue"), only the elements of the system (phonemes, letters, even contours in pictures) have to subsist mentally, while their combination is given in sedimented meanings. This also applies to "parole" (books, for instance, whether written or painted), to the extent that they consist of a certain sequence of elements taken from such systems. Whatever the truth of Sperber's conception, it certainly goes against the grain of (the family of theories of) that in cognitive science is nowadays known as the notion of extended mind.

As I have observed elsewhere, Husserl's conception of geometry as a system of connected sedimented acts of land surveying, which can be used (but not validated) in the form of sediments, as well as Lotman's idea of accumulation and Donald's notion of exogram, are all reminiscent of what has more recently been termed "the extended mind". Yet, some uses to which this and similar terms have been put are, to my mind, very doubtful, not to say paradoxical. Take the case of Otto and Inga, first broached by Andy Clark & David Chalmers (1998). As the story goes, Otto and Inga are both going to a museum. Otto has Alzheimer's disease, which is the reason why he has written down all the directions to the museum in a notebook to serve as his memory. Inga, however, is able to remember the directions using only her (un-extended) mind. The argument is that the only difference existing in these two cases is that Inga's memory is being internally processed by the brain, while Otto's memory is being served by the notebook. In other words, Otto's mind has been extended to include the notebook as the source of his memory. The notebook qualifies as such because it is constantly and immediately accessible to Otto, and it is automatically endorsed by him.

It is not entirely clear whether Clark & Chalmers really want to suggest that, in this way, the (non-extended) mind can be entirely bypassed, but something like this would seem to follow from their principle that what is functionally equivalent to processes in the mind (such as, in this case, memory) is an extension of the mind. In any case, this argument is explicitly made by Daniel Hutto, & Erik Myin (2013), in a book with the suggestive title "Radicalizing enactivism: Basic Minds Without Content". Hutto and Myin deny that any "content" (something sometimes also expressed as "any meaning") is needed at all for semiotic and/or mental acts to occur, with the possible (only sometimes mentioned) exception of language. But his cannot be: without a mind taking into account (though not necessarily reanimating, in Husserl's sense) what has been sedimented, there is nothing there. In other words, if Otto cannot read, which is a semiotic act (that is, an act of meaning or content), he cannot be doing anything at all with the notebook. The notebook is meaningless, if not actualized by a mind, just like writing in some unknown script. And that it is meaningless means that it cannot function in any way equivalent to Inga's memory, without the semiotic acts of reading and, in fact, the antecedent act of fixing the attention. Indeed, as everyone knows who has some real experience of people with dementia, Otto may very well write all the instructions down, but then he will very probably forget to look the information up.

In their more recent book, Hutto and Myin (2017) takes on mind *with* content, but they still restrict the latter to the single case of (verbal) language. There is much to be said about the semiotic inadequacy of such an account, but here we will concentrate just on their critique of the storage metaphor. Clearly, Lotman's idea of accumulation, as well as Husserl's notion of sedimentation and (perhaps less clearly) Donald's (2010) term "exogram", rely on the storage metaphor, if they are not presented as being storage quite literally. Indeed, all these notions have been forged to explain how meanings may persevere in time and be transferred in space, and how they can be available to more than one subject at a time, that is, in other terms, intersubjectively. It is difficult to see how anything of this can be accomplished without some kind of storage, however metaphorical, being involved. More specifically, without such storage, culture, let alone cultural evolution, appears to be impossible. The reason for Hutto and Myin (2017: 233ff) marking their distances to the storage metaphor seems to be that they want to think of mind as not being simply extended, but continuously so. It is no accident that they refer to Lambros Malafouris (2013: 227ff) who in his discussion of the extended mind argues in meticulous detail for there being no limit between the potter's (un-extended) mind and the potter's wheel. One is inevitably reminded of the fundamental point made by Peirce:

> A psychologist cuts out a lobe of my brain and then, when I find I cannot express myself, he says, "You see your faculty of language was localized in that lobe." No doubt it was; and so, if he had filched my inkstand, I should not have been able to continue my discussion until I had got another. Yea, the very thoughts would not come to me. So, my faculty of discussion is equally localized in my inkstand. (CP 7.366)

Another *locus classicus*, in this context, is, of course the blind man's cane, which, according to Merleau-Ponty (1945), is experienced as being a part of the blind man's body. Even if we can often experience such an umbilical link between ourselves and the inkstand (or, to modernize the example, the computer), and between ourselves in the guise of a potter and a potter's wheel, the experience of this nexus has to be reconciled with the property of storage, which is necessary for the extended mind to have any function to fulfil, not only in cultural evolution and history, but also as the intersubjective foundation, and the ongoing source of negotiation, of contemporary social life. Of course, this inkstand, computer, blind man's cane and potter's wheel, with which I am at this moment so inextricably united, will not endure, but other instances of them will. They perdure as types, but they are materially embodied as particular instances in time and space. The important thing, in any case, is that, *however extended the mind may be, it is necessarily extended from a particular pole,* that of a mind in the literal sense, which is the mind of a human (or at least animate) being.

Nevertheless, all kinds of extended mind cannot in any obvious way be considered as materially instantiated types mapped directly to the corresponding types. Geometry, logic, language, and many other systems clearly make up complex grids, only some of the members of which are materially instantiated, while depending for their meaning on their place in the mesh. Perhaps we should think of sediments not only as having attained different degrees of petrification, but also being able to shift back-and-forth in that extent. This is playing havoc with the sedimentation metaphor, but so does already Husserl's notion of reanimation. Something like this is needed to explain the continuity between the mind of the writer and his ink stand or computer, on the one hand, and the perdurance of the writing systems he uses in history on the other.[3]

---

[3] It is beyond the scope of the present paper to spell out the semiotic and/or ontological difference between the basic relation of token to type and that of token forming part of typological networks (but see Sonesson, 2020).

2 Attention to the pools of knowledge

Even though the understanding of any semiotic act may require us to conceive it from the point of the addressee, every concrete act of communication is structurally more or less biased to the addresser or the addressee, as the Tartu School has recognized, although their formulation referred to the nature of different cultures, rather than, as we will suggest, to different occasions of communication (see 2.1 below). Another question concerns the content of those overlapping bubbles of knowledge, at each specific moment of communication. We will suggest that the pool of knowledge *in situ* has to be understood in terms of the notion of relevance, as characterized by Alfred Schütz, as is can been reviewed in terms of Aron Gurwitsch's notion of the field of consciousness (see 2.2). Above (in 1.2), the notion of attention was referred to in passing, but we will not suggest that it may be considered the precursor of all semiotic acts. This can only be demonstrated, when the notion of attention is expanded to that of the field of consciousness, as will be suggested below, relying on Schutz and Gurwitsch, but also on Eco.

2.1 Orientation to the addresser – or to the addressee

It is, I think, an important modification brought to the phenomenological model employed, most directly adopted from Roman Ingarden, when Jan Mukařovský (1974) and his followers in the Prague School of semiotics set out to define the act of meaning from the point of view of the addressee, not from that of the addresser, similar in that respect to the now well-established pragmatics paradigm. Such an approach makes it understandable that traces left by an animal on the ground, or clouds harbouring rain, can be signs in equal measure to words and pictures (See Sonesson 2012 and 1.1 above). According to the Prague school model, all interpretation also takes place in accordance with a pool of knowledge, more or less shared between the addresser and the addressee, which has two main incarnations: the set of exemplary works of art and the canon, in the sense of the rules for how art works are to be made. Again, this double aspect of the pool of knowledge may be generalized from the special case of art to any artefact offered up for communication. On the one hand, there are certain exemplary artefacts, and, on the other hand, there are the schemes of interpretation.

In order to concretize this idea of a pool of knowledge, we may think of it as made up of schemes of interpretation. The notion of scheme has a history in phenomenology, particularly that of Alfred Schütz (see further II.2), as well as in cognitive psychology, from the original work on memory by Frederick Bartlett and the genetic psychology of Jean Piaget to the more recent contributions to cognitive science, where they are sometimes known as scripts, by the likes of David Rumelhart and Roger Schank. Summarizing this long and variegated tradition, Sonesson (1988: 17) describes a scheme as being "an overarching structure endowed with a particular meaning (more or less readily expressible as a label), which serves to bracket a set of, in other respects independent, units of meaning, and to relate the members of the set to each other". Bartlett talked mainly about memory schemes, and Schank notably mentions the restaurant script. The first is a scheme for mental operations, the second a scheme mainly for behaviour. Different culture may have different memory schemes, which means that a story coming from one culture is retold from memory by members of another culture using schemes prevailing in their culture. The restaurant scheme (or script) entails knowing more or less what you are expected to do while you are in a restaurant, which may also, no doubt, be different from one culture to another. More precisely, Schütz used the term "scheme of interpretation", and he claimed that it was historically constituted out of the sedimentation of earlier acts before being applied to the current act (as seen in Figure 4.3 above). Although Schütz doesn't say so, we can now add that such a process of sedimentation may be genetic or generative (see 1.2.)

According to an idea, suggested by Lotman (1976) as well as by Abraham Moles (1981), the addresser and addressee of any situation of communication start out with "codes" – or, as I would prefer to say, schemes of interpretation –, which overlap only in part, struggling to homogenise the system of interpretation as the communication proceeds. We can extend this idea by referring to the Tartu school conception that cultures may be sender-oriented and receiver-oriented (Lotman et al. 1975), transferring these properties to situations of communication. The communicative act may then be said to be sender-oriented or addresser-oriented, to the extent that it is considered to be the task of the receiver or addressee to recover that part of the system of interpretation which is not shared between the participants. It will be receiver-oriented, or addressee-oriented, to the extent that the task of recovering knowledge not held in common is assigned to the sender or addresser (see Figure 4.4). In other words, a situation of communication is addresser-oriented when it is the addressee that has to adapt to the interpretative resources at the disposal of the addresser, and the situation of communication is addressee-oriented when it is the addresser that has to adapt to the interpretative resources at the disposal of the addressee.

Art, as conceived under the regime of Modernism, has been characteristically addresser-oriented; mass media, in the entrenched sense of the term (which is not really applicable to all modern media), have been noticeably addressee-oriented.

A dialogue takes place when each of the subjects adapts his schemes of interpretation somewhat to that of the other; that is, in Piagetian terms, when there is both accommodation and assimilation. This would normally suppose there to be a large share of common ground from the beginning. On the other hand, when addresser and addressee fail to negotiate the parts of the interpretation system that they do not both possess, the resulting concretization will be a deformation. One or both of the subjects will then assimilate the message without accommodating to it. In this sense, both addresser-orientation and addressee-orientation are deformations; but they are normally deformations that are prescribed by the culture.

Although they derive from a quite different tradition, or, more exactly, from two different traditions, there are familiar names for these orientations: the adaptation to the addresser, and more generally the whole dimension going from the addressee to the addresser, can be termed *hermeneutic*: it is about the way of understanding the other and/or his works. And the adaptation to addressee, and the whole dimension going from the addresser to the addressee, can be called *rhetoric*, because it is about the way of best getting the message through to the addressee. The overall dimension, which concerns the resources at hand, is properly semiotic.

The elementary meaning-giving act, at least in the case of human beings, appears to be the act of attention. Taking my inspiration from Aron Gurwitsch's (1964) ideas about the "theme" at the centre of a "thematic field", and surrounded by "margins", later reconceived by psychologist Sven Arvidson (2006) as different approximation to the "sphere of attention", I have suggested that the gaze may function as an organizing device, transforming continuous reality into something more akin to a proposition (Sonesson 2012; 2014). It is possible to conceive of the orientation to the addresser or the addressee as part of such an act of attention. But attention clearly has a much wider scope, because is also pertains to the content of the semiotic act.

2.2 The relevance of the pool of knowledge

We have been looking from several different points of view at the partly overlapping circles of the communication model, ascribed to the addresser and the addressee, respectively – but, so far, we have not asked what the content of these circles really consists of, when considered at the precise moment at which the act of communication occurs, that is, in other words, at the moment of mutual attention between the addresser and the addressee. From early on in his scholarly carrier to the very end, Umberto Eco (1976; 1984; 1999; 2014; 2017) has been arguing, ever more persuasively, that semiotic content takes the form of an *encyclopaedia*, not a *dictionary*, where the latter is understood as a Porphyrean tree, that is, as a continuous binary subdivision of terms, in which no properties are ever encountered anew on the different branches. In contrast, Eco (2017: 40) describes the encyclopaedia as a rhizome:

> Every point of the rhizome can be connected to any other point; it is said that in the rhizome there are no points or positions, only lines; this characteristic is doubtful, however, because every intersection of two lines makes it possible to identify a point; the rhizome can be broken and reconnected at any point; the rhizome is anti-genealogical (it is not a hierarchized tree); if the rhizome had an outside, with that outside it could produce another rhizome, therefore it has neither an inside nor an outside; the rhizome can be taken to pieces and inverted; it is susceptible to modification according to the growth of our knowledge; a multidimensional network of trees, open in all directions, creates rhizomes, which means that every local section of the rhizome can be represented as a tree as long as we bear in mind that this is a fiction we indulge in for the sake of our temporary convenience; a global description of the rhizome is not possible, either in time or in space; the rhizome justifies and encourages contradictions; if every one of its nodes can be connected with every other node, from every node we can reach all the other nodes, but *loops* can also occur; only local descriptions of the rhizome are possible; in a rhizomic structure without an outside, every perspective (every point of view on the rhizome) is always obtained from an internal point in the sense that every local description tends to be a mere hypothesis about the network as a whole. Within the rhizome thinking means feeling one's way by *conjecture*.

In Sonesson (1989: 73), I followed Arthur Koestler (1978: 27ff), in suggesting that reality (at least as it is experienced by human beings) is a "multi-levelled, stratified hierarchy of sub-wholes", where each sub-whole or *holon* is, in relation to higher levels, a dependent part and, in relation to its own parts, a whole of remarkable self-sufficiency. But I also followed him in presuming that such "holarchies" can be regarded as "vertically" arborizing structures whose branches interlock with those of other hierarchies at a multiplicity of levels and form "horizontal" networks, termed reticulations. In this view, arborization and reticulation are complementary principles in the organization of meaning. Koestler's model, it seems to be, accomplishes the same task as Eco's encyclopaedia, but it does so by supposing a multiplicity of organizational networks whichintermesh at numerous points.

Patrizia Violi (2017: 234ff) takes Eco to task for claiming that, at the local level, that is, in our terms, at the specific moment that the act of communication takes place, the encyclopaedia is flattened out into a dictionary entry. I think both Eco's point and that of Violi are well-taken. To go beyond this opposition, and to get closer to grasping the nature of such a locally situated encyclopaedia, I suggest we should be exploring the notion of systems of relevancies. The most well-known proponent of a theory of relevance is no doubt nowadays Dan

Sperber (1996; 2005; & Wilson (1995 [1986]). While there are certainly things to be learnt from this approach, it is, as I tried to show elsewhere (see Sonesson 2018), basically misguided. In the present context, it is sufficient to point out that, according to Sperber & Wilson (1995 [1986]), there is only one principle of relevance, which is – relevance. Thus, meaning comes out as something completely contingent, resulting from the task of making the best of the situation at hand. The phenomenologist Alfred Schütz (1970), however, listed a series of principles, or more exactly "systems of relevancies", all broadly speaking social in nature, and having the function of guiding our interest in given situations as they occur in the Lifeworld (see Sonesson 2012). While Schütz (1970:25ff, 30ff.) did not forget about the contingencies of the present situation, the main thrust of his argument consists in imputing relevancies to the typicalities of the Lifeworld, in Husserl's sense of the term. And though he does not really define the notion of relevance either, he certainly connects it to the notion of selection, itself dependent on interest, which is operative already in perception.

What is never spell-out, however, is how what Schütz later calls systems of relevancies relates to his earlier notion of scheme. In his *Reflections*, admittedly, Schütz (1970: 2, 36, 39, 43, 170) mentions "schemes" and "schemes of interpretation" several times, and, at least on two occasions, he talks about "schemes of interpretational relevancies" (106f), which sounds as a hybrid between schemes and relevance systems. Curiously, the term scheme seems to be absent from Schütz' most important posthumous work, which abounds on the theme of relevancies, in terms of both structures and systems (Schütz and Luckmann 2003 [1979–1984]: 252ff.). Might not the system of relevancies be conceived as made up of schemes, or being equivalent to schemes, in which case we have a least something more of an account of the passive synthesis behind it, in order words, of the processes of sedimentation? We will no doubt never know what Schütz thought about this, but this idea could still be taken as a cue for developing his idea of relevance systems.

To illustrate how we might seek out those perceptions and sedimented experiences from our stock of knowledge which are relevant to the problem at hand, Schütz (1970: 4ff.) tells a story (taken from Cicero, *De divinatione*, I, XIII, XXIII) told about Carneades, the ancient Greek philosopher directing the Platonic Academy at the time when it had converted to Scepticism. In this story, Carneades enters a room which is badly lighted, not being sure whether what he sees in the corner is a pile of rope or a coiled snake. Initially, he has a roughly equally weighted motivation for believing the object to be the one or the other (see Schütz 1970: 16ff.). Carneades' point at the time, obviously, was that there is no truth available, but only verisimilitude. According to the anecdote, the man then realizes that the object is not moving, which offers him some simple evidence for taking it to be merely a coil of rope. In Carneades' terms, the first level of probability is reached, the most likely. Continuing the inspection of the object, however, the man is reminded that it is currently winter, and that snakes are torpid at this time of year. The original evidence is counter-evidenced, possibly convincing the man that extreme caution is called for. Finally, he picks up a stick, strikes the object in question, and observes that it still does not move, thereby corroborating the interpretation of it as a coil of rope. Instead of contravening evidence to the first verisimilitude, he now has confirmation of it. He has, therefore, not contented himself with gaining evidence at one level, but has sought out additional indications and counter-indications which could pertain to the situation. Thus, Schütz turns a sceptic's argument to a narrative of our progressive search for truth – which we can approach though not definitively attaining it, as both Husserl and Peirce have observed (See further Sonesson 2018).

But to give substance to the idea of relevance systems, we have to have recourse to the phenomenology of the field of consciousness developed by Aron Gurwitsch (1957; 1964; 1985), a scholar who, in Germany, was as much inspired by Husserl as Schütz, and who, in the US, become a close friend. Nevertheless, Gurwitsch and Schütz never manged to bring their different phenomenological analyses to bear on that of the other. Yet, this is exactly what we will try to do in the following. According to Gurwitsch, every perceptual situation is structured into a theme, a thematic field, and a margin. The theme is that which is most directly within the focus of attention. Both the thematic field and the margin are in contiguity with the theme, but the thematic field is, in addition, connected to the theme at a semantic level. When attending to the theme, we are easily led to change the focus to something within the same thematic field. Changing what was earlier in the margin into a theme, on the other hand, is felt to require some kind of outside incitement. In the margin is normally found some items of consciousness that always accompany us, such as our own stream of consciousness, our own body, and the extension of the Lifeworld beyond what is presently perceivable. But the margin will also contain all items that are not currently our theme, nor connected to this theme.

This is an excellent beginning for a theory of attention, as Sven Arvidson (2006) has recognized, but it is not a full-blown theory.[4] Schütz often connects his systems of relevance to such a thematic structure, though his references to Gurwitsch are rather oblique (1970, 2, 86, 161). The idea certainly originates in the work of Husserl, as well as in that of William James, but, to my mind at least, the most enlightening description was the one given by Gurwitsch, and it

---

4 For some further queries, see Sonessson 2010.

seems to inform what Schütz here writes. Interestingly, Gurwitsch (1957: 271f, 310ff; 1964: 343f, 394ff) formulates some critical remarks on Schütz' theory of relevance, with reference to the 1945 paper "On Multiple Realities" (now in Schütz 1967: 207–259). Or, to be more exact, he claims that his use of the term "relevance" is not the same as that found in Schütz' work. And he goes on to deny that Schütz' term has anything to do with the theme-thematic field-margin structure of the field of consciousness which interests him.

Clearly, at least in his later *Reflections*, Schütz (1970) took a different view. More to the point, Gurwitsch (1964: 342; his italics) observes that, to Schütz, "*a certain item is relevant to me* on the account of projects and pursuits that engage me", while to Gurwitsch himself, "a *certain item* is said to be *relevant to the theme* (which may well be a plan of action or a pursuit) and also to other items because of their relevancy to the theme". In fact, there are reasons to think that, in picking the term "relevance", Gurwitsch wanted to refer to the French verb, "relever", which, among other things, signifies something like "depending on" or "pertaining to a particular domain" (*Le Petit Robert*: "être du ressort de, dépendre de, être du domain de"). Indeed, this is precisely the meaning given to the term by Gurwitsch (1957: 270: 1964: 340, his italics): that which is relevant is not simply co-present with the theme, but it is "*of a certain concern* to the theme. They *have something to do* with it."[5]

Reading Schütz' 1945 paper, and writing the manuscript in France, before his close contact with Schütz in the US, Gurwitsch (1964: 342) observes that "though occasionally using the term in a sense close to ours", Schütz seems to understand relevance much more with reference to a given, embodied, and situated Ego. This seems to me less true about Schütz' later writings, taking into account his recourse to the Husserlean notion of typicality. In spite of Gurwitsch's critique, I think we are justified in seeing in Schütz' relevancies a kind of *thematic adumbrations*. At least Schütz' (1970: 26) *topical relevancies* could be understood in this sense: as "that by virtue of which something is constituted as problematic in the midst of the unstructuralized field of unproblematic familiarity – and therewith the field into theme and horizon." From a Gurwitschean point of view, nevertheless, one may wonder for whom something becomes problematic while other things remain familiar. The *interpretational relevancies* seem to involve the different possible interpretations of what the problematic item could turn out to be, which, in a case prominently discussed by Schütz, may be a pile of rope or a snake, and perhaps

---

5 Although written in English, Gurwitsch's (1957; 1964) book was first published in a French translation, but also the translation uses the term "relevance", not "pertinence", which at the time would have been the idiomatic translation (See Sonesson 2018).

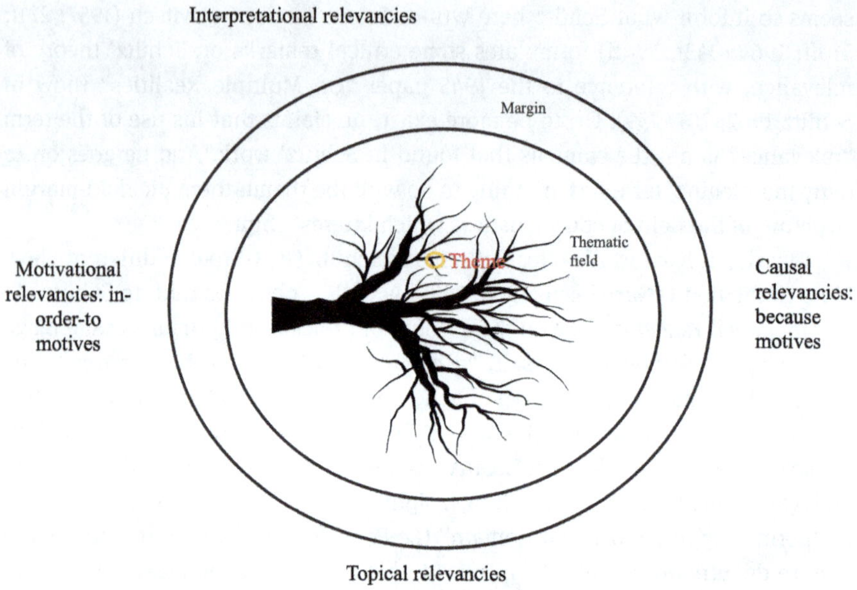

Figure 4.4: The field of consciousness, as conceived by Gurwitsch, with Schütz' systems of relevancies inscribed. Here we treat Schütz's sundry kinds of relevancies as different aspects of a system of relevancies. The figure shows arborization, in the sense of Koestler, but abstracts, in the interest of readability, from reticulation, which is what makes the difference between the dictionary and the encyclopaedia.

other things, but certainly not a table or a bed (Schütz 1970: 38ff). These interpretations seem to me to be difficult to separate from the *topical relevancies*, of which they are rather a part, somewhat like a paradigm, a set of alternatives, in relation to a syntagm, the chain of connected items. The *motivational relevancies* are more obviously beside the point in a Gurwitschean perspective, because they have to do with the motives which make us act on our interpretations (Schütz 1970: 45ff). But Schütz might have been better inspired to treat topics, interpretations, and motives as different aspects of relevance systems.

Pursuing the lead of my earlier discussion (see Sonesson 2018), I will now spell out the lineaments of a model of what may take place within the overlapping circles of the communication model, that is, the interacting minds of the addresser and the addressee (See Figure 4.4). For the sake of readability, we here abstract from reticulation, in Koestler's sense, which is what makes the difference between the encyclopaedia and the dictionary, in the sense of Eco. Interpretation, topics, and motivation are considered to be different aspects determining the choices made in the network. Following other texts by Schütz (1974 [1932]; 1962–1996), motivations are divided into in-order-to motives (motives in the ordinary language sense) and because-of motives (causes). Altogether, this model can be considered to show a particular state of the field of consciousness, with the theme being situated, as in the story of the Carneadean man, right in the corner where a configuration appears (interpretational relevance), which could be seen as a pile of rope or a coiled snake or perhaps some third thing (topical relevancies). Carneades wants to enter the room, but hesitates, because of the danger which could result from this configuration being identified as a snake (in-order to motives), where this motive itself builds on the knowledge that snakebites may be poisonous (because motive).

Thus, the concrete situation serves to prune to the savage wood of the encyclopaedia into the likeness of a Porphyrean tree. Reticulation may not have to be given up, but it retreats to the margin of consciousness, when holarchy comes to the front.

3 Conclusion

This paper started as a critique of the classical notion of communication, as defended most famously by Jakobson and Eco, along the lines of Shannon & Weaver. Instead of transport and/or recoding, we argued, it is the creation of an artefact and the setting of a task to make sense of it which is central to semiosis. As shown most clearly by what, in other traditions, is known as "natural signs", this task may be set by the addressee, rather than the addresser. While this model is most clearly inspired by Husserlean phenomenology, it could also be taken as an interpretation of the Peircean triad, but it is to the Prague School that we own the insertion of semiosis into a social framework by. Another phenomenological notion which turned out to be important here, however, is that of sedimentation, the passive mnemonic remnants of earlier semiotic acts, which form the background to the interpretation of any current act. From Husserl's posthumous writings, we learned that such sedimentation may be genetic (derived from the experience of the individual from the cradle to the grave – or almost) or generative (handed down from one generation to another, and so on indefinitely). On the other hand, we suggested that some paradoxes resulting from the notion of extended mind, in particular in an enactionist interpretation, can be resolved by distinguishing different extents of petrification of the sediments, which can be shifted back and forth. Another idea inspired by the Tartu School which we broach here is that of the orientation to the addresser or the addressee, though in terms of situations of communication rather than cultures. We go on the suggest that this is a particular application of the act of attention, which is the foundation of all semiotic acts, which then also can be applied to the content of the semiotic act. Starting out from Eco's idea of the distinction between the encyclopaedia and the dictionary, we propose a solution to the conundrum of the situated encyclopaedia, which, according to Eco, becomes a dictionary, by suggesting that each situation of communication creates its own modulation of the field of consciousness, as understood by Gurwitsch, when applied to at network resembling what Schütz called a system of relevancies. In later studies, we intend to show that this model will be productive for better understanding many semiotically relevant issues, such as signs, icons, and metaphors.

References

Arvidson, Sven. 2006. *The sphere of attention: context and margin*. London: Kluwer Academic.
Black, Max. 1962. *Models and metaphors: studies in language and philosophy*. Ithaca, N.Y.: Cornell Univ. Press.
Clark, Andy & David Chalmers. 1998. The extended mind. *Analysis* 58(1). 7–19.
Donald, Merlin. 1998. Hominid enculturation and cognitive evolution. In Colin Renfrew & Chris Scarre (eds.), *Cognition and material culture: the archaeology of symbolic storage*, 7–17. Cambridge: The McDonald Institute for Archaeological Research.
Donald, Merlin. 2010. The Exographic Revolution: Neuropsychological Sequelae. In L. Malafouris & C. Renfrew (eds.), *The Cognitive Life of Things: Recasting the boundaries of the mind*, 71–79. Cambridge, UK: McDonald Institute Monographs.
Dunér, David. 2018. Semiotics of Biosignatures. In *Southern Semiotic Review*, 9. 47–63.
Eco, Umberto. 1976. *A theory of semiotics*, Indiana U.P., Bloomington.
Eco, Umberto. 1984. *Semiotics and the philosophy of language*. London: Macmillan.
Eco, Umberto. 1999. *Kant and the platypus: essays on language and cognition*. London: Secker & Warburg.
Eco, Umberto. 2014. *From the tree to the labyrinth: historical studies on the sign and interpretation*. Massachusetts: Harvard University Press, Cambridge.
Eco, Umberto. 2017. Intellectual autobiography of Umberto Eco. In Sara Beardsworth & Randall E. Auxier (eds.), *The philosophy of Umberto Eco*, 3–66. Open Court, Chicago, Illinois.
Gurwitsch, Aron. 1957. *Théorie du champ de la conscience*. Bruges: Desclée de Brouver.
Gurwitsch, Aron. 1964. *The field of consciousness*. Pittsburgh: Duquesne University Press.
Gurwitsch, Aron. 1966. *Studies in phenomenology and psychology*. Evanston: Northwestern University Press.
Gurwitsch, Aron (1985). *Marginal Consciousness*. Athens, Ohio: Ohio University Press.
Husserl, Edmund. 1939. *Erfahrung und Urteil*. Prag: Academia Verlagsbuchhandlung.
Husserl, Edmund. 1954. *Die Krisis der europäischen Wissenschaften und die transzendentale Phänomenologie: eine Einleitung in die phänomenologische Philosophie*. Husserliana: Gesammelte Werke VI, 2nd ed. Haag: Nijhoff.
Husserl, Edmund. 1973. *Husserliana: gesammelte Werke. Bd 15, Zur Phänomenologie der Intersubjektivität, T. 3: 1929–1935: Texte ausdemNachlass*. Haag: Nijhoff.
Hutto, D. Daniel & Erik Myin. 2013. *Radicalizing enactivism: basic minds without content*. Cambridge, Mass.: MIT Press.

Hutto, D. Daniel & Erik Myin. 2017. *Evolving enactivism: basic minds meet content*. Cambridge: MIT Press.
Koestler, Arthur. 1978. *Janus: a summing up*. London: Hutchinson.
Ingarden, Roman. 1965 [1931]. *Das literarische Kunstwerk. Eine Untersuchung aus dem Grenzgebiet der Ontologie, Logik und Literaturwissenschaft*, Halle: Max Niemeyer. 3. Auflage.
Jakobson, Roman. 1960. Linguistics and poetics. In Thomas Sebeok (ed.), *Style in language*, 350–377. Cambridge, Mass.: MIT Press.
Merleau-Ponty, Maurice. 1945. *Phénoménologie de la perception*. Paris: Gallimard.
Lakoff, George & Mark Johnson. 1980. *Metaphors we live by*. Chicago: University of Chicago Press.
Lotman, Yuri. 1976. Culture and information. *Dispositio. Revista hispánica de semióticaliteraria* 3(1). 213–215.
Lotman, Yuri. 1979. Culture as collective intellect and problems of artificial intelligence. *Russian Poetics in translation* 6, 84–96. University of Essex.
Lotman, Yuri, et al. 1975. *Theses on the semiotic study of culture (as applied to Slavic texts)*. Lisse: The Peter de Ridder Press.
Malafouris, Lambros. 2013. *How things shape the mind: a theory of material engagement*. Cambridge, Massachusetts: MIT Press.
Moles, Abraham. 1981. *L'image, communication fonctionnelle*. Paris: Casterman.
Mukařovský, Jan. 1974. *Studien zur strukturalistischen Ästhetik und Poetik*. München: Hanser.
Peirce, Charles Sanders. 1931–1958. *Collected Papers I-VIII*. Hartshorne, C, Weiss, P, & Burks, A, (eds.). Cambridge, MA: Harvard University Press.
Peirce, Charles Sanders. 1998. *The Essential Peirce, volume I-II*. Ed. By the Peirce Edition Project. Bloomington and Indianapolis: Indiana University Press 1998 [EP].
Prieto, Luis. 1966. *Messages et signaux*. Paris: PUF.
Prieto, Luis. 1975a. *Pertinence et pratique. Essai de sémiologie*. Paris: Minuit.
Prieto, Luis. 1975b. *Essai de linguistique et sémiologie générale*. Genève: Droz.
Reddy, Michael. 1979. The conduit metaphor: A case of frame conflict in our language about language. In Andrew Ortony (ed.), *Metaphor and Thought*, 284–310. Cambridge: Cambridge University Press.
Schütz, Alfred. 1962–1996. *Collected papers* 1-4. The Hague: M. Nijhoff.
Schütz, Alfred. 1970. *Reflections on the problem of relevance*. New Haven: Yale University Press.
Schütz, Alfred. 1974 [1932]. *Der sinnhafte Aufbau der sozialen Welt: eine Einleitung in die verstehende Soziologie*. Frankfurt am Main: Suhrkamp.
Schütz, Alfred & Thomas Luckmann. 2003 [1979–1984]. *Strukturen der Lebenswelt*. Konstanz: UVK Verlagsgesellschaft mbH.
Shannon, E. Claude & Warren Weaver. 1949. *The mathematical theory of communication*. Urbana, Ill.: University of Illinois Press.
Sonesson, Göran (1988) *Methods and models in pictorial semiotics*. Report from the Semiotics Project, Lund University. Available at https://www.researchgate.net/publication/242511757_Methods_and_Models_in_Pictorial_Semiotics.
Sonesson, Göran (1989). *Pictorial concepts. Inquiries into the semiotic heritage and its relevance for the analysis of the visual world*. Lund: Aris/Lund University Press 1989.
Sonesson, Göran. 1999. The signs of life in society – and out if. *Sign System Studies* 27. 88–127.

Sonesson, Göran. 2002. The act of interpretation. A view from semiotics. *Galáxia*, 4, 67–99. São Paolo, EDUC.
Sonesson, Göran. 2010. Semiosis and the Elusive Final Interpretant of Understanding. *Semiotica* 179–1/4. 145–258.
Sonesson, Göran. 2012. The foundation of cognitive semiotics in the phenomenology of signs and meanings. *Intellectica* 2(58). 207–239.
Sonesson, Göran. 2014. Translation and Other Acts of Meaning. In between Cognitive Semiotics and Semiotics of Culture. *Cognitive Semiotics* 7(2). 249–280.
Sonesson, Göran. 2015a. From remembering to memory by way of culture: a study in cognitive semiotics. *Southern Journal of Semiotics* 5(1). 25–52.
Sonesson, Göran. 2015b. Phenomenology meets semiotics: two not so very strange bedfellows at the end of their Cinderella sleep. *Metodo* 3(1). 41–62.
Sonesson, Göran. 2016. Cultural Evolution: Human History as the Continuation of Evolution by (Partially) Other Means. In David Dunér & Göran Sonesson (eds.), *Human Lifeworlds: The Cognitive Semiotics of Cultural Evolution*, 301–336. Frankfurt/M.: Peter Lang.
Sonesson, Göran. 2017. Preliminaries to a Taxonomy of Intersemiosis". *Punctum* 3(1). 119–131.
Sonesson, Göran. 2018. New Reflections on the Problem(s) of Relevance. In Hisashi Nasu& Jan Straßheim (eds.), *Relevance and Irrelevance: Theories, Factors and Challenges*, 21–50. Berlin: de Gruyter.
Sonesson, Göran. 2020. Translation as culture: The example of pictorial-verbal transposition in Sahagún's *Primeros Memoriales and Codex Florentino*. *Semiotica* 2020; 232: 5–39, co-edited by Göran Sonesson.
Sperber, Dan. 1996. *Explaining culture: a naturalistic approach*. Cambridge: Blackwell.
Sperber, Dan. 2005. Modularity and relevance: How can a massively modular mind be flexible and context-sensitive? In Peter Carruthers, Stephen Laurence & Stephen P. Stich (eds.), *The Innate Mind: Structure and Content*, 53–68. New York: Oxford University Press.
Sperber, Dan & Deidre Wilson. 1995 [1986]. *Relevance: Communication and cognition*, 2nd edn. Oxford: Blackwell.
Steinbock, J. Anthony. 1995. *Home and beyond: generative phenomenology after Husserl*. Evanston IL: Northwestern University Press.
Steinbock, Anthony J (2003) Generativity and the scope of generative phenomenology. In Donn Welton (ed.), 289–325. *The New Husserl: A Critical Reader*. Indiana University Press.
Tomasello, Michael. 2008. *Origins of human communication*. Cambridge, Mass.: MIT Press.
Tomasello, Michael. 2009. *Why we cooperate*. Cambridge, Mass.: MIT Press.
Violi, Patrizia. 2017. Encyclopedia: Criticality and Actuality. In Sara Beardsworth & Randall E. Auxier (eds.), *The philosophy of Umberto Eco*, 223–250. Chicago, Illinois: Open Court.
Vološinov, N. Valentin. 1929[1973]. *Marxism and the philosophy of language*. New York: Seminar Press.
Welton, Donn. 2000. *The other Husserl: the horizons of transcendental phenomenology*. Bloomington IN: Indiana University Press.

Luis Emilio Bruni
Cultural narrative identities and the entanglement of value systems

Abstract: This chapter explores the nature and the implications of the processes of reflective construction of "cultural-selves" and "collective consciousness" mediated by the narrative function. In particular, it brings together the notions of "narrative identity" and "heterarchy of values" in order to synthesize a relational, processual and heterarchical notion of *cultural narrative identity*. For this purpose, it highlights the centrality of "values" in the determination of identities. Values may be spread throughout a web of emerging intertwined spheres and domains, encompassing inseparably the individual, the social and the cultural; in domains that go from private to public, from family to work, from local to national to regional to global, touching the many nuances of interest groups and stakeholders co-existing in a globalized civil society. Therefore the processes of identification very often confront individuals and whole cultural layers with non-transitive value scales that give place to *dynamic systems of heterarchical belonging*. The aim is to explore whether such a processual heterarchical perspective can be of utility in understanding the paradoxes and contradictions in contemporary cultural dynamics in light of the acceleration propelled by the global platform of digital technology. The approach considers *heterarchies* as the loci of competing and coexisting value systems and multiple "regimes of worth". Once we have the consideration of value-adherence in multilayer cultural processes and networks, we are bound to consider heterarchical processuality in order to be able to elucidate the rationality of the putative paradoxes and contradictions.

Keywords: cultural narrative identity, heterarchy of values, heterarchical processuality, globalization, value systems, heterarchical belonging, digital technology

1 Introduction

If you read carefully the title of this article, you will probably agree with me that it invites a lot of trouble. Every single key term (and their permutations) is problematic and practically constitutes a field of study on its own. It is hard to

Luis Emilio Bruni, The Augmented Cognition Lab – Aalborg University, Denmark

https://doi.org/10.1515/9783110662900-006

find consensus about the definition of the terms: "Culture", "Identity", "Values" and "Narrative". There are books, articles, handbooks and tons of materials trying to define these terms and review their respective histories. An additional problem is the fact that these four terms are very intuitive for us and we all have our own definitions and understanding of them.

In my attempt to bring them together, I will put into relation and intersect several established areas of research that perhaps have not been sufficiently in conversation. I will draw inspiration and knowledge from research areas such as identity studies, narrative identity, cultural identity, cognitive and cultural semiotics, and even systems theory and cybernetics. However, I would also like to invite the reader to hold as much as possible to our collective intuitive understanding of these terms, and concentrate rather on their possible interplay. This means that my strategy will be to introduce some selective aspects of these fields and terms, which are instrumental to my argument for the pertinence of synthesizing a notion of "cultural narrative identity" that entails a processual and *heterarchical* perspective, which, I claim, may be useful to tackle contemporary global cultural phenomena.

2 Short on identity

The intrinsic relations between being, unity and identity have been central in the history of ontology and philosophy. The concept of *identity* has been in many different ways considered constitutive for the definition of a being or the delimitation of a unity. The debate ranges from positions that request clear identity criteria for discriminating among existences, to positions that reject the possibility that identity – as a constitutive feature of a being – cannot be defined in an absolute and general sense, or as many seem to fear, in any "essentialist" way. Nevertheless, whether ontologically grounded or epistemologically instrumental, for practical and analytical purposes, some sort of essentialism seems to be inherent (i.e.: essential!) to human cognition. My interest here is not so much on the third person ascription of identity to "things" by an observer (i.e. a categorization or taxonomic endeavor), but rather the kinds of first person reflective, recursive and processual formation of identities, which include personal (or individual) identity, and, more specifically, cultural identity.

In their introduction to the Handbook of Identity Theory and Research, Vignoles, Schwartz and Seth (2011) list some pertinent questions that "have plagued" the literature on identity: "(1) Is identity viewed primarily as a personal, relational, or collective phenomenon? (2) Is identity viewed as relatively stable, or as fluid and constantly changing? (3) Is identity viewed as discovered, personally constructed, or socially constructed?" These questions will help us to think about the relational, processual and heterarchical nature of what we are calling *cultural narrative identity*. In their extensive review, and throughout the whole volume, Vignoles, Schwartz and Seth (2011) show the diversity and the "power" of the identity construct in a myriad of academic disciplines and fields such as psychology, sociology, anthropology, linguistics, political science, education, family studies, and public health, from different traditions, methodologies and focal levels of analysis. They conclude that on a fundamental level, identity involves people's explicit or implicit responses to the question: "Who are you?" As we will see, very often the answer to that question will take the form of a story.

3 Narrative identity

According to Ricoeur (1991: 73), narrative identity fundamentally refers to "the sort of identity to which a human being has access thanks to the mediation of the narrative function". For him, this can be the "life stories" of an individual or of a historical community. If knowledge of the self is an interpretation, this interpretation finds in the narrative mode a privileged mediation. The story that is constructed has to make sense of what it has been, what it is, and what it is desired or expected that it will be of that identity, its continuity and its permanence in time. At the same time, the story has a protagonist, the self, who brings agency into the picture.

Ricoeur (1991: 73) searches for the overlapping zone of two "modes" (or connotations) of the notion of identity: identity as "sameness", and identity as "self". These two modes refer respectively to the "what" and the "who" of the identified unity. "Sameness" can be related to permanence in time – physical and psychological continuity – while "self" can be related to agency. Narrative brings both notions together into what McIntyre (2013 [1981]) calls the "the narrative structure and unity of a human life" (Ricoeur, 1991). Personal identity as a human reflective process has been conflated with consciousness and memory, via the relation, and the continuity, between past and present. It is this temporal dimension of individual or cultural identity that makes the case for its narrative conception.

There are many psychological descriptions of what happens when identities dissipate: role-confusion, depersonalization, estrangement – the experience of not belonging to one's own psychic events, divorce and alienation from one's own internal psyche, one's own body, one's own external world, as if the natural relation of the self with these three sites is fractured. In this context, is culture to be considered part of the external world? Or, are we to say with Lotman (1990: 223) "We are within [culture], but it – all of it – is within us"? We can say therefore that cultural identity is an intrinsic aspect of personal identity, and this relation leads us to point out one of the key correlates of identity: *the sense of belonging.* The reflective processual identity that we will be referring to is not about being, but about becoming and belonging. Our becoming constantly questions our belonging.

After the seminal works on the topic of narrative identity in the late 1980's and early 1990's in different disciplines (e.g.: philosophy, Ricoeur 1984, 1991; psychology, McAdams 1988, 1996; social sciences, Sommers 1992, 1994), narrative identity became an interdisciplinary field with deep roots in psychology, therefore exploring mostly the level of individual narrative identity, with very little having said about the cultural or social levels (Sommers being an exception). Thus, at the level of the individual,

> . . . narrative identity is an internalized and evolving story of the self that provides a person's life with some semblance of unity, purpose, and meaning. Complete with setting, scenes, characters, plots, and themes, narrative identity combines a person's reconstruction of his or her personal past with an imagined future in order to provide a subjective historical account of one's own development, an instrumental explanation of a person's most important commitments in the realms of work and love, and a moral justification of who a person was, is, and will be. (McAdams, 2011: 100)

Thus according to McAdams' review, over the past 30 years the concept has evolved in many different directions, encompassing perspectives from cognitive science, life-course developmental studies, cultural psychology, sociology, and personality and social psychology, having become a central component of a "full, multi-level theory of personality" (McAdams 2011).

It can thus be argued that the construction of narrative identities has become a multidimensional and multilayer phenomenon which spans through a web of emerging intertwined spheres and domains, encompassing inseparably the individual, the social and the cultural; in domains that go from private to public, from family to work, from local to national to regional to global, touching the many nuances of interest groups and stakeholders co-existing heterarchically in a globalized civil society.

4 The cultures of culture

Even more extensive and problematic is the very notion of "culture" and the different perspectives and disciplines that study it. This is worsen by the fact that the term "culture" has been, and still is, a central weapon in intellectual political debates in senses that tend to portray culture, and/or its definitions, as ideological constructs. Another complication is the overlapping of the term culture with "sister" terms such as "society", "civilization", "tradition", and even "nation".[1]

Etymologically the term tradition stems from the Latin "traditionem" (trans = over + dare = give), which signifies delivery, surrender, a handing over, as in the Augustinian sense " . . . what they had received from the Fathers, this they delivered to the children" (Saint Augustine 430 [1957]). Therefore in the modern sense, tradition refers to "things" handed down from generation to generation, which is always implicit in the notion of culture. In Raymond Williams's terms "Whatever holds 'significance' from the 'set of meanings' received from 'the tradition' has to be valued in terms of the present experience. For this we have to return them to immediate experience." (Shashidhar 1997). According to Shashidhar (1997), what Williams attempts to show is that any hope of understanding human-social reality lies in coherently relating the significant statements received from the past instances of that "lived" reality to the "immediate" living of our present. In other words, Williams sees such social reality as a hermeneutic dialectic between the past and the present: "Somewhere, in the world of human thinking coming down to us from our predecessors, the necessary insights, the fruitful bearings, exist. But to keep them where they belong, in direct touch with our experience, is a constant struggle" (Williams 1960; Shashidhar 1997). Thus, our sense of belonging to a culture, a nation or a tradition is intrinsically related to how we

---

[1] Throughout this chapter, the reader may find problematic an apparent interchangeability of the notions of society and culture, or of social and cultural. We are by no means claiming that they are synonyms, but they are certainly mutually constitutive. There is no (human) society without culture and the there is no culture without social life. A culture may be spread in many societies, and a society may contain several cultures. This is an additional reason for pursuing a heterarchical approach to socio-cultural processes. However, I believe there is a fundamental asymmetry between the two categories. There can be social relations, but not cultural processes, without symbolic representation. Therefore I consider culture as a more human-specific and encompassing category than sociality (I say sociality here because for some "society" may be a structure of sociality exclusive of human beings). Based on this closeness between the social and the cultural, the approach advance here will be drawing inspiration and making extrapolations from both social and cultural theories, and this is what may give sometimes an impression of interchangeability, which I hope can be tolerated by the reader.

experience it temporally, i.e.: the interplay of our memories and our projections into the future, or, in narrative terms, our continuous existential synthesis of analepsis and prolepsis.

The etymological origin of the term culture as "the tilling of land" relates it to the notion of "civilization" as the passage from nomadic to sedentary modes of life: "Now Abel was a keeper of sheep, and Cain a tiller of the ground". It is with the emergence of the positivist academic disciplines that culture becomes "the intellectual side of civilization", until it ends up almost as a synonym with civilization: societies mutate through successive states of cultural or civilization *progress* in terms of knowledge, beliefs, art, morality, law, costumes, etc. However, the terms "civilization" and "progress" came about only in the XVIII century with the economists prior to the French Revolution, such as Turgot and Littré, and entered the modern dictionaries as late as 1835, under the influence of the "new ideas" of the XIX century: scientific discoveries, industrial revolution, trade, "well-being", "the age of prosperity" "indefinite progress" and the age of the "absolute civilization" (Guenon 1982 [1945]). According to Guenon, with the advent of positivism, civilization became the degree of development and perfectionism reached by the European nations in the XIX century. In that period (in 1871), Edward Tylor initiates the Modern technical definition of culture as socially patterned human thought and behavior. In 1917, Alfred Kroeber (1917), a foundational figure of cultural anthropology, made a somehow forgotten seminal contribution, emphasizing the cross-generational aspects of culture beyond its individual human carriers (the "culture bearers"). Individuals are born into and are shaped by a preexisting culture that continues to exist after they die. In this sense, Kroeber's work can be considered an antecedent to Yuri Lotman's semiotic conception of the "cultural space" (Lotman 1992). With the advent of cultural psychology comes the emphasis of culture as the production and spread of explicit representations, socially shared information that is symbolically coded, which in contemporary global society encompasses a complex merging of mass, pop and digital culture. From this complexity emerge new criteria and kinds of identity, as for example what Haug (1987) calls "commodities' identity", imaginary spaces in which individual consumers construct their own identity by comparing it with a generalized "other", where advertising is a form of "para-ideology" not on the same level of other customary cultural identifying entities such as the state, law or religion, which he claims are proper ideological powers.

The last notion that I that think worth of qualifying in this context is the notion of "nation", as it seems to have had a much stronger impact on the sense of belonging to historical collectives of people than other identity markers (such as country or civilization). The French philologist and historian Ernest

Renam delivered a conference at the Sorbonne in 1882 (Renam 1996 [1882]) with the title "What is a Nation", which provides and insightful and visionary account of the complex problem of overlapping values and identities, which emphasizes the gluing effect of the temporal experience of "having gone together through many things", in other words, what we could consider the raw material of a narrative.

Basically, Renam sees the notion of nation as a deeply-felt identity level laying above race, language, ethnicity, religion, community of interest, geography or military necessities, and therefore much above country and perhaps even culture: the essence of a nation is "that all its individuals must have many things in common but it must also have forgotten many things." He recognized very early that if racial criteria for identity should become predominant, this could lead to the destruction of European civilization, which was close to happen in the first half of the 20th century. He exemplifies how the intimate alliance between the Roman Empire and Christianity delivered a severe blow to the idea of race, excluding for centuries ethnographic criteria for the formation of identities. He supported this by pointing out how the genetic origins of humanity are tremendously anterior to the origins of culture, civilization, and language and how the primitive Aryan, Semitic, and Touranian groups had no physiological unity. According to Renam, historians – as opposed to anthropologist – understand race as a cultural construct. Therefore, shared things like reason, justice, truth, and beauty constitute more valid criteria for placing oneself within a narrative identity – things with which "those who belong" can agree upon. In other words it is values, cultural values, which more properly define narrative identities. A similar reasoning comes with language as a criterion: "Languages are historical formations, which tell us very little about the race of those who speak them" and the political importance that one attaches to languages comes from the fact that, in the past, they have been erroneously regarded as indicators of race.

With religion the issue is more complex because traditionally religion has been a transnational connector of identities, but according to Renam, with the secularization of the State – there where it has taken place – religion has become a matter of individual conscience and there are no longer single masses of people believing in a uniform faith, and whereas it is certainly a very powerful identity trait, it can no longer be considered a trait that determines the identity of (secularized) nations. "Community of interest" as an identity marker can be exemplified today with the advent of the European Union, something that Renam actually predicted. The challenge that the EU faces today is how to include the "European sentiment" in what would otherwise just be a geographically determined commercial treaty with a military alliance. Geography would be the substratum, but a "nation is a body and soul at the same time". Renam's answer to the question "what is a nation" is all about sharing a narrative identity:

> A nation is a soul, a spiritual principle. Two things, which in truth are but one, constitute this soul or spiritual principle. One lies in *the past*, one in *the present*. One is the possession in common of *a rich legacy of memories*; the other is *present-day consent*, the desire to live together, *the will to perpetuate the value of the heritage that one has received* . . . The nation, like the individual, is the culmination of a long past of endeavors, sacrifice, and devotion . . . A heroic past, great men, glory . . . this is the social capital upon which one bases a national idea. To have common glories in the past and to have a common will in the present; to have performed great deeds together, to wish to perform still more . . . One loves in proportion to the sacrifices to which one has consented, and in proportion to the ills that one has suffered. One loves the house that one has built and that one has handed down. The Spartan song – 'We are what you were; we will be what you are' – is, in its simplicity, the abridged hymn of every *patrie*. (Renam 1996 [1882], my italics)

The resulting identity in Renam's account possesses, I would claim, the fundamentals of a narrative leaning to a mythical quest: "How many trials still await you! May the spirit of wisdom guide you, in order to preserve you from the countless dangers with which your path is strewn!" (Renam 1882). Narrative identity integrates agency with foresight and hindsight at the cultural level – the individual sees him/herself in relation to the future trajectories of others and is already under the influence of the past trajectories of others. In turn, this sense of belonging feeds back to the implicit or explicit, definition of multi-agency goals, and it relates to Renam's notion of Nation as "having done and willing to do more".

5 Values in culture

There is a well-accepted mutual constituency between a culture and its individual agents and interpreters. There is a static aspect of a culture, which lays in its foundation, its origins, that which has to be handed over – Ricoeur's "identity as sameness" (idem), the permanence in time (continuity). The dynamic aspect yields development and transformation, the adapting legacy – Ricoeur's "identity as self" (ipse), which brings agency into the picture. This processual changing/permanence dialectic constitutes the narrative identity of a culture and determines the heterarchical belonging of its individuals (see below). To the old proverb that says "know where you come from to know where you are going" we could add "in order to understand where you stand right now". In cultural narrative temporal terms, this can be framed as the dialectics between the roots where you come from (idem) and the values where you stand right now (ipse). "Values" become a defining element of identity but not in a static manner. Values have been considered a powerful kind of cultural "markers" or "identifiers", however blurred at times by rigid, categorical and static considerations of value sets and systems, which approach the individual as a coherent whole subject belonging to a fix collection of such various cultural identifiers. The list of identifiers is long and diverse according to different disciplines, methodologies or frameworks. It is out of the scope of this chapter to review such methodologies and categories, but it is worth to mention some common categories such as gender, sexuality, ethnicity, history, nationality, language, religion, aesthetics, food, geography, political orientations, social class, etc. As a reaction to the fixed and static sets of identifiers, which perhaps could make a bit more sense in the pre-globalization societies, people working from the social and/or the business disciplines have suggested that an individual's social identity works more as an amalgamation of cultures across boundaries (national, organizational, professional, etc.), which fuse together to create one's overall culture. The combination would be unique to each individual (Straub et.al. 2002: 14). If this was wholly true, then there would not really be a phenomenon that we can call culture. Everything would be atomized into millions of individual cultures and their social relations. In the direction of Kroeber (1917) and Lotman (1990), which see culture as an emergent phenomena, these atomistic view of individual identities would be difficult to accommodate and a balance would be needed. A heterarchical perspective of cultural narrative identity could aid in finding such balance of dynamic categories of "shared values" that characterize culture. However, "value" is another problematic concept. The issue of "values" has been often considered in research on how to "measure" culture, where in turn the definition of culture risks becoming highly problematic by overly simplistic categorizations. In the 1980s and 1990s, the "shared values" perspective was advanced by numerous researchers (Straub et. al. 2002). With his peculiar definition of culture as a collective programming of the mind, Hofstede (1984: 18) sees values as "a broad tendency to prefer certain states of affairs over others". Once a value is learned, it becomes integrated into an organized system of values where each value has a relative priority, and therefore one could eventually hierarchize them. Such a value system would be relatively stable in nature but can change over time, reflecting changes in culture as well as in personal experience. Cultural patterns are rooted in value systems of major groups of the population and they get stabilized over long periods of history. Therefore, individuals based on their unique experiences not only differ in their value systems but also in the relative stability of these value systems (Straub et. al. 2002).

Hampden-Turner and Trompenaars (1994) emphasize the temporal dimension when they claim that members of a culture are likely to share common attitudes because they share a common history. They present a scheme of seven dimensions of culture that classify such attitudes in binary oppositions: 1) universalism/particularism, 2) individualism/collectivism, 3) neutral/affective relationships, 4) specific/diffuse relationships, 5) achievement/ascription, 6) internal/external control, and 7) perspectives on time. There are in the literature several of these "universal" schemes of categorizations based on values and/or attitudes. My interest is not how these schemes are specifically defined, but rather how the different idiosyncratic instantiations of the different categories interact dynamically in cultural processes, so eventually any scheme could potentially work in a particular domain. What many of these value-based descriptions of culture have in common is the notion of boundary e.g., the nation-state/geographic borders, organization, or profession, or the boundaries of the semiosphere in question. For example, Straub et al. (2002) make the binary distinction between core and peripheral values. Culture is primarily a manifestation of core values which influence individuals' cognitions, attitudes, and behaviors.

Many of these descriptions tend to a have a nomothetic view of culture abstracting from the particular historical idiographic instantiations of lived culture as a product of history – and thereby my insistence on the identity that results from the narrative function. The nomothetic or ahistorical perspective I call the "operative system" perspective, which sees culture as a mindset or framework in which individuals and societies interact. To this, sometimes it can be added what those individuals and societies have achieved with the given "operative system" – which perhaps brings the idiographic perspective into play. This distinction can be found in both Lotman's semiosphere and Ricouer's narrative model. Straub et al. (2002) classified the multiplicity of definitions of culture into three main groups: definitions based on shared values; definitions based on problem solving; and a third group that they call "general all-encompassing definitions". However, when these definitions do not conform exclusively to the "operative system" perspective, and take the historical uniqueness into consideration, there is a tendency to portray a geographic perspective of culture. For example, Hofstede's (1984) offers a mechanism whereby a culture value can be assigned to a particular group of people. This group is determined by a geographical boundary. Given the dynamism of historical World migration this can be very problematic, even in the pre-globalization and pre-digital era, as illustrated by Renam's account.

6 Towards a heterarchical approach

Recent understandings of narrative identity suggest that a person's life story says as much about the culture wherein a person's life finds its constituent meanings as it does about the person's life itself. In constructing self-defining life stories, people draw heavily on prevailing cultural norms and the images, metaphors, and themes that run through the many narratives they encounter in social life (McAdams, 2011). This cultural perspective still has the main focus on the individual, where culture is "just" an influence or a constraint in the development of individual narratives. On the other hand, Somers (1994) pioneered the study of identity formation through the concept of narrative regarding the "social construction of identity". Her approach to narrative identity helps to circumvent the common anti-essentialist prejudices towards any use of the notion of identity. It also seeks to avoid the hazards of misleading categorical conceptions of identity, and it makes a dynamic linkage between identity and agency. However, the main importance of her contribution, in the present context, lies in her "relational and network approach", which supports the perspective of a processual understanding of narrative identity as *a dynamic system of heterarchical belongings*, which I am proposing here. By the time of the publication of her article, the "narrative turn" in different disciplines of the humanities, social and human sciences (including anthropology, psychology and cognitive sciences) was yielding a new ontological status to narrative phenomena that transcended the customary epistemological status, which limited narrative to be a sort of qualitative method or simply a form for representation. Somers saw in this turn an opportunity "to infuse the study of identity formation with a relational and historical approach that avoids categorical rigidities by emphasizing *the embeddedness of identity in overlapping networks of relations that shift over time and space.*" (Somers, 1994: 607 my italics). This was in light of the challenges that social theory was confronting at the time, which included among others the collapse of communist regimes, the ecological crisis, the conflicts of ethnic solidarities, cultural nationalisms, a vast array of "new social movements" such as the green movements, gay and lesbian movements, feminism and multiculturalism. These challenges remain actual today while others have gained a renewed prominence, such as religious conflicts, the blurring and the exacerbation of the left/right political dichotomies, neo-populist ideologies (left and right), the political reduction of the ecological crisis to climatic change, the advancement of globalization, technological utopianism, cultural homogenization and global branding, among others. These issues and challenges today acquire a new level of complexity in light of the cultural acceleration that is propelled by the global platform of digital technology.

According to Somers, in the 1990's, the emerging social theories of "identity-politics", had shifted explanations for action from "interests" and "norms", to "identities" and "solidarities", assuming that people act on the grounds of common or shared attributes (or cultural markers) rather than on rational interest or a set of fixed learned values. In this context, Somers lucidly warned about emerging identity-categories that may end up working as new "totalizing fictions" in which a single category of experience will dominate over a set of cross-cutting simultaneous differences (for example, gender and sexual identity overruling class, ethnicity, race, age, religion, etc.). According to her, "The new identity-theories reify anew what is in fact a multiplicity of historically varying form of what are less often unified and singular and more often 'fractured identities'". Therefore, she claimed, there is a need for a new conceptual vocabulary that can enable us to plot the narrative identities which dynamically shape history and social action synchronically and diachronically. As I will try to show, a heterarchical perspective may contribute to advance in this direction by explaining how "solidarities" can overlap in seemingly contradictory non-transitive permutations of shared attributes in the cultural dynamics of the semiosphere. Social and cultural action can be better understood if we can recognize the various culturally constructed stories in which people are *emplotted* and which according to Somers consist of composed of (breakable) rules, (variable) practices, binding (and unbinding) institutions, and the multiple plots of family, nation, economic life, etc. – all of which conforms, in the view of the present work, a heterarchical entanglement of values, traits, attributes and interests in such narratives identities.

At this point it is pertinent to be more precise about the notion of heterarchy and specifically about its inherent characteristic of allowing to organize subjective values (expressed from a particular standpoint), which by being subjective do not conform to the laws of transitivity. Although Warren McCulloch (1945) introduced the notion of "heterarchy" into science more than 70 years ago, its implications and epistemological consequences have not been widespread in the scientific and academic main stream (von Goldammer, Joachim and Newbury, 2003; Bruni and Giorgi, 2015). A heterarchy is not defined in opposition to a hierarchy, but rather in a relation of complementarity. One crucial difference between both organizational principles is that hierarchies can be found in both physical and living systems, while heterarchies are to be found exclusively in the living world – where subjective, semiotic and communication processes take place. Hierarchies may be fixed and static, whereas heterarchies are by necessity processual and dynamic. Therefore, it would perhaps be more correct to speak about heterarchical processuality (Bruni and Giorgi 2015).

What McCulloch (1945) realized at the neural level (a living system), is that it is not always possible for the system to rank (hierarchically) its values with respect to the available choices. Physical processes don't deal with options or choices. Only living organisms that can sense differences act upon response-repertoires that involve two or more potential options, determining different degrees of proto-subjectivity and subjectivity in living systems (Bruni and Giorgi 2015). This is of course much more evident at the level of human subjective values. According to von Goldammer, Joachim and Newbury (2003: 2), it is precisely the process of decision itself that has to be analyzed in order to understand from a logical point of view what distinguishes a "heterarchy of values" from a kind of ranking that implies a "hierarchy of values". At whatever level of the scale of semiotic freedom in living systems, in which choices – based on assessments of the context – are enacted by the system, there is the possibility of a value anomaly between the options of the repertoire. This means that the options are not necessarily ranked hierarchically, and therefore the transitivity law is not valid. In a value system (or scale) the transitivity law would take the following form: "if A is preferred to B and B to C that means that A is preferred to C". In a physical value system – like for instance a measurement scale – transitive logical statements can always be constructed for the physical attributes and measurable physical quantities that it expresses. For example if A is taller than B, and B is taller than C, A will *always* result to be taller than C. Therefore, the three values can be ranked hierarchically from taller to lowest. On the other hand, when there is a system expressing subjective values, the values or preferences not always can be ranked in this way. If a person prefers coffee to hot chocolate, and hot chocolate to tea, that does not necessarily mean that the person prefers coffee to tea: the values are, in this sense, intransitive. Let me illustrate this point with an example from computer science. Suppose that we have a system for managing a database of books. A transitive dependency occurs only if our database relates three or more attributes. Let us say that in our case we have three distinct collections of attributes: Book (A), Author (B), and Author-Nationality (C). In these collection, the following conditions hold:

I. A → B, if we know the book we know the author (it is not the case that B → A, knowing the author does not guarantee us knowing which book).
II. B → C, if we know the author we know the author-nationality
III. Then the functional dependency A → C follows, if we know the book we know the author-nationality by the axiom of transitivity.

However, in a relational database there is not only one-to one and one-to-many relationships (like e.g. one author having only one nationality, or one author having many books) – which may yield a hierarchical model by virtue of the transitive dependency – but there can also be many-to-many relationships (suppose that a book could have many different authors and that an author can have many different nationalities, which is actually possible). This situation would require a network-like model able to exclude certain types of transitive dependencies in order to navigate the referential system. Otherwise there could emerge paradoxical loops or value anomalies, which would jeopardize the referential integrity of the database in question. In our case, the referential integrity corresponds to a coherent narrative identity. One of the implications of the transitivity law is that it cannot deal with the possibility of pondering two or more values in simultaneity. This is a very important point to understand heterarchical processuality because, as mentioned before, we need to consider the "process of decision" when "choosing" the value through which we will based our actions or our criteria for belonging. There is a circular cognitive/volitive process here: we processually "define" or choose our values in order to act, and while acting we define and actualized our values. Our options are not presented or compared one after the other, but simultaneously, and our choices between two or more potential acts are very often mutually exclusive. This presents us often with dilemmas and paradoxical or incompatible choices. Moreover, such heterarchies of values can be highly context-dependent and dynamically vary from one situation to the other (which is not possible in rigid hierarchies or categorizations). McCulloch (1945) introduced the notion of "value anomaly" (or "diallel") to refer to these logical contradictions. The term can be related to similar notions such as paradox, tautology, antinomy, contradiction, dissonance, semantic incongruence, and to Bateson's notion of double bind (Bruni and Giorgi, 2015). In normal healthy circumstances, value anomalies and vicious circles are usually resolved by recognizing a wider gestalt that may help to make meaning out of the seemingly irreducible values or criteria for belonging to a given category or collective identity. One can suspect that the condition becomes pathological when, from one or another reason, the individual or the collectivity has no access to a larger gestalt (or narrative) which could potentially put the contradiction in a new congruent perspective. A paradox of conflicting values may involve problems with self-referentiality, and therefore with identity, which can only be dealt by identifying a meta-narrative that allows inclusion into a larger or overlapping gestalt (outside of the paradoxical situation) in which the subject can alternate between seemingly different standpoints. The parallel and simultaneous logical places of each standpoint have to be mediated by some adequate narrative that will thrive to coherently accommodate the different perspectives and potential scenarios of each of the standpoints in an attempt to reconcile the dialectical contradiction through the synthesis of a new identity. Heterarchies are complex adaptive systems that interweave a multiplicity of organizing principles, becoming "the sites" of competing and coexisting value systems, which allow multiple regimes of worth (Stark, 2001).

According to Somers (1994) choosing narratives to express multiple subjectivities is a way of overcoming the apparent neutrality and objectivity typically embedded in master narratives. She provides the following example: the public narratives of working class community available in certain historical periods may have omitted women, just as many of the current feminist accounts of identity may omit class and poverty. The elaboration of counter-narratives emerges as a natural strategy when one's identity is not expressed in the dominant public ones. In Lotman's terms this would be how new meanings and narratives generated in the peripheries make it to the center of the semiosphere. The new emerging narratives may link particular spheres or domains (e.g. gender, class, nationality, background, etc.) with many other "relational complexities", which reveal "alternative values" in multiple narrative trajectories.

In a relational and heterarchical perspective, identities cannot be exclusively derived from attributes imputed to a specific social category in a particular culture at a given historical period. Rather, cultural narrative identities could be derived from the heterarchical belonging of the given groups of agents to the multiple overlapping narratives (standpoints) in which they are embedded and which they themselves identify with. What the analyst can hope for is to recognize "patterns of overlaps" of such narratives and standpoints.

A good example could be the criteria for belonging to emerging transnational political cultures in the western world, where traditional binary oppositions such as left/right, progressive/conservative, and liberal/socialist have become blurred by a plethora of overlapping values that are giving rise to new contradictions, new identities and hybrid transnational alliances. The different versions of multiculturalism and political correctness may also introduce paradoxes. Policies in these directions may intend to promote equality of opportunity, exchange across social boundaries, tolerance and "diversity" but at the same time may introduce contradictions which may facilitate discrimination or segregation by class, race, and gender (Wade, 2017). The very debate on culturalism and multiculturalism transects political identities across left and right. Stjernfelt and Eriksen (2012) claim that culturalism – the idea that individuals are wholly determined by their culture – brings contradictions across the whole political spectrum.

In line with Ricoeur's notion of semantic innovation (1984), Somers (1994) sees narratives as constellations of relationships (connected parts) embedded in time and space, constituted by causal emplotment, which preclude sense making of singular isolated phenomena. As a cognitive faculty, causal emplotment helps to discern meaning by linking multiple events in temporal and spatial relationships. Such relationships may become confused when it is impossible or illogical to integrate them into an intelligible plot. If the story and its implicit values are not clear, the capacity to act may be hindered, become incongruent or even paradoxical. Prioritizing events, like prioritizing values, entails a process of hierarchization. For example, a collectivity participating in themes such as "political correctness", "sustainability", "free market", "economic growth" "full employment" and "climate change" will have to relate to concrete examples of events from current social and cultural processes and arrange them in some order, and normatively evaluate these arrangements (Somers 1994). When there are subjectively competing or contradicting themes in the plot, events and values may not conform to the transitivity property and therefore may lead to paradoxical or unintelligible relations. The selected or predominant themes can only be arrange (or interpreted) heterarchically, and it is their consolidation in a normative frame that can freeze them into a hierarchy that may attempt to smooth contradictions and paradoxes. However, in a different time or context a different set of prevailing narratives could determine a different sense of belonging. If we rigidly place individuals or communities in fixed categories, based on common interests or values, our analysis may become blurred if we fail to consider the processual relationships and life-episodes implied in the narrative identity approach. Thus, an analytical approach should consider this sort of multilayered, processual and relational dynamics where the cultural narrative identities that emerge overlap to give place to complex heterarchical systems of values and, therefore, of belongings.

The importance of these dynamic collective narratives is their mutual constituency with agency. Individuals and collectivities adjust stories to fit their own identities, and, conversely, they will tailor "reality" to fit their stories (Somers 1994). In other words, identity and reality are mutually constitutive. The proleptic power and the normative aspect of self-fulfillment prophesies is related to this mutual constitution. For the socio-cultural level, Somers uses the term *public narratives*, which are "those narratives attached to cultural and institutional formations larger than the single individual" constituting intersubjective networks or institutions (Somers, 1994). Thus, public narratives may range from family levels, workplace, church, province or nation (curiously, Somers does not mention culture as an entity or level to which one can belong to). In this perspective, these collectivities selectively appropriate and arrange events into stories and plots with normative goals, explanatory power and inclusion/exclusion criteria.

7 Heterarchical relational clusters in the semiosphere

The narrative construction of the cultural "self" can be related to Yuri Lotman's dynamic model of the semiosphere. For Lotman (1990), the primary mechanism of semiotic individuation is the boundary between "two spheres in binary opposition", which differentiates one culture from another, and through which the culture in question divides the world in its own internal and an external space. In this sense, it is such boundary that gives place to a narrative of identity and a sense of belonging. Even though Lotman's system acknowledges the paradoxes of self-reflexivity in culture, it remains a hierarchical system organized in meta-levels, levels, and strata in relations of binary inclusion and exclusion, which do not admit grading but require either-or decisions: "something is either inside or outside, above or below; there is no in-between, nor is there a gradual transition between the two opposites" (Nöth, 2006). According to Nöth (2006) this carries the burden of the heritage of a semiotic structuralism that sought to explain semiosis in terms of oppositions even where gradations and transitions between the opposites prevail. One of Lotman's main asymmetries is the "center/periphery" asymmetry, in which the center would be the locus of stability and identity legitimization, while the periphery would be the locus of instability, creativity, transgression of norms and therefore the locus of blurring, transforming or creation of new identities. Perhaps, the picture today is not that of a static center with coherent sets of cultural values, but rather that of many centers, and many peripheries with overlapping value systems. These asymmetries, which assign value to one or the other side of the locus, do not necessarily yield value systems or sets of cultural values, which can be ranked hierarchically as in a pyramid or an onion system of concentric circles of inclusion. In Lotman's metaphor of semiotic "space", the geometrical symmetries inherent to physical space (left and right, above and below, far and near, in and out), become asymmetrical loci of cultural values (good or bad, right and wrong, beautiful and ugly). The former (space, etc.) can be organized in scales of values that conform to the transitivity law. The latter, i.e. cultural values, cannot. Nevertheless these asymmetries can still function as complex criteria for inclusion and or belonging, and therefore of identity. Thus, mapping dynamic systems of heterarchical belonging in the semiosphere can help us to account for the overlapping of subjective value systems. Different values from apparently mutually exclusive loci at different hierarchical levels can be grouped or can overlap under one narrative identity in a particular time-space and context. If cultural identities are socially constituted over time-space and through heterarchical networks, then the "other" cultures are constitutive rather than external entities. This can help to overcome dichotomist notions of (cultural) identities derived from Lotman in terms of "in and out", "us/them", "ours/theirs", giving a new meaning to the importance given by Lotman to the notion of "borders" as the most creative and productive zones of the semiosphere. Lotman also stated that cultures are oriented towards a rhetorical organization in which each step in the "increasing hierarchy of semiotic organization" produces an increase in the dimensions of the space of the semantic structure. This "hierarchy", or rather as claimed here, a heterarchy, could encompass the individual, the social and the cultural, going from the private to the public, from family to work, from local to national to regional to global, in many different domains and with multifarious semiotic resources.

According to Somers (1994), in order to make social action intelligible, the systemic typologies of our categorizations must be broken apart and their parts disaggregated and reassembled on the basis of *relational clusters*. Such relational clusters, settings or matrices, configure patterns of (heterarchical) relationships among institutions, public narratives, and social practices from which identity-formation takes shape through contested but patterned relations among narratives, people, and institutions (Somers, 1994). This cultural space of meaning making has a diachronic-synchronic development. Relational settings have history and therefore must be explored through time and space. Somers claims that spatially, "a relational setting must be conceived with a geometric rather than a mechanistic metaphor, because it is composed of a matrix of institutions linked to each other in variable patterns contingent on the interaction of all points in the matrix." (Somers 1994). This finds a congruent implementation in Lotman's spatial and geometric metaphor of the cultural space. However, as previously mentioned, Lotman's semiosphere model for cultural dynamics have been often understood as implying levels of analysis that can be ranked in terms of their inclusiveness inside each other, postulating therefore cultural levels organized in hierarchical concentric spheres. This makes it difficult to consider the heterarchical crossovers and overlaps of different cultural spheres or layers that we have been referring to. In Somers' perspective, the effects, or attributes, of different relational settings, clusters or matrices, can cross "levels" (of analysis) converging in an experienced social, geographical, cultural and symbolic narrative identity:

> A setting crosses 'levels' of analysis and brings together in one setting the effect of, say, the international market, the state's war-making policies, the local political conflicts among elites, and the community's demographic practices of a community – each of which takes social, geographical, and symbolic narrative expression.

This "cross-cutting" perspective can help us to discern complex criteria for partaking in a particular cultural narrative identity that can eventually correlate to the agency derived by such identity, by assessing how this agency is affected or constituted interactively by complex arrays of attributes coming from different relevant settings, matrices or spheres that relate to each other in a heterarchical network. Following Somers' reasoning, one could empirically disaggregate the attributes of a cluster from any presumed covarying whole, and then reconfigure them in their spatiotemporal (narrative) relationality. Different cultural layers of a "recognized cultural unity" would not be simply cast as variants of a single culture, but as different relational cluster that can overlap between each other and even share attributes with cultural layers from "another" culture. The effect of any one cluster (with its attributes or markers) could only be discerned by assessing how it is affected interactively by other relevant clusters and dimensions.

8 Conclusions

One of Somers' central question was why should we assume that an individual or a collectivity has a particular set of interests (or values) simply because one aspect of their identity fits into a pre-defined category. Furthermore, once we have place them in the given category (e.g., traditional artisan, modern-factory worker, peasant, etc.) – and therefore imputed to them a predefined set of interest or values – we proceed to explain their actions and behaviors. Even if such interests and values are considered to be somehow modulated by cultural, social or existential intervening factors, the analytical endeavor remains placing people in the right categories by identifying in them the putative interests and values of that category (Somers 1994: 623). This may apply indifferently to social or cultural analysis, or, in fact, to how much cultural dynamics is made to intersect with social action. One may argue that such fixed sets of values in given social or cultural categories perhaps were more or less homogeneous in the past. It is certainly not the case under the current cultural dynamics of a globalized digital society, in which values and interests may transect from one (cultural) category to the other in seemingly conflicting, contradictory or paradoxical ways. Somers' contention was that, epistemologically speaking, a dynamic narrative identity approach would considerably decrease the normative load implicit in the static categorizations resulting from "traditional" theories of identity. Analytically, the resulting identities should be rather considered in the context of complex relational and cultural matrices determined by empirical inquiry and not by a priori assumptions. The issue looks problematic if one considers that it is ontologically impossible to construct an analytical tool completely devoid of normativity. However, we can acknowledge that it is rather a matter of more or less, and therefore we can have as a legitimate normative goal to construct or choose our tools in such a way as to avoid (as much as possible) the kind of fixed categories that oversee the relational and "cultural matrices" in which they operate. Even the different ways of approaching the "cultural matrices", in which the identities live, may have implicit normative considerations. However, it can be a productive goal, if it implies overcoming stereotyping dichotomies (e.g. left/right wing), identity categories (e.g. socialist/capitalist/liberal/religious affiliations/ecologist/feminist, etc.), and static sets of cultural markers (e.g. western/non-western/eastern/aboriginal). Instead of prescribing avenues for agency dictated *a priori* for the given category, dynamic and context-dependent categories would allow us to discern seemingly unintelligible contradictions and paradoxes in cultural and social phenomena. Such unintelligible contradictions and paradoxes may arise in our analysis when:

1) We place an individual or a collectivity in a static cultural (or social, or political) category.
2) The identity and the belonging criteria of the category is defined by a set of values and interest that can be hierarchized (i.e. ranked) in transitive relations of dependency (e.g. if you are a right-winger, you are a conservative, therefore you are . . . ; if you are a left-winger, you are open-minded, therefore you are . . .).
3) We assume that an individual or a collectivity has a particular set of interests (or values) because one aspect of their identity fits into a given category.
4) We assume that belonging to a given category entails action based on a response-repertoire based on that category.
5) We adopt an interest approach that assumes that people act on the basis of rational means-ends preferences or by internalizing a set of values.
6) We try to make sense of social action by placing people into the right social and cultural categories by identifying their putative interests, and then by looking empirically at variations among those interests in a system of fixed hierarchical categories.

From the narrative identity perspective, people would act or express their loyalties in systems of heterarchically embedded categories such as citizenship, social class, gender, race, tradition, interests, cultural origin, etc. What can bring meaning in such seemingly disparate overlaps is the emplotting of a shared story. Looking for the story will point to the patterns that makes sense of such overlaps. The relational clusters (or settings) overlap in storylines across "levels" of analysis or categories, and the seemingly unintelligible contradictions at any level or domain can be better understood by assessing how the level or domain is affected interactively by the other relevant dimensions.

In the explosive and accelerated cultural processes mediated in the emerging digital semiosphere (Bruni 2014), the phenomenon of "fractured identities" gets an enhanced level of complexity, which diminishes any theoretical dichotomies that attempt to hierarchize forms of differences and shared values, which allegedly would allow the constitution of clear-cut categories, but which in fact may blur our understanding of current global political, social and cultural phenomena, or may give rise to new normative forms of exclusion hindering mutual cultural understanding. In such perspective, the identification of overlapping non-transitive values helps to account for the mingling of social and cultural attributes, which determine complex criteria of mixed belongings. Social roles can find analogies in different cultures and can be assumed similarly or differently in a way in which, for example, social identities can transect heterarchically across different cultures.

In spite of the *apparent* decentralization of media power entailed by the digital revolution, the mainstream (traditional) media is still able to arrange and connect events to create a "mainstream plot" that may dominate history. However, in today's atomization, these dynamics is much more complex and contradictory, including a plethora of new phenomena such as "fake news", "fact-checkers", data dredging, bubble filters, cultural narrowing, massive psychological digital profiling, social bots, etc., which may determine new forms of communicational hegemonies. Therefore, what is included in the levels of personal and public narratives cannot be seen as a hierarchic system of concentric spheres, which has the individual as a kernel surrounded by larger encompassing spheres (e.g.: individual, family, workplace, city, nation, culture). This makes Charles Taylor's term "webs of interlocution" highly relevant: intersubjective webs of relationality, which sustain and transform narratives over time (Taylor 1989, Somers, 1994). Identity can be then seen as adherence to certain shared community values (the good, the saved, the believers, the just, the tolerant), which can only make sense in such "webs of interlocution". This stresses the importance of values as determinants of narrative identities and the meaning-making process that they afford. "Contemporary selves" are "saturated" with the complex and shifting demands of social life and they have difficulties in achieving unity and purpose; instead, fragmentation and multiplicity seems to be the norm (Gergen 1991; McAdams 2011). However, as McAdams (2011: 102) points out, " . . . people living in complex, postmodern societies still feel a need to construe some modicum of unity, purpose, and integration amidst the swirl and confusion". People still seek a kind of meaning that accounts for the rapidly evolving, multi-layered, and complex social and cultural ecologies in which they are situated. Such complex digital ecologies bring a myriad of contradictory cultural tendencies that mingle in the technologically enhanced semiosphere (Bruni, 2014). Individuals and cultural collectivities identify themselves and adhere to emerging narratives that informed their actions while they attempt to conciliate cultural contradictions and dissonances. For instance, there is a myriad of cultural paradoxes in "the cause for sustainability". Today, cultural narrative identities evolve in the middle of massive information overflow, media addictions, attention deficits, and cognitive dissonances, which are reproducing many of the "thousands of cultural details" that reinforce unsustainable behaviors (Bateson 1972) – even when sustainability becomes an overt generalized normative goal. In fact, the eco-epistemological crisis can actually be re-conceptualized as a cultural identity crisis.

In concomitance, we are witnessing unsustainable cultural clashes among new emerging "paradoxical identities" (i.e.: the tensions between a given identity and its implied agency in contexts that deny the identification process). The issues propelled by the social inequalities of the globalized society, such as the massive and out-of-control migratory fluxes, exacerbate dramatically the already grave problems of inclusion/exclusion, integration, secessionism/annexationism, cultural homogenization/diversity, and the tensions between traditions and "progress". One key aspect in the conception of "narrative identity" is the anticipatory world-building practice implicit in any kind of individual or cultural identity. In the narrative approach that links identity to agency (and social action), prolepsis is all about how we incorporate hopes, expectations and goals into our stories. In this context, populist configurations of cultural narrative identities have normative effects that condition action into loops of self-fulfillment prophesies, which may eventually lead to unsustainable paths. Perhaps one of the most dominating self-fulfillment prophesies with normative power in our current narrative trajectories is that of a technological eudemonia, in which technological convergence will allegedly provide solutions to all possible harshness inherent to the human condition and to the entropic drift of life in the biosphere. This is well exemplified by the positivist and utopian techno-optimistic narratives of trans- and post-humanism, which are rhetorically implicit in many scientific and economic agendas and closely intertwined with pervasive science fiction. The new generations are submitted to a process of cultural narrowing by the pervasiveness of such narratives in a plethora of transmedia platforms and channels. These new science fiction mythologies substitute the mythologies that carry the rich cultural heritage of traditional wisdom. This is creating difficulties for the new generations to discern the ontological boundaries of their cultural narrative identities – between fiction and reality – making them vulnerable to intended or unintended rhetorical strategies for the adoption of the beliefs and value systems of what Hans Jonas calls the built-in automatic utopianism of technical progress (Jonas 1984: 21). However, if the plot advances only in terms of foresight, neglecting or even despising the cultural hindsight provided by traditions, the resulting collective agency will prospect future trajectories that are divorced from traditional sources of wisdom: "We need wisdom most when we believe in it least". (Jonas 1984: 21).

The aim of this incipient framework was to bring together different but compatible perspectives on what could be called "processual cultural narrative identities", having as gluing concepts the notion of *narrative identity* on the one hand, and the notion of *heterarchy of values* on the other. The objective was to explore whether a processual heterarchical perspective can be of utility in understanding the paradoxes and contradictions in the contemporary cultural dynamics that are shaping our reality, and which constrain the future through the negotiation of meaning-making in normative processes, which in turn are taking form in new overlapping narrative cultural identities. For this purpose it becomes important to understand the centrality of "values" in the determination of identities. Once we have the consideration of value-adherence in multilayer cultural processes and networks, we are bound to consider heterarchical processuality in order to be able to elucidate the rationality of the putative paradoxes and contradictions.

References

Bateson, Gregory. 1972. *Steps to an ecology of mind*. New York: Chandler Publishing Company.
Bruni, Luis E. 2014. Sustainability, cognitive technologies and the digital semiosphere. In: Indrek Ibrus, Wilma Clark & Peeter Torop (eds.) *Remembering and Reinventing Yuri Lotman for the Digital Age*. Special issue, International Journal of Cultural Studies (SAGE), 18(1). 103–117.
Bruni, Luis E. & Franco Giorgi. 2015. Towards a heterarchical approach to biology and cognition. *Progress Biophysics Molecular Biology*, 119(3). 481–492.
Gergen, Kenneth J. 1991. *The saturated self: Dilemmas of identity in contemporary life*. New York: Basic Books.
Guénon, René. 1982 [1945]. *Il Regno della Quantità e i Segni dei Tempi*. Milano: Adelphi Edizioni S.P.A.
Hampden-Turner, Charles & Fons Trompenaars. 1994. *The Seven Cultures of Capitalism: Value Systems for Creating Wealth in the United States, Japan, Germany, France, Britain, Sweden, and the Netherlands*. London: Piatkus.
Haug, Wolfgang Fritz. 1987. *Commodity aesthetics, ideology & culture*. New York: International General.

Hofstede, Geert. 1984. *Culture's Consequences: International Differences in Work-Related Values*. Newbury Park, CA: Sage.

Jonas, Hans. 1984. *The Imperative of Responsibility: In Search of an Ethics for the Technological Age*. Chicago: University of Chicago Press.

Kroeber, Alfred L. 1917. The Superorganic. *American Anthropologist*, New Series, 19(2). 163–213.

Lotman, Yuri. 1990. *Universe of the Mind. A semiotic Theory of Culture*. Bloomington: Indiana University Press.

Lotman, Yuri. 2005 [1984]. On the Semiosphere. *Sign Systems Studies 33*(1).

MacIntyre, Alasdair. 2011 [1981]. *After Virtue. A Study in Moral Theory*. London & New York: Bloomsbury Acedemic.

McIntyre, Alasdair C. 2013 [1981]. After Virtue: A Study in Moral Theory. London & New York: Bloomsbury.

McAdams, Dan P. 1988. *Power, intimacy, and the life story: Personological inquiries into identity*. New York: Guilford Press.

McAdams, Dan P. 2011. Narrative Identity. In Seth J. Schwartz, Luyckx Koen and Vivian L. Vignoles (eds.), *Handbook of Identity Theory and Research*, 99–115. New York: Springer.

McAdams, Dan P. 1996. Personality, modernity, and the storied self: A contemporary framework for studying persons. *Psychological Inquiry*, 7. 295–321.

McCulloch, Warren. 1945. A heterarchy of values determined by the topology of nervous nets. *Bulletin Mathematical Biophysics*, 7. 89–93.

Winfried Nöth. 2006. Yuri Lotman on metaphors and culture as self-referential semiospheres. *Semiotica* 161–1/4. 249–263.

Renan, Ernest. 1996 [1882]. What is a Nation? In Geoff Eley & Ronald Grigor Suny (eds.). *Becoming National: A Reader*, 41–55. New York and Oxford: Oxford University Press.

Ricoeur, Paul. 1984. *Time and narrative (Vol. 1)*. Chicago: University of Chicago Press.

Ricoeur, Paul. 1991. Narrative identity. *Philosophy today*, 35(1). 73–81.

Saint Augustine. 430 [1957]. *Against Julian, Book II*. Translated by Schumacher, Matthew A. New York: The Fathers of the Church Inc.

Shashidhar, R. 1997. Culture and Society: An Introduction to Raymond Williams. *Social Scientist*, 25(5/6). 33–53.

Somers, Margaret R. 1992. Narrativity, Narrative Identity, and Social Action: Rethinking English Working-class Formation. *Social Science History* 16(4). 591–630.

Somers, Margaret R. 1994. The narrative constitution of identity: A relational and network approach. *Theory and Society* 23. 605–649.

Stark, David. 2001. Ambiguous Assets for Uncertain Environments: Heterarchy in Postsocialist Firms. In Paul DiMaggio (ed.), *The Twenty-First-Century Firm: Changing Economic Organization in International Perspectivel*. Princeton and Oxford: Princeton University Press.

Stjernfelt, Frederik & Jens-Martin Eriksen. 2012. The Democratic Contradictions of Multiculturalism. New York: Telos Press.

Straub, Detmar, Karen Loch, Roberto Evaristo, Elena Karahanna & Mark Strite, 2002. Toward a theory-based measurement of culture. *Journal of Global Information Management*, 10(1). 13–23.

Taylor, Charles. 1989. *Sources of the Self*. Cambridge: Harvard University Press.

Vignoles, Vivian L., Seth J. Schwartz, & Koen Luyckx (2011). Introduction: Toward an Integrative View of Identity. In Seth J. Schwartz, Koen Luyckx & Vivian L. Vignoles (eds.), *Handbook of Identity Theory and Research*, 1–27. New York: Springer.

von Goldammer, Eberhard, Paul Joachim & Joe Newbury. 2003. Heterarchy – hierarchy: Two complementary categories of description. http://www.vordenker.de/heterarchy/a_heterarchy-e.pdf. (accessed 03 October 2019).

Wade, Peter. 2017. Liberalism and Its Contradictions: Democracy and Hierarchy in Mestizaje and Genomics in Latin America. *Latin American Research Review*, 52(4). 623–638.

Part II: **Applied Semiotics**

A **The digital age in semiotics and communication**

Kristian Bankov
Lying as a transaction of value: explorations in semiosis and communication from a new perspective

Abstract: In the first part of the paper I introduce a new, complementary approach to understanding the deep question of lying, a question that has concerned the major philosophical figures of our tradition – Plato, Aristotle, Augustine, Montaigne, Rousseau, Kant, Derrida – to mention only the most prominent. This new approach considers the lie as an act of value transfer similar to theft. In the rest of the paper I develop a basic dichotomy to distinguish the two typologies of value abused by lying (although in the real practice they most often go together): instrumental and ego values. In instrumental lying the liar's gain is commeasurable with monetary or economic value, whereas in ego lying the universal currency is recognition, according to the way Todorov defines this intriguing notion. Probably the most original contribution of this chapter is its reflection on the dynamic relation between these two types of value in our everyday life, of which the analysis of lying makes more explicit and evident but glosses only one part.

Keywords: lying, value, exchange, semiotics, ego

1 Introduction

The topic of deception, lying, and speaking untruth has always attracted thinkers. Lying refers to something easy to outline as far as it is a part of everyday experience, though at the same time it concerns every aspect of human existence, from ethics and religion, to science and rationality, from practical concerns and economy to love and identity. In the field of semiotics, lying has been consecrated by Umberto Eco's classical 1975 definition as the central distinctive feature of the discipline: in his theory, semiotics studies everything which can be used in order to lie (1976: 7). At the margins of semiotics Derrida wrote an insightful Prolegomena to the history of lying (1993/2002), wherein the whole philosophical depth of the problem is underlined. Heidegger defines Dasein, the core notion of his philosophy, as an entity that "bears within itself the possibilities for deceit and lying" (Derrida 2002: 30), and we see that the major figures of the western philosophical

Kristian Bankov, New Bulgarian University, Sichuan University

tradition (Plato, Aristotle, Augustine, Montaigne, Rousseau, Kant, to mention only the field's giants) have made their point on it.

Given all this, can an average semiotician say something original and worth the reading time about lying? We must add that Derrida never wrote the entire history of lying (although this was his intention) but left us only the prolegomenon, probably because of this difficulty. After his attempt it is clear that further scholars can contribute only small, hopefully original, insights to enrich or update the fascinating philosophical and semiotic heritage. Scholars who do this build a foundation for a different type of author to write more general theories of the post-truth age.

In this context, I would like to detail a new perspective on the outlined mechanisms of lying, *combining the semiotic sensibility in understanding the exchange of signs and meanings with the economic pragmatism of exchanged value*. I will try to represent lying as a currency exchange, or rather as a transaction that modifies the assets of the two parts. If we take the famous aphorism saying that 1) if you have an apple, and I have an apple, and we exchange these apples, then you and I will still each have one apple; but, on the other hand 2) if you have an idea, and I have an idea, and we exchange these ideas then each of us will have two ideas, here I attempt to position lying more on the side of 1) rather than 2).

In his typology of lies, Rousseau distinguished between occasions in which the act of lying does not bring any consequences. He calls this "fiction" (Derrida 2002: 32), and this observation opens an important consideration. In the semiotic tradition of Saussure and the Paris school, *lying is a fiction*: it is a textual strategy, a logical construct of veridiction, resolved inside the text and not in relation to factual reality. But while we suspend our disbelief in fiction, our critical awareness of the real value of things and time is substituted by intertextual rules of genre, where meaning is generated by differential principles and formal relations (Barthes 1986, Todorov 1981). Fictional temporality is not scarce or as finite as our primordial existential temporality, which opens the meaning of the world towards each person's unavoidable end (Bankov 2019). I propose that when lying begins to have consequences, it is because it starts to affect the primordial temporality of each of us, and that primordial temporality, as I have shown elsewhere (ibid.), finds its direct expression in our resources and self-esteem.

2 Values for the Body and Values of the Ego

This is a basic distinction that opens an important typology of human needs and the way we relate to the economy and market. The parallel with the pragmatic goals of lying is not arbitrary at all, as will become clear. The starting point is that we may analyze lying in a similar way to how people spend their time and money. Through lies people gain value and risk losses, just as they do on the market. This perspective does not contradict the dominant ethical and linguistic approaches to lying, but rather upgrades and helps with understanding the complexity of this fundamental human inclination.

The trade of lies satisfies needs and as a mechanism it closely resembles theft – getting something without paying for it. If we project the phenomenology of lying to the famous Maslow's hierarchy of human needs, we may distinguish a big variety of lies fulfilled in order to obtain, on the one hand, resources such as food, clothes, shelter, etc., which satisfy fundamental needs, and on the other lies for love, prestige and social status. We must be careful in defining the first type of lies because in some occasions these might be completely instrumental, as between people who meet occasionally and who will probably never meet again, but they might involve the ego of the participants despite obtaining the same factual result. John may need gas for his car, and do not want to beg for a full tank or buy it. He stops an occasional car on the road and he lies that he ran out of gas accidentally and he needs only a few liters to reach the next gas station. He may use this lie 10 times and get his tank full. But if he must lie to someone with whom his life projects interfere, a significant other (a neighbor, a student mate, a colleague), the lie becomes more complicated because he must preserve his ego values – the lie has to be sustainable over time because if revealed he will "lose face", and it is more credible to be taken seriously from the very beginning. In fact, most of these lies are produced for the "commerce" of Ego values.

To better understand this point we can introduce the distinction of French semiotician Jean-Marie Floch between *instrumental and the existential values* (Floch 1990: 120 ff). Kant, one of the most authoritative and categorical voices on the subject of lying, postulates the relation between instrumental and existential values in the following way:

In the kingdom of ends everything has either a price or a dignity. What has a price can be replaced by something else as its equivalent; what on the other hand is raised above all price and therefore admits of no equivalent has a dignity. (Kant 1996: n1 at 434)

Contrary to the kingdom of ends, Kant's idealized ethical world, in everyday life *dignity and price are related as connected vessels of value,* and lying is a sophisticated management of both resources for obtaining of maximum results. John might badly need gas, but still say nothing to his neighbor on the subject, particularly if there have been issues of dignity and honor between them in the past. In this case, lying will transfer value from the instrumental benefits to the existential ones: the lie will falsify the reality of the lied one in a way that John will "save" dignity (his neighbor will never know that John was in a difficult situation), but will lose the instrumental benefits of taking the gas in an easy way. One of the great insights about the fundamental role of lying, in direct confrontation with Kant's theorization, is made by his French contemporary Benjamin Constant, who claims that the categorical prohibition of the lie as the German philosopher postulated it "would make any society impossible" (Derrida 2002: 45). My proposal for the first type of lies is entirely within Constant's assumption.

3 Recognition as the currency of the Ego value and Ego lying

One notion which can give more systematic form of my proposal is that of *recognition*. It is a huge topic in the Western tradition, even bigger than the lie, but I will adopt the excellent account of it made by Todorov (2001). According to the French thinker, recognition is the basic resource of our social existence and the invisible fabric of human relations. Without recognition we may continue our biological living, but are deprived of any sense of existence. All existential values come from recognition by others and from the recognition we give to them. An instinct for recognition is already the first reaction of the newborn baby, that is, to cry when it is not the object of human attention (19, 63). This instinct never ceases, but adopts various culturally determined forms during social affirmation. Recognition is the complex phenomenology of social attention that we need and give during all our life. Kids desperately need the attention of the parent and his/her approval of their actions and cognitive achievements; teenagers change their identity because of the delegitimation of these sources of recognition they need so much; the quest for recognition marks our progress in education (grades and diplomas are formalized forms of recognition), in profession (the salary, positions), in sports (medals), in politics (the vote), in academia and in the army (various ranks, titles and prizes); rituals such as marriage consecrate us to a higher degree of social recognition; and when we get old and become functionally unproductive, the only source of recognition remains the respect of others, the love of our children, grandchildren, relatives, etc. Obviously this list could continue for several more pages.

The point here is that Todorov represents recognition as the universal currency of the existential values exchanged by the members of a society. It is similar in structure to the instrumental values and utility that the material economy models, and is universally measured through pecuniary resources and market exchange. It is a structural resemblance, which is however grounded on the categorical opposition of dignity and marketable assets. In the Kantian universe, these two things are alien to each other and all forms of lying are incompatible with them. In the real world dignity is most often compromised by the appetite for material benefits or their absence, and *lying is a necessary strategy for social survival* that allows everyone to preserve as much dignity as possible, all without losing the achieved level of material comfort.

Max Weber defines a social principle according to which social stratification is dominated by class groups in times of significant economic and technological repercussions, whereas social stratification passes gradually towards the logic of the status groups during periods of sustainable development (Weber 1978: 938). Status groups are ranked by lifestyle which is dictated by one's material welfare and expressed by conspicuous consumption. In status groups everything is about *the social estimation of honor* (932). Simmel and Veblen provide other classical sociological contributions on the same topic. Simmel describes the trickle-down effect of social prestige and the way fashion works to provide a dynamic translation of clothing, and represents the public display of a family's honor and status. Veblen describes the same status stratification expressed in its extreme by the families of the first American billionaires, whom he termed "the leisure class". In all those approaches a sociological law is derived from the translation of material consumption into dignity and honor. According to the proposed model of lying, most of this translation is rooted in the exchanging of those two types of value.

We may assume a kind of rule for the social exchange of recognition, economizing its esteem in opposition with Kant's imperative. The more recognition a person or family receives, the more its dignity (reputation, status, prestige, honor, existential value, identity) grows, and as a consequence the recognition that this person or family gives to others assumes a higher value. To Kant, the imperative of dignity prevents the value of people being compared, but society functions in the exact opposite way. People develop their social roles (some call them "masks") in order to preserve a coherent level of status, and to eventually increase this level of status in the eyes of significant others. Comparison with those slightly above them in social standing or the professional hierarchy is usually the main source of inspiration for their lifestyle choices (see Simmel's trickle-down effect). The accumulation of recognition does not work very differently from the function of private financial savings. The more one has accumulated, the more powerful his currency of recognition is to others. In this scheme our social roles function as different accounts: we sell and buy recognition through our different social masks at different exchange rates. The relationship between social masks and lying is deeper than what it first appears.

Now let's see how the commerce of lying works in all this.

Lying occurs most frequently in the name of status and honor, usually as a theft of value and recognition. Let's imagine that John builds a family. They have an average living standard, in tacit emulation of his neighbors, colleagues and friends. If his wife accidentally loses her job, for incompetence or disciplinary reasons, it is very unlikely that he will tell the truth to the others if he has the choice not to tell them. John and his wife will most probably lie that she is just changing her job because she worked enough for that company and cannot learn anything further. It is her choice. His wife will then need to start a new job, and new lies will be necessary to represent the new job as an upgrade from the old one although this will not in fact be the case.

John's daughter, a teenager, might be unhappy with her now-reduced budget and decide to earn some money on her own. She may sell pictures of herself on the internet and then invent lies for her parents, lies that serve to preserve her standard of dignity that has been shaped by the family and that the dissemination of nude pictures would seriously compromise. At the same time, she might win recognition among her close friends exactly with the fact that her pictures were requested. She may lie to them to increase the perception of her success. One and the same fact might be cashed out as recognition with the opposed sign (plus or minus), depending on the "mask" she uses. *The presentation of the self in everyday life* (Goffman) *represents a complicated semiotic value management that requires permanent lies*. In his book on Post-truth, Keyes claims that in the decade before 2004 there is a substantial increase in the amount of everyday lying done by people. In his chapter "A Whole Lotta Lyin'," he presents data from empirical research showing that young people in particular are lying more than they think they are (Keyes 2004: 3 ff). It is a kind of automatized lying that represents a normal part of the social interaction. Another researcher of lying, quoted by Keyes, arrives at the conclusion that "lying is not only a possible action, but a preferred one." (5), which would bring Kant's spirit a lot of grief.

Returning to Benjamin Constant's claim, we may conclude that this type of lying reflects the age of exaggerated communication standards where people need to adapt their lifestyle reality to the dignity expectation of the others in order to preserve the value (or recognition capital) of their identity. Through a complex management of lying, people preserve that existential value which is necessary to both sides of lying. It is always preferable that one gets recognition from someone who is rich in recognition and social capital, rather than by someone honest but undesirable in the context of the social vanity fair. In this context it is difficult to integrate people who are honest and never lie, and other people (born detectives) who are able to (and do) reveal the everyday lying of others.

4 Money, trust and instrumental lying

If present-day millennials are the generation that lies the most overall, there are still professional fields where lying and similar behaviours set the condition for business profitability. After the world financial crisis in 2009, a lot of public interest was bent on the mechanisms of the stock markets. Books, movies, journal articles, and research papers revealed a disquieting reality which was mostly a professional niche before. In this reality, the most difficult thing to understand was how *the value of money* was formed, and how the profits of the big global financial players increased while the real economy was suffering from financial deficiency, people were losing their jobs, and governments were announcing austerity policy.

I do not believe in conspiracy theories, so I doubt that there exists somewhere the mastermind of global finances who can explain definitively what happened. Thus I conducted my own research and, without any pretense of finding out the definitive truth, I discovered that today *we need semiotics to understand the nature of money* (see Bankov 2019). Here I take that work one step further: thanks to the general semiotic principle of the monetary sign, I can outline a model of the instrumental type of lies, which is the opposite of Ego lies as defined in the previous part.

In its historical development the monetary sign has gone through various phases, each of which detached it further from the tangible, valuable asset that the sign represents (Bankov 2017). The utmost expression of this logic of the monetary sign was the Gold standard. With the gold standard there were coins, paper and electronic money in public circulation, but all of these were backed up in the vault of the national bank with real gold, the value of which was established by the international market and not by the government. *The Gold standard was invented to prevent the state from lying*, as governments and central banks were inclined to print more money than what they had earned by their economies in order to pay debts towards their own population. With money backed up by gold those payments were genuine, whereas payments with printed money without golden backup were fraudulent and a serious threat to provoke inflation (money lose their value) and financial crises.

In 1971, the US Gold standard was officially suspended by president Nixon, after all the other big economic powers had given it up during the previous five decades. The reasons for this suspension are complex and hugely debated, but what is important to us is that many influential thinkers as Baudrillard (1993 [1976]), Goux (1994 [1984]), and others foresaw in this an institutionalization of the Big Lie, giving neoliberal governments a free hand to have full control on the lifeblood of the political-economic system – in other words, the value of money.

What followed the Gold standard was the system of *fiat money*. This is a system where value is determined by law, i.e. by mere political decision. Obviously this is not an arbitrary decision, and during periods of financial stability mostly external factors rule the value of money. But in times of financial crisis the arbitrariness of fiat money raises serious suspicion in the rightfulness of the whole system.

With fiat money, the monetary sign achieves its most essential form – *trust inscribed*.[1] "Trust inscribed" means that the free market financial system (i.e. not totalitarian or under absolute monarchy) may be sustainable only if the majority of its members believes in its future, in the value of forthcoming incomes, and by the purchasing power of money. As I have shown already (Bankov 2019), the wealth of an economic system, the entity of the available capital and the entity of employed productive capacity, depend on predictability in time. And predictability means trust. On the contrary, if there is no trust in a bank and people withdraw their money to the point that the bank cannot pay, this constitutes bankruptcy. The banking system relies on the predictability of the fact that no more than 15% of a person's savings will be necessary as liquidity, and the rest of this money can then be invested in lucrative long-term business projects. Present-day capitalism, more than any prior form of capitalism, is a *capitalization of trust*.

In the eyes of many analysts, the financial crisis of 2009 was provoked by the speculative and fraudulent capitalization of trust, which has been called "semiotization of money" (Bankov 2019: 346 ff). The predictability of the real economy and the capitalization of profitability prospects is something normal that made the world go around for centuries. But when the digital age began to power financial markets, pure futurity was capitalized in an uncontrolled outburst. Derivative financial products for billions of dollars were emitted by the banks, which made those who were capable of selling them rich. Selling derivatives better resembles gambling than business entrepreneurship. Practically, it is bluffing, creating trust in something which does not exist, but which becomes lucrative for the one selling it because of that stolen trust. Selling derivatives represents an economic transaction of money for lies. Today it is easier than ever to capitalize trust, and besides the real players in the market of trust, who are gaining the trust of their customers through quality and dedication, *there are also those who are capitalizing on stolen trust, i.e. on lies*.

All this to outline the general principles of instrumental lying for money. Although there are many people who had a direct experience of such fraudulent trade (*The Wolf of Wall Street*, one of the heroes of our time, was one of its

---

1 This is by far the best contemporary definition of money, given by Nail Fergusson, 2008: 32.

precursors), the real extent of this phenomenon, as with the other typology of lies, occurs on the lower level of everyday activities. One of the purest forms of capitalized lie is *fake money*. The producer of fake money sells trust in the monetary system which does not belong to him, and in this way causes a direct loss to the person being lied to. The same mechanism occurs with selling *fake branded goods*: brand means trust, sometimes representing a huge investment on the side of companies to create and sustain it, when some fraudulent producer takes the external resemblance of the original products and sells the reproduction – a trust for which he had not invested, resulting in higher profits. The interesting thing here is that often buyers know that they are purchasing fake goods, but will buy it because in certain social occasions they will display the image of the fake product for the image of the real branded one and will gain the same prestige with less investment. Obviously in this case we are talking about the Ego lying typology.

A lot of instrumental lying takes place thanks to the *abuse of professional trust*. When a person with low knowledge of vehicle mechanics has car trouble, the garage mechanic is happy because he may present a distorted reality to the person without them being able to verify the truth. The invented reality represents a bigger problem: thanks to the lie, the mechanic will take more money for less work and for changed parts. There are many laws for consumer protection, but unfortunately many businesses would not survive without this practice. The worst example of lying in this kind is when a doctor abuses professional trust. The law imposes a Hippocratic Oath to prevent doctors from lying, but this oath is not infallible. The pharmaceutical business is particularly instrumentalized, where doctors are financially encouraged to sell certain products and capitalize on the abuses of institutional trust conferred to them by law.

Another variety of instrumental lying for money, based on a different level of competence, is the tourist industry at large. Usually the very idea of a touristic adventure starts with a fantasy, stimulated by incomplete but idealized knowledge about a given place. The tourist professional creates all the valorization of the offered service thanks to this incomplete knowledge. This is not always based on explicit lying, but the result is nevertheless that the client pays for an imagined scenario with the help of the agent, which is quite different from a factual reality of the place when visited. But the real champions of instrumental lying are local touts and crooks, whose expertise comes from *a better knowledge of circumstantial reality*. Although one can hardly trust such people a priori, they create a relation of trust thanks to the verisimilitude of the description of what they can do for the tourist. The tricky part comes from the statistical constancy of a tourist's behavior. People are inclined to believe in their uniqueness and the uniqueness of newly established social contacts, this situation is reversed in the minds of touts and crooks. They capitalize on the most common weaknesses and inconsistencies of a tourist's expectations as statistically proven over the years. Long before Amazon's algorithms, these operators developed an intuitive approach towards making a tourist spend the maximum possible amount for a minimum amount or quality of received goods and services. In Echian semiotic terms this is a masterful manipulation of the empirical reader, thanks to exact identification by the author (tout) of the reader-model, based on the insufficiency of a tourist's encyclopedic competence of the tourist destination's reality.

Love and sex industries are another huge sphere of instrumentalized capitalization on fraud and deception. As a mechanism these industries are similar to touristic lying, but here these fantasies are powered by a more desublimized and direct link to the libido. Usually men are induced to believe they are in a loving or sexual relation with a younger woman, after which point they are easily manipulated into scenarios of expenditure and fake investments. Cinema and literature are full of such stories, most of the times complex and intricate, but in real life it paradoxically seems that the most common clichés are the most lucrative.

5 Conclusions

As I mentioned in the introduction, lying is a sprawling topic and our examples could continue ad infinitum. We may, however, draw some "macroeconomic" conclusions on the spread of lying in recent times. Keyes, referring to other scholars, underlines how in the sixties the political lie was an exceptional act provoking surprise in public opinion, whereas in the Reagan-Clinton-Bush era it became a habit (Keyes 2004: 12). We may consider this tendency an inflation of the lie, an *inflation of its capacity to carry both instrumental and ego value*. I would put a further accent on explaining this process through the rise of the information age and on the proportionally increased use of lies in relation to the overall increase of information and communication (in contrast with Sissela Bok, quoted by Keyes, ibid.). But the real exponential inflation of lying, in my opinion, comes with onset of the Web 2.0 era, when internet becomes dominated by user-generated content.

Web 2.0 brings about a deep cultural transformation whose outcomes are too early to generalize or describe in a theoretical way, but whose source of motivational power come from this newly established dialectic between the constant constructing of the self (where ego lying takes place), and the capitalization of trust (where instrumental lying takes place). The inflation of the value of lying is balanced by an augmented productivity of ego discourse, which in its turn is subject to capitalization as it turns out to be successful and influential. This is not only the obvious market economy of influencers, but an option for every public and private body almost independently of its intentions. This is the logic of the algorithms ruling the social media exchange, the new market of attention, of experience, of the most scarce of all resources – our time. In such a post-truth situation it seems that the only reliable reference of value is the purchasing power of money (Bankov 2019), a value on which we are inclined to convert any other existential and cultural value. In 2000, Rifkin warned us that the new digital capitalism causes "the absorption of the cultural sphere into the commercial sphere", with the consequence of a depleted cultural diversity of humanity – a diversity which represents the lifeblood of civilization (Rifkin 2000: 11–12). The widespread employment of lying, the loss of stable references for our socio-culturally constructed reality, and the universal validity of the purchasing power of money are the components of this mechanism.

References

Bankov, Kristian. 2017. Approaches to the semiotics of money and economic value. *Signs and Media* (2017, Autumn issue), 178–192. Chengdu: Sichuan University Press.

Bankov, Kristian. 2019. From Gold to Futurity: a Semiotic Overview on Trust, Legal Tender and Fiat Money. In Cheng Le & Marcel Danesi (eds.), *Social Semiotics*, Special Issue on 'Legal Discourse analyses and their advances'. 336–350.

Barthes, Roland. 1986. The reality effect. *The Rustle of Language*, Engl. trans. R. Howard, Basil Blackwell, London, 141–148.

Baudrillard, J. 1993 [1976]. *Symbolic Exchange and Death*, trans. by Iain Hamilton Grant. London: Sage.

Eco, Umberto. 1976. *A Theory of Semiotics*. Bloomington (IN): Indiana UP.

Derrida, Jacques, 2002. History of the Lie: Prolegomena. *Without Alibi*, 28–70, ed. and trans. Peggy Kamuf. Stanford: Stanford University Press.

Ferguson, Niall. 2008. *The ascent of money: A financial history of the world*. New York: Penguin.

Floch J.-M. 2001. *Semiotics, Marketing and Communication. Beneath the Signs, the Strategies*, engl. trans. R. Orr Bodkin. Pelgrave: New York.

Goux, Jean-Joseph. 1994. *The Coiners of Language*. Norman: University of Oklahoma Press.

Kant, I. 1996. Groundwork Of The Metaphysics Of Morals. In Mary J. Gregor (ed.), *The Cambridge Edition Of The Works Of Immanuel Kant – Practical Philosophy*, 37–108 at AA 429. Cambridge: Cambridge University Press.

Keyes, Ralph. 2004. *The Post-Truth Era. Dishonesty and Deception in Contemporary Life*. New York: St. Martin's Press.

Rifkin, Jeremy. 2000. *The Age of Access: The New Culture of Hypercapitalism, Where all of Life is a Paid–For Experience*. Penguin/Putnam: New York.
Todorov, Tzvetan. 1981. *Introduction to Poetics*, University of Minnesota Press, Minnesota.
Todorov, Tzvetan. 2001. *Life in Common*. University of Nebraska Press, Lincoln Nebraska.
Weber, Max. 1978. *Economy and Society: An Outline of Interpretive Sociology*, Berkeley CA: University of California Press.

Loredana Ivan, Corina Daba-Buzoianu, Ioana Bird
Effective affective campaigns? An analysis of campaigns centered on Roma

Abstract: The current paper proposes a qualitative textual analysis of Romanian campaigns centered on Roma communities, from the perspective of the semiotic interplay of cognitions and emotions, with an accent on the presence of cognitively complex self-conscious/moral emotions such as shame, guilt, embarrassment, and pride. Often, the general goal of strategic communication campaigns with social purposes is to increase self-reflective processes and people's abilities to respond to others' needs. Among the moral emotions, guilt is considered central for eliciting empathy and prosocial behaviors, including altruism. A corpus of six campaigns that ran between 2008–2019 were selected for the analysis based on online search engine results visibility. The textual analysis focuses on the messages in the communication outputs and the way the messages relate to self-conscious moral emotions.

Keywords: self-conscious emotions, moral emotions, humanitarian campaigns, communication campaigns

1 A theoretical investigation

1.1 The interplay between cognitions and emotions: social psychological perspective

The relation between feelings and thinking has been always in the center of the debate in various sciences. Traditionally, the two components of the human mind have been approached as independent entities and little attention have been payed to our emotions, whereas out cognitive life was deeply investigated. Massey (2002) argues that the long-time neglecting role of affects is due to our attempt to differentiate ourselves from our biological routes and focus on what makes us unique from other species – our cognitive life. The separation between affect and cognition lays back in the antiquity- with the separation between spirit

Loredana Ivan, Corina Daba-Buzoianu, Ioana Bird, National University of Political Studies and Public Administration (SNSPA), Romania

https://doi.org/10.1515/9783110662900-008

and feelings and the postulated superiority of the spirit (for example in Plato's conception).

It was only after 1980 that the interplay between affects and cognition witnessed a mounting interest (see Massey 2002; Forgas 2001). Early in his volume, Kevin Marinelli scrutinises the affect theory and discusses the classical rhetoric paradigm (starting from the Platonian distinction between pathos and logos); the cultural studies and the neuroscience paradigm. Besides the three paradigms, social psychology has brought important insides about the interplay between emotions and cognitions in the changing attitudes and persuasion.

Although social psychology has always considered affects in the way attitudes were conceptualized – as a tripartite construct with cognitive, affective and behavioral component (Eagly and Chaiken 1993), it was Zajonc (1980) who drew attention on the dominant force of affect in evaluative attitudes towards stimuli. Zajonc and others (see for example Niedenthal and Halberstadt 2000) have provided evidence that affective reactions are the stronger component of people's answer to different stimuli, playing an important role in the way they structure attitudes about different social experiences, and eliciting behavior.

We agree with Forgas (2012) that there is little consensus in the social psychological literature, and not only, about the distinction between terms "affect", "feelings", "emotions" or "mood". Instead, "affect" is used more as a generic label for the four concepts or the concepts are used interchangeably. In an extensive review on the importance of affects in social psychology, Forgas (2001) illustrated their role in: (1) attitudes towards the self; (2) attitudes towards others; (3) intergroup attitudes and (4) attitude change and persuasion.

Basically affects can have an *informational effect* on attitudes for example by *affect priming* or affect infusion – "the tendency for judgments, memories, thoughts, and behaviors to become more mood congruent" (Forgas and Eich 2012 63) and a *processing effect*- influencing one's judgment about an even or a situation. To integrate the theories about the role of affect in both the content and the process of cognition, Forgas (2002) suggested the *Affect Infusion Model* (AMI). The model exhibits factors that facilitate or inhibit affect infusion on attitudes (see Forgas 2010 for a review). The model specifies three contextual elements, which play a role in the way affects infuses cognition: the *task*, the *person* and the *situation*. The Affect Infusion Model might explain the interplay between affects and cognition, when dealing with persuasive messages and how affects elicit attitudes change. For example, there is evidence that people might agree more with a persuasive message when they are interviewed in a pleasant *situation*, than in an unpleasant one, also the fact that people tend to engage in more detailed processing of a message when a positive mood is induced. Similarly, the way *the task* is framed, for example by emphasizing the positive outcomes and producing a positive mood, influences the chances of favorable answers. Also, when arguments point to the negative consequences or trigger the fear of losing something or being left behind, the compliant behavior has better chance to appear (see for example Cialdini 2001). Experimental studies from the social psychology of persuasion have for a long time pointed to the fact that inducting positive affect might be effective in creating agreement with the elicited messages (Mc Guire 1985). Also fear-arousing messages proved to be effective in the attitudinal change (Petty, Cacioppo, Sedikides, and Strathman 1988). Still, as Forgas (2010) underlines, the role of *the person* involved in the information processing is relevant, as fear proved to generate attitude change, particularly when the person believes that the message indicates effectively how to avoid the negative consequences.

The role of affects on persuasion has been explored by "infusing" people with positive and negative mood and then recording the attitudinal changes. For example, Forgas and Eich (2012: 77) underlined the fact that: "in a number of studies, participants in sad moods showed greater sensitivity to message quality, and they were more persuaded by strong rather than weak arguments. In contrast, those in a happy mood were less influenced by the message quality and were equally persuaded by strong and weak arguments." Generally speaking, the experimental studies conducted in the social psychology to investigate the way affects infuse cognition and generate attitudes change have elicited positive or negative moods and did not investigate the role of more complex emotions, as for example the moral emotions.

Humanitarian communication and subsequently social campaigns share the appeal to a specific emotion for persuasive purposes, most of the time the so-called *moral emotions*, Haidt (2003) or *self-conscious emotions* (Lewis 1995), as shame, guilt, embarrassment, and pride. The psychology of emotions, dominated by the so-called "basic" emotions (Ekman 1992) – happiness, sadness, fear, anger, disgust, surprise, has turned the attention to more complex self-conscious emotions, acknowledging their value in a range of phenomena, from altruism to nationalism (this happened particularly after the 1995, when the volume, *Self-Conscious Emotions: The Psychology of Shame, Guilt, Embarrassment, and Pride* by Fischer and Tangney was launched).

1.2 Self-conscious moral emotions

The humanitarian discourse often elicits emotions as guilt, pride or shame, as these are "cognition-dependent" emotions, (Izard, Ackerman, and Schultz, 1999) involving self-evaluative processes. Michael Lewis (see Lewis and Brooks 1978) coined the term self-conscious emotions, as they occur after a process of self-evaluation. Currently it is acknowledged that self-conscious emotions involve not only self-reflections, but also perceiving the fact that another person is expressing an emotion about one's self (Fessler 2007). Thus they are *other-oriented* emotions and involve some level of empathy and cognitive dissonance (an aspect that we will discuss further in this section). Their reflective value both on self and potential others is important as this is probably uniquely human and created the premise for self-conscious emotions to be evaluated as *cognitively complex* (Lewis 2000). Often, the general goal of strategic communication campaigns with social purposes is to increase self-reflective processes and people's abilities to orient to others' needs.

Moreover, self-conscious emotions are related to the promotion an attainment of complex *social goals*, as for example the enhancement of social status or the prevention of group rejection (Tracey and Robins 2007). While the primary emotions are more related to the attainment of survival goals, self-conscious emotions are social embedded in a way that they promote socially valued behaviors. The self-conscious emotions occur once people are able to form some stable representations towards self and reflect consciously about what is desirable/non-desirable in certain situations. Furthermore, self-conscious emotions appeal to social values shared by societies/communities – and that is why they are labeled moral emotions: "shame, guilt, embarrassment, and pride function as an emotional moral barometer, providing immediate and salient feedback on our social and moral acceptability" (Tangney, Stuewig, and Mashek 2007: 22). The self-conscious moral emotions do not only emerge as a result of people's actual behavior, but also as a feedback of anticipated behavior. Thus, we can talk about *anticipatory* shame, guilt, or pride and *consequential* shame, guilt, or pride. Anticipatory self-conscious emotions are particularly interested when considering humanitarian communication, because they can reinforce moral choices once people consider behavioral alternatives. The humanitarian discourse does not necessary trigger actual affective responses but anticipatory moral emotions. Nevertheless, the anticipatory social emotions are inferred from past emotions on similar behaviors and the awareness of the moral standards in that particular situation (as resulted through self-reflective processes). In addition, moral standards vary from individual to individual, for particular class of behaviors, though there is some degree of universality or social consensus regarding of what is expected in a particular situation.

1.3 Guilt in the center of the humanitarian discourse

Several humanitarian campaigns appeal to the moral value of guilt. Moreover, in the current literature, guilt is considered central for eliciting empathy and prosocial behaviors – including altruism (Batson 1987; Tangney, Stuewig, and Mashek 2007).

Even though people are not aware of the events that are more prone to guilt or prone to shame and often threat the two as indistinctive, "the emotional signals from others that help generate shame are not the same as those that generate guilt" (Campos 2007: xii). In addition, guilt triggers reparatory acts by influencing people to reflect on "the thing" that they have done wrong to others, following for example by apology or confession after a social trespass (Tangney and Dearing 2002). On contrary, shame is considered maladaptive in several contexts, causing people to withdraw from the situation and reflect on "themselves" as subjects of the transgressed behavior. Guilt is also related to people's capacity of making internal causal attribution (as a core aspect of self-reflective processes explained above). More specific, internal, unstable, and controllable attributions are likely to produce guilt and generate a reparatory answer (Tracy and Robins 2007). Following this research line, a humanitarian message should be conceived so that people would be prone to think that "they have the power" to interfere, they were "not aware before about the situation", and that their interference "would make a difference".

Providing the fact that elicited guilt is not extreme and causes a state of humiliation, experiencing guilt might have better chances of reactions to humanitarian appeal than other self-conscious emotions. Still there is evidence that anger inhibits the generation of guilt: For example when a person already feels angry on the "chaos" created by the homeless situation in her neighborhood, it might be difficult to experience guilt while presenting with a humanitarian message about homeless people.

There are at least three arguments for including guilt in the center of the humanitarian discourse. (1) When feeling guilty, people do not need to defend their global self, but to consider their behaviors and the consequences. This would cause them remorse and regret over the "things" they did wrong. Guilty feelings appear as consequences of internal, unstable and controllable attribution (For example: "I did not try all the possibilities to solve the problem") and claim for reparatory actions; (2) Guilt is considered a typical moral emotion (Tangney, Stuewig, and Mashek 2007) basically because it has the ability of forming emphatic connections with others (other-oriented empathy). Practically any humanitarian discourse elicits from the audience an increased orientation for other people's needs. Also other-oriented empathy moderates, along with guilty feelings, the link between moral standards and moral behavior (Leary, Tate, Adams, Allen, and Hancock 2007); (3) Guilt is responsible for inducing cognitive dissonance and generate attitudinal chance or behavioral compliance. Modern interpretations of cognitive dissonance (Festinger 1957) suggest the fact that the discomfort people feel in dissonant situations is not due to inconsistency between cognitions, but rather due to feeling personally responsible for producing negative consequences (see for example Cooper and Fazio 1984). In experimental induced dissonant situations, people reported negative affect similarly to guilt, which were further reduced following attitudinal change (or behavioral compliance). The action-based model of dissonance has in the center the role of affective state – people would engage in reparatory acts because they "would feel bad" otherwise. Though guilt is not explicitly described in the action-based model of dissonance (Harmon-Jones 2000), it is most probably a factor in creating change.

2 A qualitative analysis of humanitarian communication centered on Roma communities in Romania

2.1 Methodology

The present study will conduct a qualitative textual analysis, investigating a corpus made out of outputs of communication (messages produced by communicators) in the form of communication campaigns centered around Roma issues. Textual analysis is the method communication researchers use to describe and interpret the characteristics of a recorded or visual message. The purpose of textual analysis is to describe the content, structure, and functions of the messages contained in texts (Frey, Botan, and Kreps 1999). Textual analysis involves understanding language, symbols, and/or pictures present in texts to gain information regarding how people make sense of and communicate life and life experiences. The messages can be understood as influenced by larger social structures and historical, cultural, or political factors (Allen 2017). In this current analysis we will focus on the messages in the communication (with an accent on the interplay between cognition and emotion).

2.1.1 Research objectives

The objective of this study is to analyze a corpus of communication campaigns centered on Roma communities in Romania, to both a) analyze the messages of the communication outputs and b) investigate how they relate to self-conscious moral emotions. To structure the analysis, the following research questions will be employed:

RQ1: What are the primary messages, visual signs and symbols used in the campaigns?

RQ2: What are the self-conscious/moral emotions triggered in the analyzed campaigns?

2.1.2 Corpus for analysis

The corpus for this study was selected based on an online search using keywords, and by selecting the campaigns with most visibility in the google search results. This method of selecting the corpus allows us to identify the campaigns that have had most visibility online, which would lead to more endurance over time in the search results. We must firstly note the paucity of such campaigns and communication outputs, which has. The most visible campaigns that were selected for this study are divided into two: campaigns on a national level or run by NGOs in Bucharest, Romania's capital, and campaigns for smaller, local communities.

National campaigns tackle broader issues

National campaigns or campaigns run by NGOs in Bucharest that were among the first results retrieved in a browser search centered on communication objectives such as attitudinal change

(A) regarding Roma individuals, particularly regarding their professional potential, raising awareness and of the systemic prejudice that can hinder the access of Roma individuals to the workforce. Elements of behavior change (B) objectives are also present through concrete calls to action.

'**A generation of Roma specialists in the medical field**' (*O generație de specialiști romi in domeniul medical*) ran between December 2011 and March 2012 and was implemented by the ActiveWatch human rights foundation. Each of the visuals used in the campaign contains the photograph of a child, followed by messages such as 'When I grow up I want to be a beggar' (Când mă fac mare vreau să mă fac cerșetoare), 'When I grow up I want to be a thief' (Când mă fac mare vreau să mă fac ciorditor). These messages are doubled by the tag line of the campaign: 'Roma children dream what we allow them to dream' (Copiii romi visează ce-i lăsăm noi să viseze) followed by the line 'Now they can dream about becoming doctors. 400 scholarships in the medical field'. The suggestion is that stereotypes of Roma children have a significant influence on both their self and public perception. The campaign was covered by online media and was promoted on public institutions' website, announcing the activities and the outcomes.

Figure 7.1: Image from the campaign 'A generation of Roma specialists in the medical field'. https://www.paginademedia.ro/2012/01/campania-antidiscriminare-cu-mesaje-socante/.

'Like Rom' is a campaign run by Agenția Împreună NGO with the goal of setting up role models for the young generation of Roma. 100 Roma professionals from a great variety of fields tell the story of their education path and professional life. The campaign fosters models through a discourse triggering pride and stating that it shapes the community of the Roma professionals. By underlining that it echoes the story of successful Roma people, Like Rom is organized like a book with stories worth telling and invites the audience to see how prejudice, discrimination and poverty can be replaced with good education, well ranked and paid jobs and social recognition. Most of the campaign is based online, social media terminology is being used to create a familiar online environment.

'Stop preconception about Roma people' (S.P.E.R.) is a governmental program run by the General Secretary of the Government and the first national campaign aiming to inform people about ethnic discrimination of the Roma people, fighting against prejudice and discrimination. The program was launched in 2007, but reached new editions targeting the adult population in Romania with the headline: Discrimination of Roma is taught at home. Know them before judging them! (Discriminarea romilor se învață acasă. Cunoaște-i înainte să-i judeci). Workshops, seminars, press materials, posts and news. Roma population is also targeted by the program with two media campaigns *Jan Angle, Romale!* and *Rom European,* radio and TV show and a newsletter promoted in Roma communities.

Figure 7.2: Image from the campaign 'Discrimination of Roma is taught at home. Know them before judging them!' https://www.galasocietatiicivile.ro/resurse/campanii-sociale/apararea-drepturilor-individuale-colective/discriminarea-romilor-se-invata-acasa-cunoaste-i-inainte-sa-i-judeci-9008.html.

'Young Roma, beyond stereotypes' is a campaign run by NGO Agentia Impreuna presenting pictures of young Roma women and men addressing the audience with messages aiming to brake ethnic stereotypes and to increase trust and positive behavior. A collection of several pictures representing Roma young people with messages triggering the stereotypes of Roma are spread in online media. Both text and pictures portrayal young Roma as any other young Romanian in terms of profession, hobbies, lifestyle and physical appearance. The six images convey different types of topics like: language proficiency, interest in art and patriotism. One particular topic aims to underline that the color of the hair is not a particularity of the Roma people (e.g. 'Brunette or blonde, we are both Roma').

Local campaigns and projects
'S.O.S.-Roma children', organized by the local chamber of Partida Romilor Pro Europa. The aim of this local project is to convince Roma children to stay in school, by explaining the role education has in their personal and professional life. Through seminars and discussions with Roma children, 'S.O.S.-Roma children' intends to increase awareness of Roma children on educations' importance for their future and better life. Unlike other campaigns and national projects, 'S.O.S.-Roma children' is directly addressing Roma children, without targeting other audiences. Traditional Roma clothing can be observed in the pictures about the campaign, both in the media coverage and in the news promoted by organizers.

'Need for quality and equality in education' is a local campaign run between 2009 and 2011aimimg to increase children's enrollment in school in four counties in Romania. With the help of an intercultural curricula and an intervention model, 800 Roma children were supported to adapt to kindergarten and to enroll in school. The intervention model developed helped to identify children at risk and to stop them from leaving school. The project also involved a large group of school teachers and school workers in order to reach the objectives.

2.1.3 Results

The communication outputs of the campaigns put under scrutiny in the current paper use images, frames and key words widely encountered in representations of Roma people. The idea of poverty and discrimination prevail over the other elements composing the image of Roma. With the overall goal of raising awareness on Roma conditions and on changing behavior among Romanians, these campaigns have used symbolic elements in order to trigger the representation of Roma and inequality within the educational and social system. The visual stereotypes of Roma children have the role to guide the audience throughout the story told by the picture and text and to activate empathy. In addition, the campaigns use the dominant discourse on Roma to envisage a collective guilt over the general conditions of Roma people, with an emphasize on children and young adults. Both the visual stereotypes and the use of the dominant discourse on Roma people have the role of setting a common representation ground with the audience.

Although some differences can be observed between national campaigns and local initiatives, it is obvious that the children in the pictures can be easily identified as Roma due to the portrayal techniques. Darker skin, black hair, along with a poor appearance are visual elements used to announce the presence of a Roma child in the picture. Although the color of skin and hair and poor clothing are not to be seen exclusively among Roma children, they are parts of the visual stereotypes of Roma children. When having both Roma and Romanian children in the picture, the Roma is pictured in shadow, low light and looking up to the other. Unsurprisingly, images portrayal inequalities among children and the negative attitude that adults have on this matter. As indicated in one campaign, adults drag children away from Roma in order to avoid contact. Inequality, poverty and low access to education are the elements that comprise the dominant discourse on Roma people. The responsibility for this is put on the adults and society at large, as both images and texts show that Roma people deserve a change in life and that this starts with education.

Drawing on the campaigns on Roma communities in Romania, the textual analysis conducted revealed several themes covered by the campaigns, together with some portrayal techniques aiming to fight stereotypes triggering moral emotions. Two types of audiences emerge from the communication campaigns, as both the general Romanian and Roma publics are addressed. Pictures and texts convey moral emotions to shed a light on the situation of Roma people, mainly children and young adults, by pointing on the stereotypes they have to fight and the lack of resources and support. National campaigns aim to portrayal Roma as similar to Romanians to ensure inclusion and assimilation, by triggering shame due to egocentric and discriminatory attitude, whereas local campaigns are more anchored in community life and propose interventions in schools to avoid abandon. Local projects and campaigns represent Roma in traditional and poor clothing and aim to stop Roma children from leaving school.

Education enhances success
This theme is conveyed in informative messages, aiming to increase awareness of the huge importance education has. Although no call to action is present, some hidden advices can be observed in texts, as an important element of the campaign is training of Romanian educators in order to help them assist Roma children in school ('Need for quality and equality in education'). By telling that without education Roma children can't dream of becoming something else expect beggar and poor, the campaign sheds a light on a collective guilt, making the audience being responsible of non-action. The headline 'Roma children dream what we allow them to dream' along with the picture of a Roma girl triggers a moral emotion and emphasizes the idea that it is the responsibility of the society that these children have equal chances in life and the possibility to hope for a better future. The guilt is addressed through two dimensions: the responsibility of non-

action and the regret and sympathy that due to their belonging to an ethnic community, Roma children are sentenced to poverty and low quality of life. For the responsibility of non-action, the solution is suggested to be in the hands of each Romanian family as 'Discrimination of Roma is taught at home. Know them before judging them!' The regrets and sympathy for Roma children's situation is constructed through an increase of self-reflective processes and by making use of people's abilities to respond to others' needs. Ignorance and negative attitude towards the Roma people are seen as part of the social process that is sentencing Roma children to no education, lack of professional perspective and good life quality.

Similar, not different
Through storytelling techniques, the portal LikeRom.ro changes the discourse about Roma personal and professional path. By underlining the importance that education had in their development and success, LikeRom.ro aims to show that Roma people have professions and hobbies just as the Romanians do. By underlining the aspects of their personal and professional development, the portal is actually pointing towards the similarities with any other successful Romanian. The shame of acting as superior and egocentric is aimed to be felt by the audience, in an attempt to use self-conscious emotions in diminishing the differences. As 'Young Roma, beyond stereotypes' campaign states, when compared with Romanians, there aren't differences in the lifestyle, professional path and physical appearance of the Roma people. Equally important is the fact that these similarities are represented in opposition to the wide stereotypes of the Roma. Through a word-play in Romanian language, the headline of one of the pictures states 'I don't eat letters, I study them' ('Nu mănânc litere, le studiez'). In Romanian, eating letters means being unable to use properly the language and make grammar mistakes, Romas being considered to speak poor Romanian. Another word-play headline mentions 'I have canvas on my wall, not carpets' is presenting a young woman interested in art instead of preserving Roma traditions (Roma and people living in traditional house in country side use carpets to decorate walls). Similarities are also underlined in the physical appearance, underlining that the color of the hair is not characteristic for Roma: 'Brunette or blonde, we are both Roma'.

Stereotypes and preconception lead to poverty and underdevelopment
While the theme related to success and education is conveyed in informative messages, the poverty and underdevelopment of Roma is represented as a consequence of the stereotypes and prejudice. By stating that Roma children are not

allowed by us to dream of becoming anything else but beggars, it is shown what stereotypes and discrimination can lead to and the negative lifelong impact they might have. Pictures and texts shading a light on the consequences of stereotypes and preconception represent children being ignored, labeled, put aside in poor conditions. By picturing Roma children facing poverty, dirty and wearing poor clothing in a general context of ignorance, judgement and inequality, self-conscious emotions are triggered. Thus, the reader is intended to feel a mix of guilt and shame for rejection and non-action. While guilt is linked to rejection and stigmatization, shame accompanies the passivity and lack of action to enhance Roma children access to education and welfare.

Headlines like 'Roma children deserve a chance', 'A chance for education and better future' and 'S.O.S. Roma children – A gate for education' are in line with the usual discourse fighting prejudgment and stereotypes and are highlighted by the key word chance for an emotional dimension of the message.

Models of Roma

While guilt and shame are moral emotions used in addressing the general public in Romania, some campaigns show Roma as models, successful people with highly valued professions like lawyer, manager, medical doctor, social worker, journalist, and PR practitioner. With the aim of motivating Roma youth to stay in school and in showing that education and strong will are the key to success, some programs tell the story of Roma's personal and professional life. These models are constructed to inspire other Roma and to trigger pride among the Roma community.

2.1.4 Discussion

This study on affective communication campaigns reveals not only some interesting ways of articulating moral emotions/self-conscious emotions, but also different ways of constructing the image of Roma through similarity. The presence of visual stereotypes and of the dominant discourse of Roma people in the campaigns is aimed to reach a common representational ground with the audience. In this line, the campaigns emphasize on the idea that the differences between Roma and Romanians are due to poverty and low education, all these being the result of social inequality and discrimination. Due to low access to education and poor life conditions, Roma people are rejected by society at large. Equally important is the fact that when looking at national and local projects and campaigns on Roma important differences can be reported. While local projects and campaigns mainly address the Roma community and propose intervention models in schools and kindergartens in order to help children continue their education, national projects, programs and campaigns engage a larger audience and try to reach a general public. Thus, the local initiatives present concrete cases of Roma children and talk about their situation, whereas national ones try to determine behavior change through triggering moral emotions.

In this light, our study points out a new element in the discourse on Roma, as campaigns highlight the similarities between Roma and other groups in terms of language proficiency, interest in art and leisure. This may be considered an important change in the foremost discourse on Roma people and does not trigger guilt and shame, but rather admiration and pride. In a very interesting way, national campaigns and projects seem to address the idea of similarity with the large majority of Romanians, pointing out that, when educated and with equal chances in life, Roma people succeed as any other person. Pity, empathy and guilt is replaced with moral emotions that have not be previously addressed in the Roma discourse. By pointing out that their personal and professional path is similar to any other Romanian and that they have jobs, preoccupations and hobbies that are not culturally rooted, these campaigns convey new representations of Roma, constructed in terms of similarity.

References

Allen, Mike (ed.). 2017. *The sage encyclopedia of communication research methods*, Thousand Oaks, CA: SAGE Publications.

Batson, C. Daniel. 1987. Prosocial motivation: Is it ever truly altruistic? *Advances in experimental social psychology 20*. 65–122.

Campos, Joseph J. 2007. Foreward. In Jessica L. Tracy, Richard W. Robins & June Price Tangney (eds.), *The self-conscious emotions: Theory and research*, ix–xii. New York: Guilford Press.

Cialdini, Robert B. 2001. Harnessing the science of persuasion. *Harvard Business Review* 79 (9). 72–81.

Cooper, Joel & Russell H. Fazio. 1984. A new look at dissonance theory. *Advances in experimental social psychology* 17. 229–266.

Eagly, Alice H. & Shelly Chaiken. 1993. *The Psychology of Attitudes*. Harcourt Brace Jovanovich College Publishers, Orlando, FL.

Ekman, Paul. 1992. Are there basic emotions? *Psychological Review 99*(3). 550–553.

Fessler, Daniel. 2007. From appeasement to conformity: evolutionary and cultural perspectives on shame, competition, and cooperation. Jessica L. Tracy, Richard W. Robins & June Price Tangney (eds.), *The self-conscious emotions: Theory and research* Guilford Press, New York. 3–20.

Festinger, Leon. 1957. A theory of Cognitive Dissonance. Stanford CA: Stanford University Press.
Fischer, Kurt W. & June Price Tangney. 1995. *Self-conscious emotions: The psychology of shame, guilt, embarrassment, and pride*. New York: Guilford Press.
Forgas, Joseph P. & Eric Eich. 2012. Affective influences on cognition: Mood congruence, mood dependence, and mood effects on processing strategies. *Experimental Psychology* 4. 61–82.
Forgas, Joseph P. 2002. Feeling and doing: Affective influences on interpersonal behavior. *Psychological Inquiry* 13(1).1–28.
Forgas, Joseph P. 2010. Affective influences on the formation, expression and change of attitudes. In Joseph P. Forgas, Joel Cooper & William D. Crano (eds.), *The Psychology of Attitudes and Attitude Change*, 141–161. New York: Psychology Press.
Forgas, Joseph P. 2012. Feeling and speaking: Affective influences on communication strategies and language use. In Joseph P. Forgas, Orsolya Vincze & János László (eds.), *Social Cognition and Communication*, 63–81. New York: Psychology Press.
Forgas, Joseph P. (ed.). 2001. *Feeling and thinking: The role of affect in social cognition*. Cambridge: Cambridge University Press.
Frey, Lawrence R., Carl H. Botan, & G. Kreps. 1999. Investigating communication: An introduction to research methods, 2nd edn. Boston: Allyn & Bacon.
Haidt, Jonathan. 2003. Elevation and the positive psychology of morality. In Corey L.M. Keyes & Jonathan Haidt (eds.), *Flourishing: Positive psychology and the life well-lived*, 275–289. Washington, DC: APA.
Harmon-Jones, Eddie. 2000. An update on dissonance theory, with a focus on the self. In Abraham Tesser, Richard B. Felson & Jerry M. Suls. *Psychological perspectives on self and identity*, 119–144. APA, Washington, DC: American Psychological Association.
Izard, Carroll E., Brian P. Ackerman & David Schultz. 1999. Independent emotions and consciousness: Self-consciousness and dependent emotions. In Jefferson A. Singer & Peter Salovey (eds.), *At play in the fields of consciousness: Essays in honor of Jerome L. Singer*, 83–102. NJ: Erlbaum, Mahwah.
Leary, Mark R., Eleanor B. Tate, Claire E. Adams, Ashley Batts Allen & Jessica Hancock. 2007. Self-compassion and reactions to unpleasant self-relevant events: the implications of treating oneself kindly. *Journal of Personality and Social Psychology* 92(5). 887–904.
Lewis, Michael & Jeanne Brooks. 1978. Self-knowledge and emotional development. In M. Lewis & L.A. Rosenblum, *The development of affect, vol.1*, 205–226. New York: Springer Science and Business Media.
Lewis, Michael. 1995. Self-conscious emotions. *American Scientist* 83(1). 68–78.
Lewis, Michael. 2000. Self-conscious emotions: Embarrassment, pride, shame, and guilt. In M. Lewis & J. M. Haviland-Jones, eds. *Handbook of emotions*, 623–636. New York: Guilford.
Massey, Douglas S. 2002. A brief history of human society: The origin and role of emotion in social life. *American Sociological Review* 67. 1–29.
McGuire, William J. 1985. Attitudes and attitude change. In Gardner Lindzey & Elliot Aronson (eds.), *The Handbook of Social Psychology*, 3rd edn., 233–346. New York: Random House.
Niedenthal, Paula M. & Jamin B. Halberstadt. 2000. Emotional response as conceptual coherence. In Eric Eich, John F. Kihlstrom, Gordon H. Bower, Joseph P. Forgas & Paula M. Niedenthal (eds.), *Cognition and Emotion*, 169–203. New York: Oxford University Press.

Petty, Richard E., John T. Cacioppo, Constantine Sedikides & Alan J. Strathman. 1988. Affect and persuasion: A contemporary perspective. *American Behavioral Scientist* 31(3). 355–371.

Tangney, June Price & Ronda L. Dearing. 2002. *Shame and Guilt*. New York: Guilford Press.

Tangney, June Price, Jeffrey Stuewig & Debra J. Mashek. 2007. What's moral about the self-conscious emotions. In J. L. Tracy, R .R. Robins & J. P. Tangney (eds.), *The self-conscious emotions: Theory and research*, 21–37. New York: Guilford Press.

Tracey, Jessica L. & Richard W. Robins. 2007. The self in self-conscious emotions: A cognitive appraisal approach. In J. L. Tracy, R .R. Robins & J. P. Tangney (eds.), *The self-conscious emotions: Theory and research*, 3–20. New York: Guilford Press.

Zajonc, Robert B. 1980. Feeling and thinking: Preferences need no inferences." *American Psychologist* 35(2). 151–175.

Zajonc, Robert B. 2008. Feeling and thinking: Closing the debate over the independence of affect. In Joseph P. Forgas (ed.), *Studies in emotion and social interaction, second series. Feeling and thinking: The role of affect in social cognition*, 31–58. New York: Cambridge University Press.

B **Political semiotics and communication**

Massimo Leone
The semiotics of extremism

Abstract: "Extremism" is a term and a concept that does not designate a precise, determined, and definitive location in the semiosphere (the abstract area in which a community produces, handles, and circulates meaning). As the etymology of the word itself suggests, its meaning is intrinsically topological and relational. Something can be extreme only by situating itself at the periphery of a spectrum, at the border of an area of potentialities. The term and the corresponding concept, moreover, signal that this spectrum, as well as this area, are not neutrally arranged but contain at least one dialectics and, as a consequence, one polarization. In other words, "extremism" implicitly refers to an axiology, which includes also the impossibility not to adopt a perspective on it, a point of view.

Thus, when something is qualified as an expression of "extremism", such qualification inherently points out that: 1) this something, be it a statement or a behavior, is comparable and commensurable with other similar events in the semiosphere; 2) all these events can be arranged along a spectrum, in relation to the pertinence that determines their commensurability; 3) those who deem the statement or behavior as extreme consider that they are entitled to position themselves at the center of the spectrum, and simultaneously situate what they judge at one or the other extreme of it.

The article semiotically investigates the meaning of "extremism" in present-day communication.

Keywords: Extremism, communication, populism, cultural semiotics, interpretive community, hermeneutic reasonableness

> Much of junk culture has a core of crisis – shoot-outs, conflagrations, bodies weltering in blood, naked embracers or rapist-stranglers. The sounds of junk culture are heard over a ground bass of extremism. Our entertainments swarm with specters of world crisis. Nothing moderate can have any claim to our attention.
>
> (Saul Bellow, *A Second Half Life*, 1991, p. 326)

Massimo Leone, University of Turin, Shanghai University

https://doi.org/10.1515/9783110662900-009

1 Public hermeneutics endangered

Book IV of Jonathan Swift's *Gulliver's Travels* (1726) describes the fictional race of the Houyhnhnms, a breed of intelligent horses whose perfect rationality starkly contrasts with the beastly manners of the humanoid Yahoos. Houyhnhnms are endowed with a philosophy and, above all, with a language that are completely void of any political and ethical nonsense. Their language, for instance, does not contain any word for "lie", to the point that, in order to refer to it, Houyhnhnms must use a circumlocution: "to say a think that is not".

Eliminating all imperfection from thought and all ambiguity from language has long been a human dream. On the one hand, the quest for the perfect language attests the persistence of such utopia. On the other hand, analytical philosophy seems to prolong it into a modern, philosophical version with Wittgenstein's attempt at purifying language from all meaningless contradiction. In reality, although these projects of conceptual and linguistic purification might contribute to a better understanding of both thought and language, they, the projects, are intrinsically doomed to fail. The reason is simple: a code, and especially that socially institutionalized code that is a grammar, can regulate the syntactic construction of a language, introducing a complex and efficacious system for either the approval or the disapproval of its acceptable or inacceptable forms. No code, however, can explicitly determine the semantics of a language, since this inevitably consists in a content, in meaning, in the invisible and mental reality to which access is, indeed, gained only through the shared signifiers of language.

For instance, we can collect empirical evidence about the fact that a journalist might or might not have abided by the grammatical rule prescribing the socially approved usage of the subjunctive form of a verb. We might not, however, empirically verify whether the usage of such subjunctive actually corresponds to the epistemic attitude that it, the subjunctive, is grammatically supposed to express, that is, uncertainty. We can correct the syntax of a non-grammatical journalistic discourse, but we cannot correct its semantics. We must, to a certain extent, trust it. In the same way, we must trust that the meaning of the words that we use is shared, at least to a certain extent, by our interlocutors and, more generally, by the other speakers of our same linguistic and, most importantly, hermeneutic community.

A society thrives when its members have the impression that they share a solid nucleus of common sense and cognitive encyclopedia with their fellows. It thrives, that is, when they, the members, believe in the existence of a public hermeneutics that, like the safety net of the acrobats, allows the most daring linguistic and conceptual summersaults to be performed precisely for the net of language itself underlies them, like a silent, unperceived, and yet indispensable fundament. On the contrary, a society starts to disintegrate not only when its public hermeneutics is threatened from the outside (the arrival of an invading civilization, with another language and different manners, for example), but also when it is endangered from the inside.

A society's language is not, indeed, a perfectly integer safety net but one that has been at risk of breaking many times, and that nevertheless the society in question has managed to mend. Entire wars were fought in Europe and elsewhere because citizens of the same country, the same language, and sometimes even of the same family would not agree any longer on the meaning of the word "God". Catholics and Protestants would both employ this word, and share the same grammar in arranging its syntax, yet the semantics that they would attach to it was completely at odds with each other's. Years of violence elapsed before the safety net of language would be recomposed, albeit not with the same integrity that it previously enjoyed but with the scars that the past conflict had entailed. Europeans learned, for instance, that they had to mostly expel the word "God" from their political discourse in order to be able to recover the sentiment of a common public hermeneutics.

More generally, a community's language is full of the scars of past semantic conflicts, scars that sometimes are still fresh and, as a consequence, keep being at risk of bleeding again. Linguists and semioticians, as well as political philosophers, cannot prevent the public opinion and the common language from producing variety and differentiation (Sunstein 2009). The blossoming of various interpretations is consubstantial with the functioning of the human language and cannot and must not be thwarted in any way. It must, on the contrary, welcome as an outcome of- and a tribute to- human creativity, the same that has made the human species so adaptive throughout the natural evolution and through cultural changes.

Scholars focusing on the construction of a public hermeneutics, however, should ensure that the flourishing of interpretations does not give rise to sociocultural and socio-political mechanisms whose outcome is not simply the introduction of variation in the common sense, but the disruption of any possibility of reaching social consensus and a shared encyclopedia (Merkl and Weinberg 2003; Mulloy 2004; Midlarsky 2011). On the one hand, each citizen is a player who, in front of a chessboard, can move her or his pieces in an infinite number of different combinations. On the other hand, those who do not limit themselves to add variety and creativity to the game, but actually disrupt it by disrespecting its basic rules, or even by breaking the chessboard, should neither be welcome nor tolerated, exactly for their action is not stimulating the adaptive creativity of the human society but, on the contrary, introducing stiffness and aridity through the dismantlement of any potential for a public hermeneutic.

The article that follows introduces a reflection leading to understanding some of the linguistic and semiotic mechanisms through which a logic of extremism is introduced in the semiosphere of a society. Fuelled by technological development, which in turn brings about the cultural infrastructure for an explosion of discursive productions (Lotman 2009) – oscillating more and more rapidly between opposing value polarities, and leaving less and less space for mediation, compromise, and equilibrium – extremism is an obstacle to any possibility of establishing a long-lasting public hermeneutics, for the former is conducive to a semantic topology in which the safety net of language is stretched to the utmost, and sometimes breaks with no possibility of mending.

2 Topological relativism

"Extremism" is a term and a concept that does not designate a precise, determined, and definitive location in the semiosphere, the abstract area in which a community produces, handles, and circulates meaning (Berger 2018). As the etymology of the word itself suggests, its meaning is intrinsically topological and relational. Something can be extreme only by situating itself at the periphery of a spectrum, at the border of an area of potentialities. The term and the corresponding concept, moreover, signal that this spectrum, as well as this area, are not neutrally arranged but contain at least one dialectics and, as a consequence, one polarization. In other words, "extremism" implicitly refers to an axiology, which includes also the impossibility not to adopt a perspective on it, a point of view.

Thus, when something is qualified as an expression of "extremism", such qualification inherently points out that: 1) this something, be it a statement or a behavior, is comparable and commensurable with other similar events in the semiosphere; 2) all these events can be arranged along a spectrum, in relation to the pertinence that determines their commensurability; 3) those who deem the statement or behavior as extreme consider that they are entitled to position themselves more at the center of the spectrum, and simultaneously situate what they judge at one or the other extreme of it.

For instance, when a nutritional ideology and its consequent behavior such as fruitarianism is labeled as extreme, that results from the fact that 1) this diet and its ideological underpinnings are implicitly considered as comparable and contrastable with similar nutritional behaviors, not only in general, but also in the sub-spectrum of those that enjoin restrictions in relation to the mainstream; 2) such comparison and contrast allow one to situate fruitarianism along a conceptual spectrum, which essentially measures the extent of restrictions that are imposed on the mainstream diet; 3) a perspective is assumed that places the center of this spectrum at a remote distance from the ideal position occupied by fruitarianism, which is judged, therefore, as an "extreme" practice, resulting from an "extreme" ideology.

Conceiving culture as a semiosphere, indeed, is conducive to a form of relativism that is not a moral one but a structural one: depending on that which occupies the center of the semiosphere, one is able to perceive and stigmatize as "extreme" fragments of textualities, including eating behaviors and ideologies, which are situated at the margins of the semiosphere or even linger at the threshold between what is meaningful and what is meaningless. Taking the example of fruitarianism again, it appears as evident that adopting such topological and dynamic understanding of culture allows one to realize the following features.

3 Extreme features

First, extremism is relational: it is only because diets including a larger variety of nutrients, such as meat or fish or cereals, are posited more at the imaginary center of the semiosphere of a sociocultural community, that fruitarianism can be considered as "extreme".

Second, extremism is gradual: given an extreme behavior or, more generally, sociocultural position in the semiosphere, one can always imagine nuances that differentiate it from less or even more extreme behavior, as well as internal articulations based on a more or less extreme adhesion to a principle or set of principles; within fruitarianism, for instance, one can distinguish between those who eat only fruits from those who eat only fruits that spread their seeds when they are eaten; the former will be inclined to consider the latter as 'extreme', although perhaps not with the same vehemence by which the former are, in turn, labeled as 'extremists' by mainstream eaters.

Third, extremism is dynamic: the semiosphere is a conceptual representation of semiosis as it produces meaning within a sociocultural community; as a consequence, it can be static only in the fiction of the analysis, for in reality the mutual position of the center and the borders of a semiosphere continuously evolve and the content of the label 'extremism' with them (Wintrobe 2006). Being a vegetarian, for instance, is rarely considered as "extreme" nowadays, for the simple fact that this diet and its ideology, once at the border of the semiosphere, have progressively moved toward its center while more restrictive diets would appear at the margins of the semiosphere and as more mainstream diets would increasingly lose their centrality; nowadays, being a vegetarian is not usually judged as an extreme stance anymore both because veganism appeared and multiplied at the periphery of the semiosphere, and because eating meat progressively ceased to be considered as a neutral behavior and, as such, constantly immune from social stigmatization. In the eating community that shares the primary modeling system of the Italian language, for example, the change has been so radical and visible that offering a suckling lamb for dinner to friends might be a social risk, and probably doomed to be stigmatized as inappropriate and a manifestation of cruelty.

The same goes with furs: wearing them was a seldom contested behavior up to the 1990s, whereas they became an 'extreme' fashion item in the following decades, to the point that only someone who is totally unaware of such evolution or a provocateur could don it at the present time. Being dropped from the center of a semiosphere, indeed, means also losing neutrality: the more a meaning producing practice moves far from the barycenter of a semiosphere, the more it is likely to appear as salient, in the sense that the practice ceases to be a habit for the majority and increasingly turns into the driver of a semiosic process. Being a mainstream omnivore at a dinner does not stir any curiosity; being a fruitarian, instead, is likely to prompt questions, objections, and reactions, exactly because it embodies an ideology, and the ensuing practices, that cannot be perceived as habits but as provocations to the mainstream.

4 Deontic meta-discourses

The cultural semiosphere constantly produces a deontic meta-discourse, which explicitly or implicitly suggest what meaning producing practices should be situated at the center of the semiosphere, which ones at the periphery, and which ones even out of it, through a process that usually entails not only sociocultural stigmatization but also a formalized system of legal interdiction and sanction. In Italy, eating lasagna is not a particularly salient behavior, meaning that it is performed daily by thousands of mainstream eaters; eating a vegetarian lasagna is not as mainstream, but is not at the periphery of the semiosphere anymore; presently, it could hardly be labeled as "extreme"; eating a vegan lasagna, or even more eating a fruitarian lasagna, will probably incur in curiosity on the side of mainstream eaters, but also in social stigmatization, ranging along a spectrum that goes from blandly poking fun at vegan friends to attacking them as dangerous extremists; eating a lasagna made of human body parts, finally, will incur not only in sociocultural stigmatization but also, probably, in police investigations, meaning that such behavior is situated outside of that which the semiosphere, and the community that it represents, are ready to consider as acceptable and meaningful.

As regards the relation between extremism and communication, such topological understanding of the former should stimulate the following questions: 1) What determines that a meaning-producing practice is situated at the center of the semiosphere, at its periphery, or even in the semiotic darkness outside of it? 2) What agencies promote or demote a meaning producing practice toward the center or the periphery of a semiosphere, and with what modalities? 3) What semiospheres are more likely to produce extremisms, and how should they cope with them? 4) Is communication between the center and the periphery of the semiosphere possible, is it desirable, and in what forms? 5) Can there be moderation in such communication involving the extremities of a semiosphere? What agencies should take responsibility for it, and in what ways? 6) Finally, what is the impact of the evolution of the technologies of communication on the arrangements of such modalities? Has the formation, evolution, and communication involving extremisms received a significant impact from the mushrooming of digital social networks and their communication practices? (Aly 2016)

5 Extreme rationales

As regards the first question, concerning the rationale for the distribution of mainstream and extremities in a semiosphere, one is tempted to immediately embrace a statistical and, therefore, quantitative answer: certain semiosic events are mainstream because they occur in greater quantities than those that are situated at the outskirts of the semiosphere. According to this perspective, for instance, veganism is not mainstream for mere statistical reasons: were there more vegan than meat-eating people, the former behavior would be located at the center of the semiosphere. Such explanation would, however, be simplistic. It would neglect, in particular, that the topology of the semiosphere is never bi-dimensional but, rather, multi-dimensional, stemming from a complex intertwining of different levels of discourse each representing an instance of meta-discourse for the others. According to Lotman, and understandably, the verbal language occupies and exerts a special role in structuring and nurturing the sinews of this imbricated superposition of discursive layers: a vegan meal signifies through the non-verbal substance of its food and eating style but its meaning is also carved by the position that a myriad of verbal characterizations have prepared for it in the semiosphere.

Competing medical discourses on the benefits and the nuisances of a vegan diet; a whole *imaginaire*, made of words and other signs, concerning animal rights; even jokes: the statistical weight of a meaning producing practice is important, and in the long term even determinant, but its position in the semiosphere is ultimately brought about by the complex and multilayer discursive topology that connotes such practice in relation to similar practices and in relation to the holistic structure of the semiosphere, wherein eating behaviors, like any other semiosic event, are never compartmentalized but receive their sociocultural position from a multitude of meta-levels. This dynamic is further complicated by the fact that, in many sociocultural communities, such imbricated meta-discourse is hierarchized, meaning that some genres, formats, and streams of meta-discourse weigh more than others and usually not for mere quantitative reasons but because of the ways in which the dynamic transmission of meaning as cultural memory of the semiosphere is arranged.

Eating behaviors might quantitatively and statistically evolve, yet this numeric evolution is never independent from the qualitative inertia introduced by the religious discourse, or by the discourse of tradition, which tends to maintain the centrality of meat eating despite its quantitative progressive decline. Similarly, furs might have practically disappeared from the fashion arena of most present-day western societies, yet fur still holds a place as a decorative vehicle of prestige in some settings, such as academic rituals, for instance (Italian rectors in many universities, for example, still don a toga with ermine hedge on special ritual occasions, such as the awarding of a *honoris causa* degree).

6 The semiotic danger of extremism

The statistical weight of a meaning-producing practice is not determinant in begetting its position in the semiosphere for another important reason as well: although the cultural semiotician might adopt a bird-view perspective on culture, and talk about it in terms of "texts", "practices", and "artifacts", these cultural units are not homogenous, but characterized by an internal structure that determines their specific way of producing meaning in society. That which a semiosphere situates as "extreme", thus, is not only the minority. Extremists are usually a minority, but a minority is not necessarily extremist. That is the case for a semiosphere relegates some semiosic events at its extremes also and perhaps above all because their internal structure is seen as particularly subversive of the semiosic structure of one or more mainstream behaviors. Fruitarianism, for instance, is not extreme only because it is embraced by a minority and not only because it implies more dietary restrictions than vegetarianism or veganism; it is extreme because it inherently subverts the hierarchy of values that usually underlies the preparation and consumption of food in most societies, replacing aesthetic rejoice or health benefits with such alternative ideological principles as the preservation of life as an abstract and generalized concept or the defense of the environment perceived as a living whole. In a nutshell, what is extreme in a semiosphere is not only what is statistically lesser, but what challenges, by its own existence and development, the semiotic structure of those texts and practices that occupy the center of the semiosphere, as well as the whole hierarchic articulation of meta-discourses that co-determines such central position.

This topological, non-strictly numeric understanding of the dialectics between center and periphery, mainstream and extremes is meant to lead to a better comprehension of the agencies that turn this static equilibrium into an ever-changing configuration. Indeed, were the explanation of such dialectics purely quantitative, the nature of the agencies that transform it would be easy to grasp: multiplication of a semiosic event pushes it toward the center of the semiosphere, whereas, symmetrically, texts and practices that are not reproduced as in the past tend to be demoted from the center to the periphery of a culture. Such explanation would, however, be tautological: it would not be able to clarify what brings about the multiplication of a cultural artifact, what casts it into oblivion, why some behaviors are stigmatized for a certain period of time as "extreme" and then they are promoted into the kernel of a semiosphere, or else they are rejected out of it into the realm of the intolerable and even into that of the illegal.

A better explanation would, on the contrary, stem from the qualitative understanding of the semiospheric dynamics exposed above: when a practice is labeled as 'extreme', its occurrence and reproduction is seen as subversive of an entire arrangement of texts and meta-texts, which defines the identity of a semiosphere and its members. Fascist discourses, for instance, are relegated into the peripheries of most European semiospheres, and stigmatized as extreme, not only because they are the expression of a minority, and not only because their tenets are contrary to those of most post-WWII European constitutions (to the point of being illegal in same states, that is, rejected out of the semiosphere *tout court*), but also and above all because, were these fascist discourses promoted toward the center of present-day western semiospheres, such promotion would be tantamount to the progressive and inexorable dismantlement of the whole structure of discourses and meta-discourses that confer an identity to these semiosic communities and its members. The entire worldview of a present-day citizen should change, if fascist or fundamentalist ideologies gained the core of contemporary European semiospheres (Pratt 2018).

7 Extreme agencies

That leads to the question of what semiospheres are more inclined to have extremisms proliferate and even be promoted to the rank of central and mainstream contents of a sociocultural community. That is not an easy question to answer, for it implies the determination of an agency or a set of agencies in the semiosphere. If one excludes the possibility of non-semiotic causes that drive internal change in the semiosphere, as it is the case in the Marxist theory of culture, for instance – which tends to explain cultural change in terms of economic changes – then the mechanism for the doing and undoing of centrality and extremism in the semiosphere must be found in its internal structure, in its inherent articulations. If the quantitative argument is discarded too, then the reason for the promotion of extremist cultural contents must be found in a sort of cognitive economy of the semiosphere as a whole. The concept of "cognitive economy" has been long introduced and fruitful in psychology, where it contributes to explain resistance to psychological change in terms of the effort that it would entail as regards the reorganization of ideas, emotions, and plans of action. It is sometimes very difficult, for instance, to admit to oneself that one has been cheated by one's spouse, for such belief, although supported by uncontroversial evidence, would imply, when accepted, a restructuration of one's entire body of cognitions, emotions, and pragmatic schemes, leading to a reconfiguration of one's identity as a whole.

The concept of "cognitive economy" may be applied to the semiosphere as well: there are texts, practices, semiosic events, and so on that a culture cannot admit, or must keep at its margins, lest they proliferate and disrupt the entire semiotic structure of the semiosphere. Going back to political extremism, a semiosphere such as that of post-WWII democratic countries cannot let the concept and the practice of apartheid seep into their core, for that would entail an entire restructuration of the set of ideas, emotions, and other discursive and meta-discursive configurations providing the main semantic skeleton of the semiosphere itself. Given this homeostatic definition of a semiosphere's cultural economy, the usual perspective on cultural change should be perhaps reversed, so as to provide a better vintage point on this crucial phenomenon. The cultural semiotician should maybe inquire not only and not predominantly into how a semiosphere yields to extremism, but rather into how it resists to it, and especially into the conditions given which this resistance is dropped.

8 The conundrum of cultural change

Like seismologists, culture analysts, and therefore also semiologists, observe these changes under the microscope, and see clues and traces. They are therefore able to describe how language and culture modify but remain mostly silent as regards a second, more ambitious question, which wonders not only how, but also why this change takes place. Why does language change? And why does culture change? Moved by what forces, by what agencies? Ferdinand de Saussure does not answer these questions, and for the most part he does not even ask them. Intent on building up a linguistics with scientific ambitions, purging it from the subjectivity residues of the romantic philosophy of language, he avoids any consideration that does not remain confined to a systemic conception of language, in which no external force is seen as being able to exercise a pressure on it.

The romantic and post-romantic philosophy of history, on the other hand, formulates hypotheses on the etiology of cultural change, culminating in the theory that, more than any other, marks and influences the twentieth-century thought on this topic. In Marx, culture evolves in an ancillary manner, as a superstructure whose change is not just autonomous but linked to the deeper and 'more real' change of the economic structure. For example, clothing fashions in this framework do not have any phenomenological dignity and autonomy on their own but result from 'rockier' changes in the system of value production and in the formation of socio-economic classes.

Fascinating as it may be in its simplicity, however, this perspective – albeit nuanced by cultural interpretations *à la Althusser* – is increasingly problematic in the eyes of the student of cultures, to whom it is more and more evident that the economy itself, including some of its fundamentals as the very concept of value or surplus value, do not remain unchanged throughout the ages but are, instead, subject to alterations that shape them rather as linguistic phenomena or, better still, as semiotic configurations. While in the twentieth-century production of value, mostly linked to heavy industrial activity, one could be tempted to explain the attachment of a Fiat worker to the Juventus football team as a superstructural phenomenon in which a socio-economically hegemonic class instilled and inculcated a sport passion in the proletariat in order to bridle its working force, today, as the turnover of the imaginary industry equals and in some cases exceeds that of the automotive one, the argument must perhaps be reversed, in the sense that sport attachment produces passions and values that guide also more traditionally structured sectors of value production.

The semiotic critique of romantic or post-romantic philosophy of history consists, more generally, in recognizing that cultural change has nothing outside of itself and, therefore, cannot be explained with reference to an external agency, be it economic, social, or religious. Even natural change, which seems to rest on a hard and immovable basis, out of the reach of language, on closer inspection cannot be received by the human species if not through the filter of culture. The fact of living in an extremely telluric territory, for example, undoubtedly affects the religious ideology of Japanese culture, yet this incidence is never direct but it is expressed through the filter of cultural re-elaboration, a re-elaboration that is so insistent that it ends taking on the foreground of the scene of change, replacing it as a second cultural nature that overshadows first nature. In fact, there are many telluric territories inhabited by the human species but not all of them show the same cultural traits, which is evidence of the fact that the nature of the geological conformations or climatic areas provides the framework but not the definition of the cultural concretions that manifest themselves therein, and that end up leading individuals to sometimes diametrically opposed readings of the same 'naturalistic' data.

9 Internal dynamics of semiospheric changes

But what remains if, as in the model of semiotics of culture proposed by Jurij M. Lotman, nothing, neither economics nor nature, is external to the semiosphere, and instead everything remains inexorably enclosed, confined in the homogeneous space of signification, with the need therefore to be explained in terms internal to it, without recourse to the causal link that anchors the cultural effect to a heterologous origin? Although Lotman takes his inspiration from Saussure but detaches himself from him in an attempt to "set structuralism in motion" – an indispensable operation, given the transition from a semiotics of language to one of culture – the Russian scholar, like the Swiss one, seems to ask himself *how* cultures change, but not *why*. The sophisticated meta-language of the semiotics of culture captures the minute processes of cultural change with precision, underlining, among other things, the need to think of the cultural frontier as a dynamic device of translation and, therefore, generation of meaning. This metalanguage, however, neither explain nor formulate hypotheses about the origin of the wind that, lightly blowing at the edge of the semiosphere, takes then strength as a cultural meme is reproduced, pushing it vehemently towards the center and the heart of a culture, a wind that moves and removes the universe of meaning; it does not explain, therefore, from which interplay of cold and warm masses of air it spurts, and why at a certain point it rises, or else dwindles away.

10 Reversing the big question

Perhaps it is not the answer that is a problem, but the question. Perhaps, the distinction between how culture changes and why it changes is not so relevant. It is so if a causal epistemological framework is projected onto it, which is the one that we spontaneously adopt when we try to decipher our lives. Nevertheless, the more we embrace with our gaze not so much the minute events that concern us, and that we can read, according to this scheme – against a narrative background of causes and effects – as the great lines of our existence, the more the reason for this or that revolution begins to elude us as in a nebula, where we lose the subtle and often invisible threads that have led us to a certain mark and we finally struggle to grasp the deep rationale of things, often despairing about it.

But if we change perspective, if we radically change perspective, and consider, instead, that this condition of not being able to see the causes of what happens to us is not the exception in the universe but the norm, and that our recognizing causes and effects is itself an illusion – the product of a cognitive scheme that is without doubt useful but relegated to the local area of our cognition – then perhaps we would understand that the world and its cultures, down to the natural language we speak, are too vast areas to be underlain by a rationale, by a cause driving changes, and that everything manifests itself on the basis of mostly random interactions, which become causes only if they are seen within a narrow and limited scope of observation, such as the psychology of interactions, for example.

The philosophical dream of translating casualness into causality should then be reversed into its opposite, that is, into a sort of philosophical quietism that considers causes as a limited optical illusion within an essentially casual universe, where directions of both natural and cultural sense emerge, but as result of mostly unintentional interactions, at least not underpinned by the precise project of creating 'that' sense. New stars, new planets, new species, but also new pictorial styles, new words, new stories, would emerge therefore not by virtue of a determined agency but by impulse of a diffused agency, where micro-agents follow mostly local logics, or what they consider to be such by virtue of a perspective illusion, but then, since their actions cumulate on a very large scale, give rise to structural alterations of global scope, whose outcome cannot be reduced to the single micro-intentionalities that have led to them.

11 The metaphysics of fashion

There is, then, only one word that captures this passage from microscopic intentionality to macroscopic emergence, from the apparent causality of local interactions to long-term structural randomness and, therefore, to the emergence of trend lines in the diachronic development of reality, unexplained in terms of a purely causal logic. This word is "fashion". The universe follows fashion. Culture follows fashion. Language follows fashion. It is therefore not necessary to project the causal scheme that underlies our local reading of the world on a universal scale, as proposed by the theory of intelligent design but, instead, to lean towards the idea that the pockets of causality that we see in the universe are nothing but concretions, limited in time and space; they are local phenomena on which a causal logic manages to have a grip only if it considers them as isolated from the very long period in which they are located, and with respect to which, instead, it would be more correct to adopt a theory of *ignorant design*, that is of a blind, casual, but nevertheless effective development of the universe. In fact, it is not difficult to explain Picasso causally, but it is rather difficult to accept to explain him casually, as a product that emerges from the extremely long time of culture exactly like a new animal species evolving through the eons of biological evolution. Fashion, therefore, would be the key concept not in order to explain, but in order to accept not to explain the folds of culture.

11.1 Oscillation trends

By a sort of law of large numbers, the force of chance that is exerted throughout history and cultures produces polarizations and, therefore, dialectics. But it would be misleading, and a symptom of a surreptitious return to a causal logic, to interpret these dialectics in a Hegelian way, as oppositions destined to produce a final synthesis, a balance. Instead, as it seems that the universe expands starting from an initial big bang only to then subsequently contract in a diametrically opposite direction, and so on to infinity without any possibility of storytelling that is not that of the mere observation of such movement, so cultures also give rise to dialectics, but these do not recompose themselves in a state of equilibrium but, rather, translate into mere oscillations: cultures tend randomly towards the exaltation of a certain characteristic, then reach a saturation juncture, and subsequently fold away on themselves to pursue timidly at first, then in an ever more decisive way, an opposing cultural direction, until the slowdown and stasis on the other side of the spectrum is reached.

11.2 A case study: transparency and opacity

One of the characteristics that the semiotic metalanguage finds in semiospheres is precisely the tendency to a certain degree of opacity, as opposed to the diametrically symmetrical pole of transparency (Lozano 2013). Opacity and transparency are relevant in fashion *stricto sensu*, meaning that fashion is essentially based also on the decision to show or conceal one's body, but they are also relevant in fashion in a broader sense, meaning that all cultures, and not only those of clothing, must regulate, within themselves, that which the Greimassian semiotics would call "the actant observer", that is to say, the way in which all texts, and not just textiles and vestments, decide to favor or disfavor the circulation of information. It is only by virtue of this concept, applied to the polarization between transparency and opacity, that one can grasp the structural relationships that exist between various discursive regimes of ostentation and concealment, within textual configurations that are, nevertheless, very distant in terms of media, genres, formats, styles, and contexts. Judgment on the extent of transparency or opacity that characterizes a society's culture at a given stage of its development, therefore, depends on one's topological and axiological position in relation to the spectrum of this polarization, without the possibility to establish measures or even absolute standards. A culture is more or less transparent compared to its past or future stage, and never absolutely so. The semiotics of cultures, however, must draw its materials from multiple spheres of discourse, trying to carry on relevant and representative samplings but then striving to reconnect results in order to mature a holistic understanding of the semiosphere, of the extent to which it happens to be in relation to the structural polarizations that characterize it.

11.3 The circuit of glasnost

It is quite evident that, since the disintegration of the Soviet Union, transparency has become an essential element of any rhetoric of political change. Along a line that has very ancient roots but that has found its essential coagulation, at least as far as current times are concerned, in Mikhail Gorbachev's decision to adopt "glasnost" — starting from 1986 — as the keyword of his attempt at reforming Soviet politics and society, transparency, which is one of the possible translations of "glasnost", has quickly gained the heart of the global semiosphere and become, for this reason, an object of fashion (Nove 1991; Gibbs 1999). Not only in the countries of the former Soviet bloc, but also in Western Europe, accumulating political consensus requires a systematic reference to this 'magic word', whose reverberations inside the semiosphere's ganglia are, then, multiple, meaning that they do not concern only the activity of political hierarchies or bureaucratic apparatus but begin to 'infect' other spheres of the cultural production, starting with those related to the political-bureaucratic systems.

When, ten years after the fall of the Berlin Wall that ensued the Soviet policy of glasnost, Germany decided to adopt a new Reichstag, its architectural manifestation seemed to embody, in the structure of the building that can be visited today in the German capital — and that is one of its greatest attractions — the transition from a fashion trend of opacity to one of transparency, not only in politics, but also in other areas of cultural production, first of all that of the architecture that translates politics into institutional buildings. On the one hand, in 1995, artists Christo and Jeanne-Claude 'packaged' the building as it was usual in their visual poetics, so as to hint at what German politics had been before the fall of the Wall: a place of extreme opacity, impenetrable to any external look, a building that would keep the imposing size and the austere form of a palace of power, but that would offer nothing to satisfy a desire for transparency (Figure 8.1).

Figure 8.1: Christo and Jeanne-Claude: Wrapped Reichstag, Berlin 1971–95. Photo: Wolfgang Volz. ©1995 Christo + Wolfgang Volz (Wikimedia Commons)

On the other hand, when the restructuring work of the Reichstag according to the project of the English archistar Norman Foster was completed in 1999, it seemed inevitable to add a detail that was not actually contained in the initial project but that later became its more recognizable element and ended up incarnating the new political and architectural identity of the Berlin landmark: a splendid, immense glass dome. There where absolute opacity had predominated, denounced by Christo's wrapping, absolute transparency now prevailed, enbodying the total visibility of politics exposed to the gaze and control of citizens (Figure 8.2).

Figure 8.2: Berlin, Reichstag building at night 2013 (Wikimedia Commons).

As it is often the case in semiospheres, the paths of structural polarizations must be translated into material manifestations, in which the choice of the specific matter adopted is not at all indifferent but translates into the cultural configuration of materials. It is evident that a political rhetoric based on transparency must be incarnated through a diaphanous material, such as glass, which embodies this rhetoric and at the same time is impregnated by it (Leone 2018).

The progression of the structural polarizations of the semiosphere, however, is such that they not only tend to saturate a fashion trend before slowly turning back on themselves and give rise to the opposite tendency; they also entail a continuous overflowing that, contagiously, affects discursive domains that are parallel to that in which the initial trend started. After that "glasnost" became a watchword in politics and ended up characterizing even the architectural shape of institutional buildings, transparency slowly began to pervade also discursive domains traditionally devoted to discretion, secrecy or even lies, such as diplomacy, for example. In other cultural eras it would not have been questioned that, so as to defend the national interests of a country, its ambassadors would resort to low profile, understatement, confidentiality, or even spread rumor and lies in order to damage hostile governments. The advent of Wikileaks, which a historical but above all a rhetorical thread links to the Soviet turn of Glasnost, revolutionized the realm of secrecy for the sake of transparency: it ended up seeming normal to everyone, at a certain point, that diplomatic dispatches would be revealed and spread, just as it seemed legitimate, a few years later, that the military secrets of a superpower would be exposed through the work of whistleblowers Snowden and Manning (Lozano 2012).

Interpreting these phenomena as signs of pure progress of humanity in its destiny of continuous approach to an ideal democracy means misrepresenting them or, at least, neglecting their aesthetic component, which is equally fundamental. Transparency takes hold of global discourse, from politics to architecture, from diplomacy to emotions, not because it is right in itself but because it travels and swells on the crest of a wave of fashion. Desecrating as it may sound, not grasping the fashionable aspect of certain phenomena of political aesthetics means deluding oneself with respect to the very probable fact that, for instance, this tendency towards ever greater transparency is not destined to continue indefinitely but to reach an extreme beyond which it is inevitable that it will be configured as a problem rather than as a resource, with the result of giving rise to cultural phenomena of rupture in the sense of opacity rather than transparency.

11.4 The circuit of the veil

These counter-tendencies already manifest themselves in the clothing trends that collect and institutionalize in marketing, advertising, and, of course, design, a global trend that essentially has its roots in the same event that marked, from a political point of view, the break of the tendency towards ever greater geopolitical transparency inaugurated with Glasnost in 1986. This event undoubtedly is September 11, 2001. On the one hand, it marked the end of the post-Soviet American pax: starting from this date, the West would increasingly have to revise its transparency standards in the need to situate and defend itself in relation to new hostile forces, reconsidering, for example, the work of national intelligence services and their stealthy monitoring of the 'lives of others'; on the other hand, this event also marked the arrival, on the scene of global history and culture, of minority or marginal cultural and aesthetic forces, which, taking shape within the western semiospheres, would seek to impose new political as well as aesthetic standards of transparency, thus forcing the semiospheres of the West to an action that is either of contrast or integration.

The Islamic veil, for example, a device of concealment and opacity, first manifested itself as a stumbling block of the Western aesthetics of bodily transparency, which started to come about in the West with the beginning of secularization but which exploded with the rediscovery of the nude in the late 1960s; the Islamic veil in the West represents a problem because it undermines this now established conception of the female body and its erotic-sexual charge. Following the shock, however, as the cultural novelty has become more and more established, it has given rise to semiotic phenomena of accommodation, such as the one manifested in the recent advertising campaign of the Italian fashion brand Versace (Figure 8.3).

Figure 8.3: Shanghai Versace Advertisement of Shanghai, Jinan Temple Area (personal archive).

From central accessory of the Islamic dressing code, impregnated with the value of opacity projected onto the female body, the veil is normalized as a fashion garment, revisited in a vintage key through recovering the western woman's head handkerchief, through hinting at the ways of dressing of men in the Arabian peninsula, and above all, in a central position, valorized precisely as hijab, as garment whose polarity, however, Versace reverses from concealing textile surface to exhibited fabric, thus accentuating the erotic paradox already inherent in the Islamic veil (attracting the gaze to that which is meant to hide).

12 Counterbalancing trends

Cultures evolve through fashions. Micro-interactions among individuals give rise to macro-changes in the semiosphere. Little can be done to govern this chaotic phenomenon. Its causal rationality is mostly apparent, bound to local areas, whereas it becomes random as the scale of observation is enlarged. At the macro-level, causes and effects become a narrative illusion and are replaced by the ontology of long oscillations between polarities, without any progress in one sense or the other. Evil, from this point of view, is the result of the inexplicable injustice of being born at the wrong place, in the wrong time. It would be reasonable, for humans, to avoid pain, yet pain too is the byproduct of the inexorable motility of cultures, of their inability to freeze into peaceful states. On the other hand, though, this motility also works for the opposite trend, the one that frees increasingly larger groups of human beings from suffering.

Promoting an agenda of changing the inner structure of a semiosphere is a legitimate ethical stance, which can give rise to an equally viable political plan. Yet, semiotics must distinguish between how humans would like their world to be and the general tendencies of development of the semiosphere, wherein the origin and development of intentionalities and agencies are less clear. Extremism, for instance, for as much as we might dislike it, grows in culture exactly like a predator species grows in nature: we might be hurt by it, but it is there according to laws that largely exceed causal comprehension and, even more, corrective action. That does not mean that, in a sort of philosophical quietism, humans should accept whatever befall them; that means, on the contrary, that one should be cautious about the power of communication in changing the minds of other human beings or groups. We might want to persuade others to moderation, but we might well also hit the impenetrable wall presented by a fashion trend. Once a wave of extremism has gained momentum in a society, it is exceedingly hard to stop it; unfortunately, such wave will dwindle and even turn into a back-wave of tolerance only after extremism has reached its most intolerable extremes, with unbearable pain being brought about in the process.

Given the impossibility of blocking cultural oscillations among extremes, the only reasonable task intellectuals, and semioticians among them, can carry on is to constantly warn their communities against the lures of fuelling a trend without counter-balancing it by reference to opposite movements. For instance, as far as transparency is concerned, and as much fierce advocates of its sociocultural implementation we might be, it is certainly commendable to require more transparency from political institutions, but not to the extreme that such transparency impairs or even paralyzes negotiations among diverging sociopolitical forces. From this point of view, and following the metaphor of fashion as theoretical framework to explain cultural trends, intellectuals should always be dandy, meaning that they should always expose the possible drift of cultural trends and propose some counter-moves without ever yielding to narcissistic contrarianism.

13 Conclusions

The wind that moves the universe, that moves the subatomic particles, that incites the stars to their revolutions, that baptizes new planets, that wind which blows around a new species of plants, or into the unusual shape of a fish's fin, and that also blows in theaters, those of the Greek invention, or in the new patterns of Byzantine mosaics, or in the emergence of perspective, up to the myriad of interactions that crown a meme as a champion of virality, and below all this, from when life is meaning, it pushes forward language, the systems of signs, and above all the natural language that everyone speaks and that everyone changes, but without realizing the subtle lines that animate the change, well this wind is not a real wind, it does not emanate from a distant engine, from an epochal contrast of cold and heat, and it is only for a moment, through the instant of our lives, when by a mystery the universe pretends to align itself with the will of our thought and action, that we seem to feel it, while then it diverges in a catastrophic and hilarious manner, so much so that if it had a face it would continually laugh at the thought of humans, and their playing with the empty simulacra of causes and effects, for this wind is an apparent wind, it is the wind that comes alive from the same animation of things, which are not moved by the wind but which are wind, we are wind, and you who read these words, and these words themselves, which seem to you to have some sense, but are just a lump of dust of some beauty moving in a senseless swirl, through darkness.

References

Aly, Anne, *et al.* 2016. *Violent Extremism Online: New Perspectives on Terrorism and the Internet*. Abingdon, Oxon; New York, NY: Routledge.
Berger, John M. 2018. *Extremism*. Cambridge, MA: The MIT Press.
Gibbs, Joseph. 1999. *Gorbachev's Glasnost: The Soviet Media in the First Phase of Perestroika*. Texas: Texas A. & M. University Press.
Leone, Massimo. 2018. Symmetries in the semiosphere: a typology. *Theoretical Studies in Literature and Art* 38(1). 168–181.
Lotman, Jurij M. 2009. *Culture and Explosion*, ed. Marina Grishakova, Engl. trans. Wilma Clark. Berlin and Boston: Walter de Gruyter.

Lozano, Jorge (ed.) 2012. *Secreto*. Monographic issue of the *Revista de Occidente*, 374–375. Madrid: Fundación Ortega y Gasset.

Lozano, Jorge (ed.) 2013. *Transparencia*. Monographic issue of the *Revista de Occidente*, 386–387. Madrid: Fundación Ortega y Gasset.

Midlarsky, Manus I. 2011. *Origins of Political Extremism: Mass Violence in the Twentieth Century and Beyond*. Cambridge, UK and New York, NY: Cambridge University Press.

Mulloy, Darren J. 2004. *American Extremism: History, Politics and the Militia Movement*. London and New York: Routledge.

Merkl, Peter H. & Leonard Weinberg (eds.). 2003. *Right-Wing Extremism in the Twenty-First Century*. London and Portland, Or.: Frank Cass.

Nove, Alec. 1991. *Glasnost in Action: Cultural Renaissance in Russia*. London: Routledge.

Pratt, Douglas. 2018. *Religion and Extremism: Rejecting Diversity*. London, UK and New York, NY: Bloomsbury Academic.

Sunstein, Cass R. 2009. *Going to Extremes: How Like Minds Unite and Divide*. Oxford, UK and New York, NY: Oxford University Press.

Wintrobe, Ronald. 2006. *Rational Extremism: The Political Economy of Radicalism*. Cambridge, UK and New York, NY: Cambridge University Press.

Nicolae-Sorin Drăgan
The dynamic aspect of the semiotic behavior of political actors in TV debates

Abstract: In this paper we propose a formalized mathematical model to analyze the dynamic aspect of the semiotic behavior of political actors in dialogic forms of interaction, such are presidential TV debates, from an interdisciplinary perspective that brings together: *positioning theory* (Davies and Harré, 1990; Harré and Moghaddam 2016), *multimodality* (Maricchiolo et al. 2013; Gnisci et al. 2013; Navarretta and Paggio 2013) and *functional theory* of political campaign discourse (Benoit and Wells 1996; Benoit 2014, 2016). We are interested in how the political actors involved in such forms of communication manage the *persuasive potential* (Benoit 2014) of discourse to position themselves in a credible way, as a *political persona* in front of the public, through various strategies of complementarity of the semiotic resources.

Keywords: presidential debate, multimodal analysis, political semiotics

> A prince, therefore, need not actually have all the qualities [...], but it is absolutely necessary that he seem to have them [...]. Generally men judge more with their eyes than with their hands because everybody can see, but few can feel. Everybody sees what you seem to be, few can touch what you are. (Machiavelli, *The Prince* [1512] 2008: 283–285)

1 Introduction

The ancient Greeks seem to have realised one crucial aspect of human nature: people place more value on *appearances* and *reputation* than reality. Sophists were the first "merchants of words" who understood and applied this principle in a pragmatic manner. Politicians also understood that in order to convince citizens of their ideas and political vision, they needed to adopt the practice of oratory – *ars rhetorica* –, learn to manage a public debate skilfully and try to shape the *perception of reality*. Later on, Machiavelli also lucidly noted that "men judge generally more by the eye than by the hand" (187). *Appearances* prevail over *facts*, and *faces* (reality) become *masks* (images). Starting with the

---

Nicolae-Sorin Drăgan, National University of Political Studies and Public Administration (SNSPA), Bucharest, Romania

Florentine secretary, politics began to be considered a show, and political actors had to give the impression of truly performing before the public – a phenomenon that later become known and acknowledged as the *theatralisation* of political life. In the subtle play between appearance and reality in today's politics, the focus shifts to the candidate's individual characteristics, to their image (the phenomenon of *personalisation* of politics). The image and reputation of a politician build upon and rely on a genuine system of illusions, an entire constellation of signs that make up the *political character*. According to Paul Lichterman and Daniel Cefaï (2006: 403), this is a political culture of performance type, of dramatic performance, in which "dramatic conventions shape political communication". Taking advantage of the *effet de réel* (reality effect) of television, described by Pierre Bourdieu (2007: 29–30) or the "television realism", in terms of John Fiske and John Hartley (2002: 258–264), politicians have now become political actors.

Basically, they propose to the public certain representations of reality, a certain "moral order, with offenders, victims, heroes, witnesses, and experts" (Lichterman and Cefaï 2006: 403). In this way, "the social dramas enacted by institutional actors shape a public's perception of social problems" (Lichterman and Cefaï 2006: 403). On other occasions we have shown that a certain form of discursive interaction, such are final TV debates for presidential elections, can be understood as "an ideal opportunity for candidates to establish an interpretation of reality, to impose control and symbolic representations of the situation in the political field" (Drăgan 2019a: 196). Through this real exercise of image and communication with the public, political actors have the opportunity to perform in front of the public in the role of president. Which of the candidates interprets this role more consistently and creates an impression of stronger preferability in the president posture?

In this article we analyze the *dynamics of the positioning acts* of political actors in such dialogic forms of interaction from the perspective of *positioning theory* (Davies and Harré 1990; Harré and van Langenhove 1998; Harré and Moghaddam 2010, 2016), *multimodality* (Allwood et al. 2007; Colletta et al. 2009; Maricchiolo et al. 2012; 2013; Gnisci et al. 2013; Navarretta and Paggio 2013) and *functional theory of political campaign discourse* (Benoit and Wells 1996; Benoit, McKinney, and Holbert 2001; Benoit, Hansen, and Verser 2003; Benoit 2014, 2016).

Thus, at the level of the verbal discourse, the strategic action of the candidates was captured using the functional theory of political campaign discourse (FT). However, as any discourse including political discourse, is "inherently multimodal, not monomodal" (LeVine and Scollon 2004: 3) political actors employ various methods of modes of sign production as a way of constructing meaning (Jewitt 2009). According to Michael Lempert și Michael Silverstein (2012) the political message may be assessed as a "multimodal discursive interaction" (27). Based on the opinion of the two authors, in TV debates the messages are intentionally theatrical, in the sense of constructing a certain moral profile of the candidate. Therefore, the political message becomes a *personal brand*, granting political actors a certain "biographical aura" (Lempert and Silverstein 2012: 100). The effort of creating a *political persona* implies the mobilization of an entire semiotic system. This is the reason why we will use an interdisciplinary approach – analysing the positioning complementarily with a professional framework for multimodality research, ELAN (Eudico Linguistic Annotator) – to capture the dynamics of the discursive exchange in the episode we considered. We have presented, in short, the research design in the third section, research methodology.

We will focus especially on presenting a formalized mathematical model to analyze the dynamics of the political actors' positioning game in a specific communication situation, model pre-tested in a previous study (Drăgan 2019a). We are interested in how the political actors involved in the debates manage the "persuasive potential" (Benoit 2014) of discursive functions to position itself through various strategies of complementarity of the semiotic resources.

2 The theoretical and conceptual framework of the research

2.1 Elements of functional theory of political campaign discourse

As we mentioned earlier, a functional approach to political campaigns discourse allows recovery the *strategic dimension* of TV debates (Beciu 2015: 260). The FT is based on five axioms presented in detail with other occasions (Drăgan 2016, 2017a, 2017b). We only mention here that according to the FT candidates are positioned on the preference scale (of the public) by three discursive functions: *acclamation*, *attacks* and *defenses*. Acclamations are positive statements aiming to promote self-image, and to increase the social desirability of the candidate. Attacks are discursive interventions targeting weaknesses and limitations of the opponent, used to reduce the candidate's social desirability. Defenses are statements which reject the opponent's attacks and which could influence the candidate's level of preference (Benoit 2014). The three discursive functions are mutually stimulated and conditioned (Benoit and Wells 1996: 112).

Benoit (2014: 14) argues that the three discursive functions generate a certain *persuasion potential* in favor of one or other of the candidates. Basically, voters may exercise their (apparent) *preferability* or *impression of preferability* (Benoit et al. 2003: 19; Benoit 2014: 14; Benoit and Benoit-Bryan 2014; Benoit 2016, 2017) towards one of the speakers based on a comparative act (Drăgan 2017a: 115). The manner in which political actors use discursive functions strategically during discursive interaction may stimulate the impression of preferability in favor of one candidate or the other (Drăgan 2017a: 115–116). According to Manuel Castells (2015), during electoral debates, "the audience have a tendency to appoint their preferred candidate as the winner, and not to vote for the more convincing debater" (238). Therefore, the (apparent) preferability or impression of preferability is built based on a comparative assessment of the candidates, substantiated both by their actions (policies), as well as by their personality traits (character or image) (Benoit 2014; Benoit and Glantz 2017).

2.2 Elements of positioning theory

2.2.1 The roots of the concept of positioning in marketing, brand strategies and advertising

During the 70's, Al Ries and Jack Trout (1972) announced the very beginning of a "Positioning Era". According to Al Ries and Jack Trout (1981) "The basic approach of positioning is not to create something new and different but to manipulate what's already up there in the mind, to retie the connections that already exist" (5). In semiotic terms, positioning actions activate a certain system of interpretants (interpretation scheme) "already existing" in the mind of the consumer: "position is something that exists in the mind" (Ries 2017). Thus, positioning becomes the semiotic practice that triggers certain values in the mind of the consumer, certain consumer predispositions. The effects of such a significant practice (positioning) are observed in consumption practices and determine a *business logic*. Moreover, Jack Trouth and Steve Rivkin (2009) capture the *cognitive aspects* of positioning actions and define positioning as the "body of work on how the mind works in the process of communication" (10). According to Al Ries (2017), in marketing logic, positioning is the cognitive mechanism that determines correlations between certain cognitive categories (values) and brand: "Every category in the mind is either filled with a brand name, or it's not. If the category is not filled, then it's an open hole or position which your brand can easily occupy." (Ries 2017). From the perspective of this article we note that positioning is a type of semiotic practice based at all times on communication strategies.

2.2.2 The discursive approach to positioning

From a discursive psychological approach to positioning, positioning acts appear as discursive practice of social actors in the effort to build identity. We return to a few aspects of the positioning concept and of the positioning theory also discussed on other occasions (Drăgan 2019a, 2019b). In communication acts, social actors communicate not just content, but also the way they relate to that content (positioning). According to positioning theory (Davies and Harré 1990; Harré and Gillet 1994; Harré and van Langenhove 1998; Moghaddam et al. 2008; Moghaddam and Harré 2010; Harré and Moghaddam 2016), identity and self are discursive constructions. Positioning can be reproduced in social practices, bargained in various moments, discursive or ideological rules can be attached thereto, which make them socially recognisable.

Although, at first, the concept of positioning has been introduced to replace the concept of *role* (from dramaturgical perspective of communication), considered static and less flexible in understanding the discursive behaviour of social actors (Harré and Gillet 1994: 33–36; Harré and van Langenhove 1998: 14–17), later, the authors of the positioning theory argued that positioning theory must be understood in complementarity with the theory of role. Additionally, in certain situations, positioning acts can be "the birthplace of roles" (Harré et al. 2008: 9). We suggested there that, at symbolic level, the role can be regarded as the *continuous* aspect of the discursive performance of a social actor, and positioning is the *discreet* aspect of the discursive exchange (Drăgan 2019a: 197). We mention the fragile balance between the continuous and the discreet aspect of communication is maintained by the very conflictual relationship between "seeing" and "understanding" (Marcus 2011: 183). We see the role more easily, we perceive its moments of discontinuity, of inadequacy, but we understand better the conjunctural discursive position.

In addition, for a better understanding of the concept of discursive positioning, we have analyzed two other concepts directly related to it. The first concept is that of *footing* and *footing acts*, or interlocutors' roles which are just one aspect of positioning. The second concept is the concept of *places report* or *system of places* (Flahault 1978).

Below we will present the three-dimensional conceptual model on which the positioning theory is built (see Figure 9.1). Such a pattern of positioning captures the relational aspects and the significance conditions of any dialogue interaction. The three (de)construction conditions of the significance of the discursive exchange determine each other. There is a dynamic relationship between them.

The first dimension of the model refers to positions and positioning. The way that various categories of people engage in the discursive exchanges (*positioning*

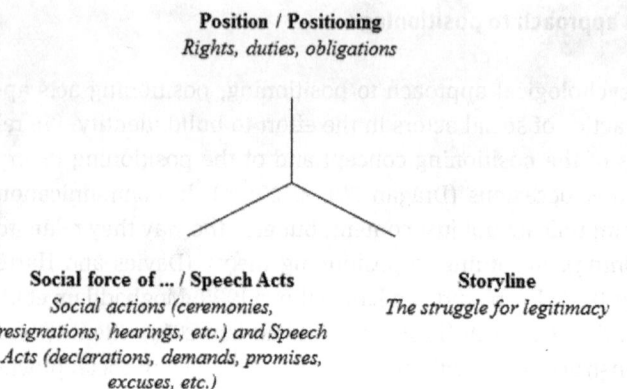

Figure 9.1: The three-dimensional positioning model (Drăgan 2019a:198; adapted after Harré and Langenhove 1999: 18; Harré et. Al. 2008: 12; Moghaddam and Harré 2010: 50).

acts), the rights and obligations they assume in their communication acts are "positions" (Harré and Gillet 1994: 34). They are predefined, by default. The rights, responsibilities and obligations associated to positions are set in the sociodiscursive imaginary of the community, and they follow a certain moral order established within the society at a certain point: "Positioning Theory is the tool for exploring the relation between what is possible and what is permitted" (Harré et al. 2008: 13). Moreover, positioning analysis allows us to capture how "rights and duties are distributed among the actors in the course of complex discursive interactions" (Moghaddam and Harré 2010: 6). I highlighted this with an example of the latest French presidential election debates, on May 3, 2017 (Drăgan 2019a: 199). When French presidential candidate Marine Le Pen portrayed her opponent as "out of touch and elitist" and "Mr. Macron is the candidate of savage globalization" (BBC May 4, 2017) has not just made an offensive statement (attack) against the candidate Emmanuel Macron. She has put into practice a certain positioning strategy. Practically, she indexed a certain type of social relationships built on the semiotic model of binary oppositions: elite / ordinary people, republican citizens / followers of globalization, etc.

We mentioned that "positions" are not socially regulated, they are not institutionalised. They are merely socially recognisable. Political actors have "the practical sense" (Bourdieu 2012: 191) of political game that allows them to recognize certain positions in the *experiential matrix of positioning*, or in the "semiotic web" of other discourses. This enables them to build positioning strategies in discursive interactions. This is the reason why positioning can be regarded as a *semiotic configuration* with a certain stability in the communicational practices of the society (Drăgan 2019a: 199).

The following dimension refers to the discursive ways in which certain positions are activated during communication. We introduce ourselves to the others, we build our discursive Ego with individual narratives (*storylines*) that we employ in verbal interactions: "Positioning can be understood as the discursive construction of personal stories that make a person's actions intelligible and relatively determinate as social acts and within which the members of the conversation have specific locations" (Harré and Langenhove 1991: 395). It is with these stories that we make our entrance in the context, we present ourselves to the others.

The last aspect of a positioning act activates the concepts of "illocutionary force" and "speech acts". As expected, these concepts were taken from *the speech act theory*, in order to describe the social significance of a discourse, gesture or social action (Harré et al. 2008: 10). Any conversational exchange has an illocutionary value by default, a certain social strength. According to Rom Harré and Grant Gillett, "Each speech-act has an illocutionary force, its social power as uttered in a certain context." (1994: 32). Therefore, positioning acts have the capacity to create social significance (Moghaddam and Harré 2010: 71).

3 Research methodology

3.1 The multimodal analysis model of positioning acts in discursive interactions

The multimodal analysis of the positioning acts of political actors in the TV presidential debates in Romania, on December 3, 2009, is the starting point for this article. In the first part of the research design we presented the *modeling of positioning analysis* according to the positioning theory authors. In short, a positioning act entails a two-step procedure. The first is *prepositioning*, an act that occurs by default when the interlocutors' attributes are known (or presumed). It is the moment when you guess "who you are talking to". We can speak of reflexive prepositioning (biographic elements, personal narratives, etc.) and prepositioning actions of the other (statements about the interlocutor's character, etc.). The next step is *positioning itself* in the dynamics of discursive exchange, and performing the assumed role. When analysing the positioning, we are looking for answers to some research questions (Harré and Rossetti 2010: 114). Which are the narratives that each social actor brings into the context to legitimise himself? Which is the social significance of speech acts (illocutionary force), following the performance of individual narratives? Which rights and obligations have appeared in the discourse and how were they assigned, refused, accepted or contested by the speakers?

Therefore, the positioning analysis entails a few steps (Harré and Rossetti 2010: 115–122):
1) Identifying prepositioning moves (reflexive prepositioning and the prepositioning of the Other)
2) Identifying and analysing individual narratives
3) Confronting positional moves with the ideological profile of the social actors.

Practically, the first step of the positioning analysis is to evaluate the communication situation, how "the target person has, or lacks, a desirable attribute for the ascription of a right or a duty relevant to the situation in hand" (Harré and Rossetti 2010: 115). The second step of the analysis helps us to identify and analyze the narratives through which political actors build the presidential character. The last step, which aims at confronting the positioning movements with the candidate's ideological profile, allows the leader's autonomy to be observed in relation to the political family that supports him, the degree of freedom to the reference. Summarizing, by analysing the positioning, we can capture the dynamic relationship between the three dimensions of the positioning of social actors during a discursive exchange (individual narratives, social significance and position).

In the second part of the research design we presented a *multimodal analysis model* that allowed us to capture on the one hand, the dynamics of the discursive exchange in the episode we considered, and on the other hand to disambiguate the positioning (and the interpretation thereof) of the political actors involved in the debate. We suggested there that this model is basically the foundation on which we apply the positioning analysis (Drăgan 2019a: 202). We have built a model of multimodal analysis that capitalizes on previous experiences of multimodal research of discursive interactions specific to political discourse. Basically, we were interested in analyzing and understanding how inter-semiotic relationships (or inter-modal) are established between different *semiotic modes*, or different semiotic modes work together for the construction of meaning (O'Halloran 2011: 121; Siefkes 2015: 114). We recall the steps of this multimodal model of analysis:
1) The annotation of semiotic resources (speech and gestures), using ELAN (EUDICO Linguistic Annotator).
2) Identifying semiotic types of gestures.
3) Attributing function to gesture (Colletta et al. 2009: 61–62).
4) Identifying the relationship of gestures with corresponding verbal discourse (semantic relations).

5) Identifying the semiotic resources that political actors use to build their individual narratives and discursive position.
6) Comparative analysis of movements and positioning strategies.

The data were annotated and analyzed using ELAN (EUDICO Linguistic Annotator), a multimedia annotation tool developed at the Max Planck Institute for Psycholinguistics in Nijmegen, The Netherlands. To study the gestures, we used The MUMIN coding scheme as the starting point, developed in the Nordic Network on Multimodal Interfaces (Allwood et al. 2007). In addition to the MUMIN coding scheme, for the annotation of hand gestures, we used a coding system used in the analysis of the various forms of political communication (Maricchiolo et al. 2012: 408; 2013:117; Gnisci et al. 2013: 881). In order to identify the "relational meanings" (Rovența-Frumușani 1999: 193) that appear in the interactions between the semiotic modes – especially the correlations between gestures and corresponding verbal discourse – we used the taxonomy of semantic relations proposed by Colletta et al. (2009: 62–63). It is noteworthy that for every communication situation that has been the subject of our analysis, the gesture-word correlation has been disambiguated in the context by comparing the informative content of the verbal discourse to that conveyed by the gestures performed by that political actor. We have detailed this model of multimodal analysis on several occasions (Drăgan 2018, 2019a, 2019b).

Therefore, an interdisciplinary approach observing the complementarity of the functional theory of political campaign discourses, positioning theory and multimodality, allows the recovery of the strategy dimension of the positioning concept.

To illustrate the dynamics of positioning acts of political actors and how they structure their discursive functions from the perspective of semiotic resources, we analyze a single communicational sequence with a duration of approximately 5 minutes and 40 seconds, during the final presidential debate on December 3, 2009. This debate was organized by the Institute for Public Policies and moderated by journalist Robert Turcescu. The protagonists of this TV debate were Mircea Geoană, the political left-wing representative (PSD+PC) and Traian Băsescu, the political right-wing representative (PDL), the current President of Romania at that time.

The communication sequence analyzed capture an exchange of replies, from the middle of the debate, which capture the correlative positioning of the two candidates towards on two of the most important subjects of the political-media agenda. This is the theme of "Justice" and the theme of "fight against corruption". This sequence was chosen based on criteria that take into account: the homogeneity criterion of selection (important moments in the debates); the common thematic criterion (the sequences considered capture the position of both political actors involved in the debate on a common theme) and criteria for selecting strategic messages.

Figure 9.2 shows a screenshot illustrating the distribution of semiotic resources for each discursive function, for the first statement made by incumbent Traian Băsescu in the discursive sequence analyzed. Within each type of enunciation (discursive function) we have marked (in an oval framing) the moments when the political actor performs gestures that can influence the "persuasive potential" of the utterance. The other discursive sequences have been assessed in a similar manner.

For example, we identified nine such moments during the statements made by the candidate Traian Băsescu in the analyzed sequence. The first four for a positive type statement – acclaim (A1) –, marked in the figure with numbers from 1 to 4 (see Figure 9.2), the next five for a positive statement as well (A1), also marked with numbers from 1 to 5. Regarding the statements made by Mircea Geoană, we have identified ten such moments in the analyzed sequence. The first five for an offensive type statement (attack, A2), the next two for a positive statement (acclamation, A1), and the last three for an offensive statement (attack, A2). We will not detail here the multimodal evaluation of the positioning actions of the political actors during the selected communication sequence. We presented such an analysis in a recent study (Drăgan 2019a).

3.2 The dynamics of positioning acts. Formalized mathematical model

As we mentioned earlier, in this study we are interested in how the political actors involved in the debates manage the "persuasive potential" (Benoit 2014) of discursive functions to position itself through various strategies of complementarity of the semiotic resources. In practice, we analyze how semiotic resources used by political actors at different times within selected sequences contribute to maximizing or minimizing the "persuasive potential" of the statements (discursive functions) in the play of discursive positions.

In order to illustrate the dynamics of this mechanism, we present things in a mathematically formalized manner Figure 9.3 shows the result of applying this model for the first of the two statements made by the incumbent Traian Băsescu in the analyzed sequence. Figure 9.4 shows the evaluation result for all five statements made by both candidates, in a comparative way. Both figures are structured according to a matrix logic: the types of statements (discursive functions) are marked on the columns – *acclaims* (A1), *attacks* (A2) or *defenses* (A3) –, while the lines distribute the occurrence of *moments* when relevant

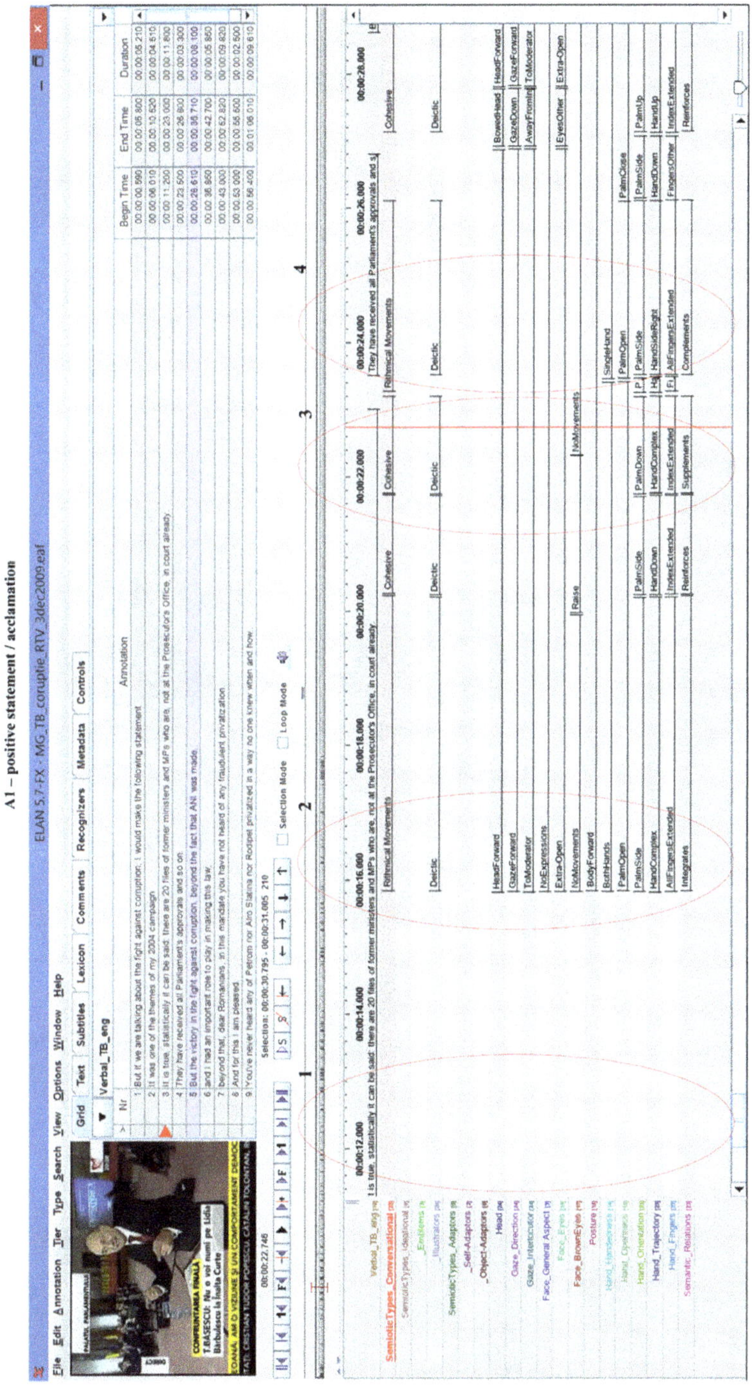

Figure 9.2: Screenshot of the ELAN user interface – First sequence, final statements, RealitateaTV station, December 3, 2009.

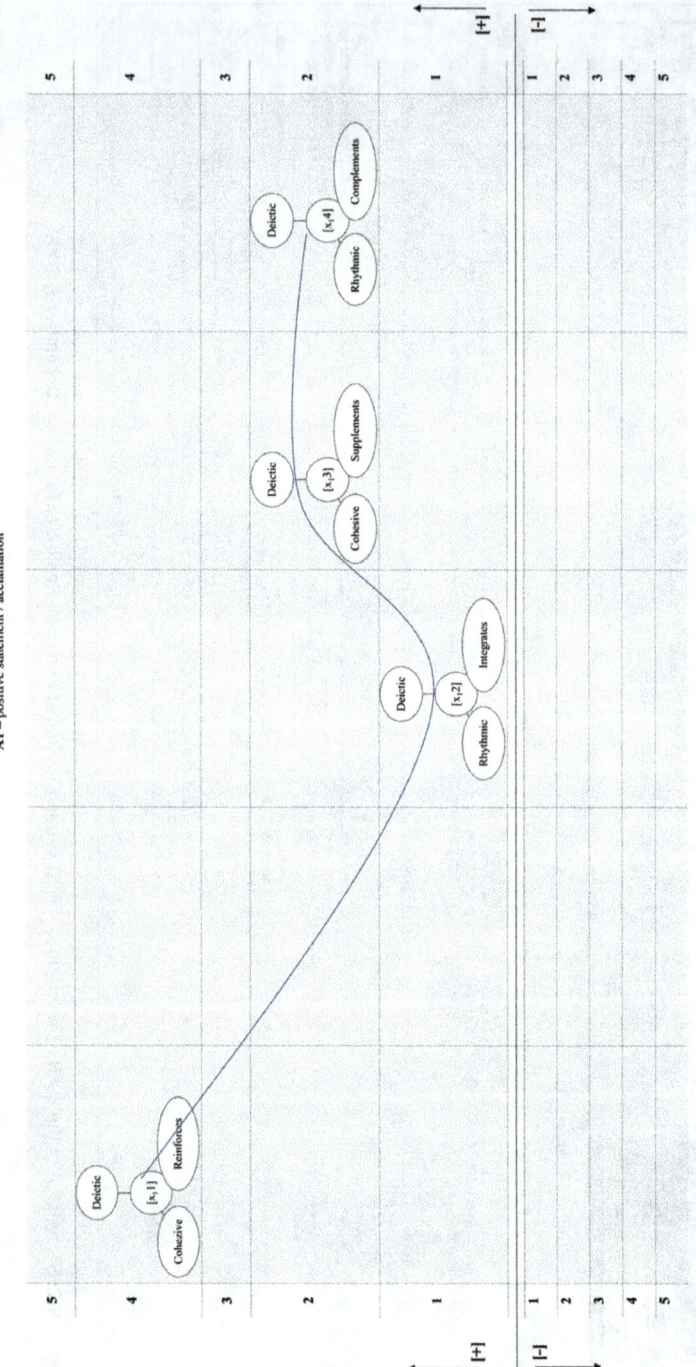

Figure 9.3: The dynamics of discursive functions-gesture correlations in the case of Traian Băsescu (incumbent) – the first statement in the analyzed sequence, positive statement (acclaim, A1).

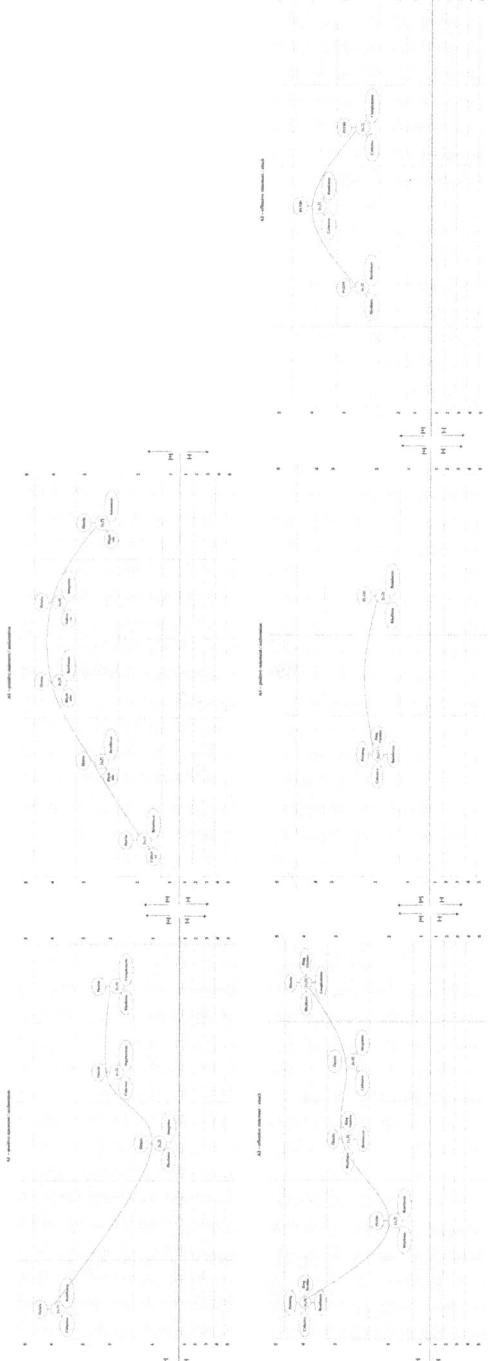

Figure 9.4: The dynamics of discursive functions-gesture correlations in the case of both candidates: on the left side for the incumbent Traian Băsescu, respectively on the right side for the challenger Mircea Geoană.

gestures that can positively [+], or negatively [-] influence the dynamics of the "persuasive potential" of statements are performed. Each such moment incorporates a certain meaning potential.

In mathematical terms, the meaning potential of gestures acts as a *multi-variable function* (type of gesture, semiotic value of the gesture, relational significance, type of interaction, etc.), which we will note with the letter F. In the case of the incumbent Traian Băsescu it will be marked with $F(x_1)$, respectively $F(y_1)$ – due to the fact that we have identified two types of statements, both of positive type (acclaims, A1). Similarly, in the case of challenger Mircea Geoană the meaning potential of gestures will be marked with $F(x_2)$, $F(y_2)$, respectively $F(z_2)$ – due to the fact that these are three types of statements (following the sequence A2-A1-A2).

Conventionally, the values that these functions can take vary on a scale from -5 ÷ +5 (from minus 5 to plus 5), values that can be appreciated for example using a Likert scale. The value of function F is correlated (corresponds) with the level of persuasive potential of the moments when the political actors perform a relevant gesture in the economy of debate. Basically, the x_ij, y_ij, z_ij, i=1 ÷ n, j=1 ÷ m, variables express the relevant gesture attributes along with the meanings attached (where "n" represents the number of candidates involved in the discursive interaction and "m" the number of relevant gestures performed by each candidate). The value of the respective function is graphically marked, together with the variables that determine it. For example, the value of the function $F(x_11)$ – which represents the level of persuasive potential corresponding to the first relevant gesture performed by the incumbent Traian Băsescu during his first statement (A1) –, is given by the multi-variable, or vector (x_11). As we suggested earlier, the multi-variable or vector x_11 has the components: u – type of gesture (*deictic*); v – semiotic value of the gesture (*cohesive*), t – relational significance (*reinforce*). The way in which the incumbent Traian Băsescu integrates these components during the performance of this gesture was appreciated in the positive area of the graph, with approximation somewhere in the area of values 3 and 4, as can be seen in Figure 9.3. Due to the fact that we follow the dynamics of the discursive exchange of the political actors in the selected communicative sequence, we are interested in appreciating approximately the value of the function $F(x_ij)$ and not in determining its exact values.

The persuasive potential of the text performed by any of the candidates is the result of the interaction between the persuasive potential of the verbal utterance – P(Ai, i=1÷3) – and the significant potentials of (relevant) gestures performed throughout the utterance – $F(x_ij)$, $F(y_ij)$, $F(z_ij)$, etc. Basically, it is a complex function thereof:

$$P_{TB\_s1}(A1) = P(A1) \circ F(x_11) \circ F(x_12) \circ F(x_13) \circ F(x_14) \qquad (1)$$

$$P_{TB\_s2}(A1) = P(A1) \circ F(y_11) \circ F(y_12) \circ F(y_13) \circ F(y_14) \circ F(y_15) \qquad (2)$$

$$P_{MG\_s1}(A2) = P(A2) \circ F(x_21) \circ F(x_22) \circ F(x_23) \circ F(x_24) \circ F(x_25) \qquad (3)$$

$$P_{MG\_s2}(A1) = P(A1) \circ F(y_21) \circ F(y_22) \qquad (4)$$

$$P_{MG\_s3}(A2) = P(A2) \circ F(z_21) \circ F(z_22) \circ F(z_23) \qquad (5)$$

The first two formulas, (1) and (2) point to the persuasive potential of Traian Băsescu's statements, the following three point to the persuasive potential of Mircea Geoană' statements from the communicative sequence analyzed (RealitateaTV, December 3, 2009).

For example, P_{TB_s1} (A1) represents the persuasive potential of the first statement with a positive value, P(A1), which integrates the persuasive potential of each of the four moments when the incumbent Traian Băsescu performs certain relevant gestures: $F(x_11)$, $F(x_12)$, $F(x_13)$ și $F(x_14)$. Similarly, P_{TB_s2} (A1) represents the persuasive potential of the second statement with positive value, P(A1), which integrates the persuasive potential of the five moments when Traian Băsescu performs relevant gestures: $F(y_11)$, $F(y_12)$, $F(y_13)$, $F(y_14)$, and $F(y_15)$.

P_{MG_s1} (A2) represents the persuasive potential of the first statement with an offensive function, P(A2), performed by the challenger Mircea Geoană, which integrates the persuasive potential of the five moments when the challenger performs certain relevant gestures: $F(x_21)$, $F(x_22)$, $F(x_23)$, $F(x_24)$, and $F(x_25)$. P_{MG_s2} (A1) represents the persuasive potential of the statement with a positive value, P(A1), which integrates the persuasive potential of each of the two moments when the challenger performs relevant gestures: $F(y_21)$, $F(y_22)$. Finally, P_{MG_s3} (A2) represents the persuasive potential of the statement with an offensive function, P(A2), which integrates the persuasive potential of the three moments in which the challenger Mircea Geoană performs relevant gestures: $F(z_21)$, $F(z_22)$, and $F(z_23)$.

Let us consider, for instance, the first moment when Traian Băsescu performs a significant gesture (x_11) within the first statement (TB_s1). During this sequence, the incumbent party's candidate builds an positive type utterance (acclamation, A1) which, at the level of the verbal utterance, implies a certain persuasive potential, P(A1). The (dominant) variables that contribute to the outlining of the semiotic potential of the gesture performed, $F(x_11)$, are: the type of gesture (the *pointing gesture* family), the semiotic value of the gesture (deictic), the type/aspect of the interaction (cohesive), respectively the relational meaning (the gesture reinforces the informational content of the verbal utterance). The manner in which the political actor integrates these variables during the performance of the gesture determines the semiotic value or the semiotic potential of the gesture. Such a performance may positively or negatively influence the construction of the persuasive potential of the statement as a whole. In our case, it provides a positive contribution due to the semiotic properties of the gesture and also to the performance of the political actor. The incumbent Traian Băsescu, when talking about the fight against corruption, looks at the moderator, the mediating court that represents the public (it is practically addressed directly to the viewers), seeming involved in the interaction, yet the gesture seems to be performed authentically. The statement accompanying this gesture expresses ideas with positive valence, which refers to the achievements of the incumbent: "But if we are talking about the fight against corruption, I would make the following statement. It was one of the themes of my 2004 campaign". As a result, the moment is market in Figure 9.3, in the positive influence zone of the discursive function – gesture relation.

Similarly, the semiotic potential for each (relevant) moment of the communicational sequences has been established for each individual type of utterance of each political actor.

This process was followed by the marking of each such moment: in Figure 9.3, the moments corresponding to Traian Băsescu's statements, while Figure 9.4 shows the dynamics of discursive functions-gesture correlations in the case of both candidates in a comparative manner. In both Figures the variations (the dynamic) of the semiotic behavior of the candidates are captured, by drawing the approximation curves of the relations (correlations) established between the discursive functions and the gestures.

Unlike the model pre-tested in a previous study (Drăgan 2019b), the dynamics of these relationships (variation of the approximation curve) in the situation of both candidates is similar. There are no significant differences regarding the semiotic behavior of the two competitors. It seems that the two manage fairly balanced relations between the discursive functions and the gestures. Indeed, the entire debate was considered by journalists quite balanced, with the exception of a single discursive sequence. The respective sequence was managed with skill by the incumbent Traian Băsescu who used a series of offensive-type statements against the challenger Mircea Geoană on the theme of corruption, focusing in particular on the *episode of nocturnal visit* of the challenger to Sorin Ovidiu Vântu, the main shareholder at the time of the RealitateaTV station. During the final televised debates for presidential elections in Romania in recent years, the way in which political actors manage such "key moments" can decide the winner of the debate (Corbu and Boțan 2011: 101).

The audience has the ability to observe imbalances occurring in the nonverbal behavior of political actors, moments of inadequacy occurring between the verbal discourse and the performance of a gesture. In addition, as mentioned earlier, voters build an impression on preferentiality towards a candidate in the electoral race via a comparative action. Furthermore, certain types of gestures – for instance, the "precision gesture" family (Calbris 2003; Lempert 2011: 243) – can contribute to the creation of favorable impressions in terms of a political actor's performance. For example, challenger Mircea Geoană performed such a gesture even at the beginning of his first offensive type statement – see Figure 9.4, left, the first performance gesture, $x_2 1$.

4 Conclusions

The general purpose of this study was to assess the manner in which the political actors involved in debates use various semiotic resources in order to construct their communicational strategies. We are interested in how the political actors involved in the debates manage the "persuasive potential" (Benoit 2014) of discursive functions to position itself through various strategies of complementarity of the semiotic resources. We have analyzed the way in which discursive functions are structured from the perspective of semiotic resources, the correlations occurring between the types of statements (discursive functions) and semiotic resources. In order to observe the dynamic of these relations, we have provided an example based on a mathematically formalized manner, applied to the communicational sequences analysed. Such formalization allows a better understanding of how semiotic resources used by political actors at different times within selected sequences contribute to maximizing or minimizing the "persuasive potential" of the statements (discursive functions) in the play of discursive positions. Intuitively, this model can be understood as a "diagnosis" of the discursive behavior of political actors during a discursive interaction, similar to an electrocardiogram (EKG or ECG) that measures the electrical activity of the heartbeat and allows the doctor to identify certain problems.

Beforehand, in order to determine and analyzing the correlations established between the types of statements and semiotic resources, we have identified the semiotic value and the functions of relevant gestures performed by the political actors, the relational meanings occurring within the gesture – verbal discourse interaction.

A first conclusion relates to the interdisciplinary approach of this study, as well as the use of a professional framework for multimodality research, ELAN (Eudico Linguistic Annotator). Such methodological framework allowed us to capture on the one hand, the dynamics of the discursive exchange in the episode we considered, and on the other hand to disambiguate the positioning (and the interpretation thereof) of the political actors involved in the debate. We have discussed this aspect in detail during a recent study (Drăgan 2019a).

Another conclusion refers to the dynamic of the non-verbal behavior of political actors. The formalized mathematical manner in which we presented the analysis allowed us to easily observe the variations. As mentioned earlier, compared to the communication situation analyzed in a previous study (Drăgan 2019b), there are no significant differences regarding the semiotic behavior of the two competitors. Such a result suggests a certain balance in the strategic construction of the messages, as well as in the discursive positioning actions in the situation of both candidates. Of course, the analysis of a single communication sequence – with a duration of approximately 5 minutes and 40 seconds – is not relevant for the analysis of the semiotic behavior of the political actors throughout the whole TV debate, which lasted about an hour and a half (December 3, 2009). Our model tested here can be extended to the whole debate to capture all the "key moments" that we mentioned earlier.

In our opinion, the political actor whose non-verbal behavior seems more predictable, whose meanings are congruent with the meanings of the verbal discourse, and who will exhibit a dynamic of the relation between the discursive function and semiotic resources with as few peaks and lows as possible during important moments, will create a more significant *impression of preferentiality* or *apparent preferentiality*. Basically, he will seem more convincing to the audience.

References

Allwood, Jens, Cerrato, Loredana, Jokinen, Kristiina, Navarretta, Costanza & Patrizia Paggio. 2007. The MUMIN Coding Scheme for the Annotation of Feedback, Turn Management and Sequencing Phenomena. *Language Re-sources and Evaluation* 41 (3/4). 273–287.
BBC News. 2017. French election gets dirty: Insults that marked a fierce de-bate. BBC News, 4 May 2017. http://www.bbc.com/news/world-europe-39803333. (accesed 7 May 2017).
Beciu, Camelia. 2015. Dezbaterile Electorale și Rolul Mediei în Campania Prezidențială 2014 din România [The Electoral Debates and the Role of the Media in the 2014 Presidential Campaign in Romania]. *Revista Română De Sociologie* 26(3). 253–278.
Benoit, L. William. 2014. *Political Election Debates: Informing Voters about Policy and Character*. Lanham U.K.: Lexington Books.
Benoit, L. William. 2016. A Functional Analysis of the 2012 London Mayor Debate. *J Mass Communicat Journalism* 6(2). 1–6.
Benoit, L. William. 2017. A Functional Analysis of 2016 Direct Mail Advertising in Ohio. *American Behavioral Scientist*. 1–12.

Benoit, L. William & William T. Wells. 1996. *Candidates in conflict: Persuasive attack and defense in the 1992 presidential debates*. Tuscaloosa: University of Alabama Press.

Benoit, L. William, Mitchell S. McKinney & Lance R. Holbert. 2001. Beyond learning and persona: Extending the scope of presidential debate effects. *Communication Monographs* 68. 259–273.

Benoit, L. William, Glenn J. Hansen & Rebecca M. Verser. 2003. A meta-analysis of the effects of viewing U.S. presidential debates. *Communication Monographs* 70(4). 335–350.

Benoit, L. William & Jennifer M. Benoit-Bryan. 2014. A functional analysis of the 2010 Australian Prime Minister debate. *Journal of Argumentation in Context* 3(2). 153–168.

Benoit, L. William & Mark J. Glantz. 2017. *Persuasive Attacks on Donald Trump in the 2016 Presidential Primary*. Lanham UK: Lexington Books.

Bourdieu, Pierre. 2012 [1996]. *Limbaj și Putere Simbolică [Language and Symbolic Power]*. București: ART.

Bourdieu, Pierre. 2007 [1996]. *Despre televiziune [Sur la télévision*, in French; *On Television*, in English]. București: Editura ART.

Castells, Manuel. 2015. *Comunicare și Putere [Communication and Power]*. București: Comunicare.ro.

Calbris, Geneviève. 2003. *L'expression gestuelle de la pensée d'un homme politique [The gestural expression of the thought of a politician]*. Paris: CNRS Editions.

Colletta, Jean-Marc, Kunene, N. Ramona, Venouil, Aurélie, Kaufmann, Virginie & Jean-Pascal Simon. 2009. Multi-track Annotation of Child Language and Gestures. In Michael Kipp, Jean-Claude Martin, Patrizia Paggio & Dirk Heylen (eds.), *Multimodal Corpora. From Models of Natural Interaction to Systems and Applications*. 54–72. Berlin, Heidelberg: Springer-Verlag.

Corbu, Nicoleta & Mădălina Boțan. 2011. *Telepreședinții. Radiografii unei campanii electorale [Telepresidents. Radiography of an electoral campaign]*. București: Comunicare.ro.

Davies, Bronwyn, and Rom Harré. 1990. Positioning: the discursive production of selves, *Journal for the Theory of Social Behaviour* 20(1). 43–63.

Drăgan, Nicolae-Sorin. 2016. Strategic positioning of social actors in the semiotic act of TV debate. *Bulletin of the Transilvania University of Brasov*, vol. 9 (58) no.2. 37–48.

Drăgan, Nicolae-Sorin. 2017a. From Narcissus to the Pygmalion Effect in Political Debates: A Semio-Functional Analysis of the Romanian Presidential Debates of 2014. In Evripides Zantides (ed.), *Semiotics and Visual Communication II: Culture of Seduction*. 108–125. Cambridge: Cambridge Scholars Publishing.

Drăgan, Nicolae-Sorin. 2017b. A semio-functional analysis of TV debates for presidential elections in Romania, from November 2014, December 2009 and December 2004. *INTERSTUDIA, Discursive Forms. Dream and Reality* 21. 108–119.

Drăgan, Nicolae-Sorin. 2018. Left/Right Polarity in Gestures and Politics. The Multimodal Behavior of Political Actors in TV Debates. *Romanian Journal of Communication and Public Relations* (RJCPR), Vol 20, No. 3(45). 53–71.

Drăgan, Nicolae-Sorin. 2019a. Semiotic Practices in TV Debates. In Alin Olteanu, Andrew Stables & Dumitru Borțun (eds.), *Meanings & Co.: the Interdisciplinarity of Communication, Semiotics and Multimodality, Numanities – Arts and Humanities in Progress* series, 193–211. Switzerland: Springer International Publishing AG.

Drăgan, Nicolae-Sorin. 2019b. A semiotic approach to the dynamics of positioning and position concepts. In Evripides Zantides (ed.), *Semiotics and Visual Communication III: Cultures of Branding*, 23–41. Cambridge: Cambridge Scholars Publishing.

Eco, Umberto. 2008 [1976]. *O teorie a semioticii* [*A Theory of Semiotics*] (translated by Cezar Radu & Costin Popescu). București: Trei Publishing House.

Fiske, John & John Hartley. 2002 [1978]. *Semnele Televiziunii* [Reading Television] (trans. by Daniela Rusu). Iași: Institutul European.

Flahault, François. 1978. *La parole intermédiaire* [*La Parole intermediate*]. Paris: Seuil.

Gnisci, Augusto; Maricchiolo, Fridanna & Marino Bonaiuto. 2013. Reliability and validity of coding systems for bodily forms of communication, In Cornelia Müller, Alan Cienki, Ellen Fricke, Silva H. Ladewig, David McNeill & Jana Bressem (eds.), *Body – Language – Communication: An International Handbook on Multimodality in Human Interaction*, Handbooks of Linguistics and Communication Science 38.1., Volume 1. 689–706. Berlin: De Gruyter Mouton.

Goffman, Erving. 1981. *Forms of Talk*. Philadelphia: University of Pennsylvania Press.

Harré, H. Rom & Luk van Langenhove. 1991. Varieties of positioning. *Journal for the Theory of Social Behaviour* 21(4). 393–407.

Harré, H. Rom & Grant Gillet. 1994. *The discursive mind*. London: Sage.

Harré, H. Rom & Luk van Langenhove (eds.). 1998. *Positioning Theory: Moral Contexts of Intentional Action*. Oxford: Blackwell Publishing.

Harré, H. Rom & Mark Rossetti. 2010. In Fathali Moghaddam & Rom H. Harré (eds.), *Words of Conflict Words of War. How the Language We Use in Political Processes Sparks Fighting,*. 111–123. Santa Barbara, CA: Praeger.

Harré, H. Rom & Fathali M. Moghaddam (eds.). 2016. *Questioning causality: scientific explorations of cause and consequence across social contexts*. Santa Barbara, CA: Praeger.

Iedema, Rick. 2003. Multimodality, resemiotization: extending the analysis of discourse as multi-semiotic practice. *Visual Communication* 2(10). 29–57.

Jewitt, Carey. 2009. An introduction to multimodality. In Carey Jewitt (ed.), *The Routledge handbook of multimodal* analysis. 14–27. New York: Routledge.

Jones, H. Rodney. 2013. Positioning in the analysis of discourse and interaction. In Carol A. Chapelle (ed.), *The encyclopedia of applied linguistics*. 4474–4479. Oxford: Wiley Blackwell.

Kress, Gunther & Theo van Leeuwen. 2001. *Multimodal discourses: The modes and media of contemporary communication*. New York: Oxford University Press.

Kress, Gunther & Theo van Leeuwen. 2010 [1996]. *Reading images: The grammar of visual design*. London: Routledge.

Kristeva, Julia. 1980. Problemele structurării textului [Text structuring problems]. *Pentru o teorie a textului (Antologie Tel Quel 1960–1971)* [*For a Theory of Text (Antology Tel Quel 1960–1971)*]. 250–272. București: Univers.

Lempert, Michael. 2011. Indexical order in the pragmatics of precision-grip gesture. *Gesture*, 11 (3). 241–270.

Lempert, Michael & Michael Silverstein. 2012. *Creatures of Politics: Media, Message, and the American Presidency*. Bloomington: Indiana University Press.

LeVine, Philip & Ron Scollon. 2004. Multimodal Discourse Analysis as the Confluence of Discourse and Technology. In Philip LeVine & Ron Scollon (eds.), *Discourse and Technology. Multimodal Discourse Analysis*. 1–6. Washington, D.C.: Georgetown University Press.

Lichterman, Paul & Daniel Cefaï. 2006. The idea of political culture. In Robert E. Goodin & Charles Tilly (eds.), *Contextual Political Analysis*. 392–414. Oxford, U.K.: Oxford University Press.

Machiavelli, Nicolo. 2008. *The Prince* (Translated, with Introduction and Notes, by James B. Atkinson). Indianapolis/Cambridge: Hackett Publishing Company, Inc.

Marcus, Solomon. 2011. *Paradigme Universale [Universal Paradigms]*. Bucureşti: Paralela 45 Publishing House.

Maricchiolo, Fridanna; Gnisci, Augusto & Marino Bonaiuto. 2012. Coding Hand Gestures: A Reliable Taxonomy and a Multi-media Support. In Anna Esposito, Antonietta M. Esposito, Alessandro Vinciarelli, Rüdiger Hoffmann & Vincent C. Müller (eds.), *Cognitive Behavioural Systems 2011, LNCS 7403*. 405–416. Berlin, Heidelberg: Springer-Verlag.

Maricchiolo, Fridanna; Gnisci, Augusto & Marino Bonaiuto. 2013. Political Leaders' Communicative Style and Audience Evaluation in an Italian General Election Debate. In Isabella Poggi, Francesca D'Errico, Laura Vincze & Alessandro Vinciarelli (eds.), *Multimodal Communication in Political Speech. Shaping Minds and Social Action*, International Workshop, Political Speech 2010. 114–132. Berlin, Heidelberg: Springer-Verlag.

Moghaddam, M. Fathali, Harré, H. Rom & Naomi Lee (eds.). 2008. *Global Conflict Resolution through Positioning Analysis*. New York: Springer.

Moghaddam, M. Fathali & Rom H. Harré (eds.). 2010. *Words of Conflict Words of War*. Santa Barbara, CA: Praeger.

Navarretta, Constanza & Patrizia Paggio. 2013. Multimodal Behaviour and Interlocutor Identification in Political Debates. In Isabella Poggi, Francesca D'Errico, Laura Vincze & Alessandro Vinciarelli (eds.), *Multimodal Communication in Political Speech. Shaping Minds and Social Action*, International Workshop, Political Speech 2010. 99–113. Berlin, Heidelberg: Springer-Verlag.

O'Halloran, L. Kay. 2011. Multimodal Discourse Analysis. In Ken Hyland & Brian Paltridge (eds.), *Companion to Discourse Analysis*. 120–137. London: Continuum.

Popescu, Cristian Tudor. 2014. Magistrul Loază şi ucenicul Forfotă [Master Scoundrel and apprentice Fuss]. *Gândul*, November 12, 2014. http://www.gandul.info/puterea-gandului/magistrul-loaza-si-ucenicul-forfota-13537738/. (accessed 13 November 2014).

Ries, Al & Jack Trout. 1972. The Positioning Era Cometh. *Advertising Age*, April 24, 1972, 35–38.

Ries, Al & Jack Trout. 1981. *Positioning: The battle for your mind*. New York: McGraw-Hill.

Ries, Al & Jack Trout. 2000 [1981]. *Positioning: The Battle for Your Mind* (2nd edition). New York: McGraw-Hill.

Ries, Al. 2017. A few words about Jack Trout and positioning. In *AdAge publication*, published on June 09, 2017. Available at: http://adage.com/article/al-ries/a-words-jack-trout-positioning/309341/. (accesed on: 19 January, 2018).

Rovenţa-Frumuşani, Daniela. 1999. *Semiotică, Societate, Cultură [Semiotics, Society, Culture]*. Iaşi: Editura Institutului European.

Siefkes, Martin. 2015. How Semiotic Modes Work Together in Multimodal Texts: Defining and Representing Intermodal Relations. *10plus1: Living Linguistics* 1. 113–131.

Trout, Jack & Steve Rivkin. 2009. *Repositioning: Marketing in an Era of Competition, Change and Crisis*. New York: McGraw-Hill.

Van Langenhove, Luk & Rom Harré. 1999. Introducing positioning theory. In Rom Harré & Luk van Langenhove (eds.), *Positioning theory: Moral contexts of intentional action*. 14–31. Oxford: Blackwell Publishers.

C **Communication, semiotics and multimodality**

Dario Martinelli
The different rifles of audiovisual communication: a semiotics of foreshadowing and the case of Roberto Benigni

Abstract: The present essay attempts to construct a prolegomenon to a semiotic theory of foreshadowing in audiovisuality, with a particular focus to the following topics:
1) A semiotic definition of foreshadowing (particularly the symbolic form and the so-called "Chekhov's rifle", or "Chekhov's gun"), in relation to the need for "essentiality" in storytelling and what will be here called a "subtractive reality".
2) An articulation of foreshadowing in four operative steps: preparation, reinforcement, variation, delivery – probably the most innovative part of this study.
3) The case study of Roberto Benigni's filmography as concrete application of the reflections proposed.

Keywords: Foreshadowing, Audiovisual communication, Storytelling, Semiotics, Roberto Benigni

1 Introduction

"Remove everything that has no relevance to the story. If you say in the first chapter that a rifle is hanging on the wall, in the second or third chapter it must go off. If it's not going to be fired, it shouldn't be hanging there" (Shchukin 1911: 44). With these words, Anton Chekhov delivered what is probably the best-known plea for essentiality in a story. Since the Russian playwright used that example, a common phrasing in any form of storytelling is the expression "Chekhov's rifle (or Chekhov's gun)", to underscore the fact that something "not essential" has been introduced in a story.

Let us imagine a woman sitting at her desk, working at the computer. She types something, then gazes at a book, then sips from a coffee mug. At some point

Dario Martinelli, Kaunas University of Technology

the phone rings, she picks it up and says "Hello? . . . Ok, see you later!" She turns off the phone and keeps on working. We may agree that this is an absolutely normal situation, featuring four actions that probably happen hundreds of times. We can all relate to a situation like this, and nothing strange has happened – has it?

Let us now imagine the exact same sequence, this time featured in the beginning of a movie. The woman writes at the computer – fine; she reads from a book – fine; she sips from a cup – fine; she talks at the phone . . . wait!!! All of a sudden a little bell rings in our mind: what was that call about? Who was she talking to? She said "see you later": what is going to happen "later"? Is she meeting a lover? A person she will kill? A simple and – let us repeat it – extremely normal gesture, when placed in a fictional text, has brought us on an alert mode, as if one of the four actions is not so "normal" anymore. Most of all, we now expect something to happen. Creating a set of expectations of different sorts and in different moments is a crucial task in storytelling, because, at the end of the day, the engaging quality of a story relies very much on expectations that are met and expectations that are disappointed, clues that go somewhere and clues that do not.

We call *foreshadowing* the communicative device in which an expectation is created and then satisfied (that is, what eventually happens was indeed anticipated by that particular clue); while we call *sideshadowing* the device that, instead, leads to the disappointment of that expectation (that is, what eventually happens is not connected to that clue, or specifically contradicts that clue).

The notions are abundantly defined and explained in most introductions and guides to storytelling and fiction writing (e.g., Hockrow 2015 for a specific application to audiovisuality). This essay, more specifically, will attempt to semiotically structure the notion of foreshadowing, in the awareness that – especially in audiovisual texts – the "rifles" may have different configurations that are not just limited to their appearance at an early stage of the text, and a "shot" towards the end of it. A deeper and more comprehensive analysis of this topic will appear in the forthcoming Martinelli 2020.

2 On foreshadowing

Strictly speaking, foreshadowing and sideshadowing refer to just *any* clue that gets satisfied or not, without necessarily stirring our attention in the way the woman typing at the computer and receiving a call does. In a most general sense, *all* the actions of the woman in our example can be clues, not only the phone call. In fact, it is not unlikely that the author will use a marginal clue that did *not* particularly catch our attention. This way the text creates more ambiguity and requires a more attentive cooperation from the reader. In other words, it would be like saying that, instead of the phone call, the actual clue thrown in the scene of the woman at the computer would be the fact that she is drinking coffee, rather than receiving the call. Usually, the author does that not just because they enjoy making things more complicated (ok, sometimes that may be the case too), but because they intend to provide a symbolic value to that particular instance, something that would refer to a deeper meaning than just the narrative construction. For this reason, the expression "symbolic foreshadowing" is often employed to describe this particular use of the device.

An example of very subtle, symbolic foreshadowing appears for instance in Stanley Kubrick's *The Shining*. After about 27 minutes we see the main character, the writer Jack Torrance, playing with a tennis ball in the lobby of the Overlook Hotel. His son Danny is playing in the same area and we see a couple of toys on the floor: among them, a small black teddy bear with its torso clothed in red. After about 1 hour and 45 minutes, Jack, who has by now gone mad and murderous, brutally kills Mr. Halloran by repeatedly slashing his torso with an axe: his body, red-blooded, falls down and ends up in the same position as the teddy bear. The latter was thus foreshadowing the death of Halloran, but the clue was extremely subtle, and almost unnoticeable. Once a kid is shown playing, a couple of toys on the floor seems a very normal condition, especially when presented in such a casual manner as Kubrick does in that sequence. The camera, indeed, follows Jack playing with the ball, and Danny's toys are simply part of the environment: there is nothing that particularly attract our attention towards the teddy bear.

In a case like this, the director is demanding close cooperation from us. He wants, so to speak, undivided attention: we have to be attentive to practically everything happening in a scene, considering each element as potentially meaningful, and of course try to appreciate the symbolic value of a child's toy that predicts someone's death (you will remember that Danny has premonitions and visions of the horrors that will follow). This is more energy-consuming than a situation in which – say – Jack crosses the lobby with his ball, leaves it, and the camera zooms in on the teddy bear and stays on it for a couple of seconds.

Having said that, foreshadowing occurs more often in the manner described by Chekhov: an author, in one way or another, "points the finger" at something, usually in such a way that the emphasis is clear but not trivial, in order not to cheapen their story. Let us instead take a second-rate movie, or a cartoon for children, where a character reveals themselves as the villain, exhibiting that typical evil grin behind the hero's back. In Chekhov's metaphor, the evil grin would not simply correspond to hang a rifle somewhere in the scenery, but it would also imply that a character in the play said something like "Oh! There is a rifle there, I wonder if someone will shoot!".

We reach a golden middle between the evil grin and Kubrick's teddy bear when we are actually able to create a certain situation of suspicion in the reader, a vague "Mmm . . . I wonder what that was for . . . " type of discomfort. This way, an "expectation" is created for something to happen, which can or cannot be met.

2.1 On essentiality in storytelling

In order to bring this question into a more specifically-semiotic realm, we may ask, what did Chekhov exactly mean by "essentiality"? What is, and what is not, essential in a story, once we have understood that the concept does not overlap with "normality"? We first of all need to understand that the normality of the real world is something larger and more complex than the normality we see in a fictional text, even when that text is doing its best to meticulously imitate reality.

The real normality, that is, the real life, is a "text" full of meaningful variables and virtually unlimited in its events, contexts, etc. Moreover, the signifying potential of these variables is neither unidirectional nor simple. Every single signifier in real life leads to endless signifieds, each changing depending on who reads the sign, to whom, where, when, how . . . Nothing occurs in relation to a single set of contents. In a fictional text we witness the exact opposite. Signifiers are expressively displayed in order to contribute to one or few signifieds, and signifieds have to be planned in advance with the specific purpose to inform the readers of certain particular occurrences, and not others (which, instead, in real life would be still possible). In practice, if sense and meaning are in real life a mathematical process of additions and multiplications, in fictional texts they are a process of subtraction.

Not only. A text operates also in a Gestaltic manner in respect to the management of the available information. In order to cope with the chaos and disorder of the surrounding reality, the mind "informs" what the eye sees by making sense of a series of items. This way, certain things end up being more important than others, and form a sort of foreground in our perception, leaving the rest in the background. When we sit at home watching a TV series, at that particular moment the "foreground" is our TV (with all that happens "inside" it, in the diegesis of the text), and possibly the snack and the drink we have prepared to better enjoy the show. Our eyes potentially can see everything (the wall behind the TV, the sofa on one side, our body . . .), but what really worry about is the TV show and the food. All the rest now is "background": if by any chance – say – we finish our beer, then for a few seconds the fridge will become foreground, in order for us to go and pick another bottle.

However, while we focus on the TV series and on the food, the wall, the table, our body and all the rest do not *cease to exist*. These things are still there, and we can switch the focus at any point, without doing anything "abnormal". Now: in a fictional text, save few exceptions, we just cannot do that. If, say, a man in a movie is watching a TV show and then stares at the sofa, we start wondering why. Why would he do that? What happened on that sofa? In sum, a person who has the misfortune to live in a fictional text can only do one thing at a time: if they add more activities they have to offer a valid justification to the readers.

Moreover, even that single activity is subject to restrictions. In our example, the man cannot just watch the TV show anyway he likes: that, too, has to be done in a way that does not raise suspicion. For example, if a mosquito bit his neck, he cannot really scratch it more than once, otherwise that bite has to mean something in the story. So, unless we are talking Peter Parker bitten by a spider, or something like that, the poor man has to resist the urge to scratch his neck. If he has not understood a sequence of the program he is watching, he cannot really reel it back to watch it again: that, too, requires a justification. And so on. Reality and normality, the way they are constructed fictionally, have to limit the quality, the quantity and the articulation of every event. If they do not do that, it means that other events are bound to appear, or have already appeared, around which another set of meanings must be constructed. Fictional reality, thus, is a type of minimalistic reality, and that, indeed, is what Chekhov means by essentiality. Of course, the degree of essentiality may change, and usually the artistic greatness of a text can be also measured by a greater degree of essentiality (a topic that reminds us of how Eco underlines the importance of ambiguity in art). Directors such as Federico Fellini were absolute masters in expanding essentiality to many more elements than one would normally expect: you may think about the richness of moments like the eating out sequence in *Roma*, with so many characters and things happening almost all at once.

2.2 Preparation, reinforcement, variation, delivery

The other important aspect is that foreshadowing and sideshadowing are "meta-signs", that is signifiers that, besides conveying a signified that appears there and then, are also referring to another signified, somewhere else, later in the text. The meta-sign appears at some point (usually at an early stage of the text) in a configuration signifier-signified that we shall call here *preparation* (FP, where F stands for foreshadowing), then it appears again in a different configuration, that we shall call *delivery* (FD). The relation between FP and FD, that is, the sense of anticipation created by the former towards the latter, is suggested by the text and established by the reader, in a "lector in fabula" regime of cooperation (see Eco 1979). And finally, unlike FD, which usually appears one time only, FP may be proposed and re-proposed on more than one occasion (which we shall call FR, as in *reinforcement*), also with a slightly different nuance (which we may call FV, *variation*), without however altering its "anticipatory" function.

Let us take Peter Weir's *The Truman Show* as example (for the full screenplay and useful notes, please check Niccol 1998). In this dystopic scenario where a man named Truman Burbank is the unaware protagonist and prisoner of a massively-organized reality show, the various actors tend to interact with him in rather redundant manners, in order to create a controllable routine around Truman: there is his best friend who always comes in with a pack of beer cans, there are the two twins who practically nail him on the particular wall where one sponsor appears, there is the newspaper kiosk seller from whom Truman always buys the same magazines, and so forth. Among these regular encounters, there is one that occurs every morning, as soon as Truman comes out from his house to go to work. It is the Afro-American neighbor family, always smiling at the other side of the street. They wish him a good day, and Truman replies: "Oh! In case I don't see you: good afternoon, good evening and good night!", after which he indulges in a hearty laugh, reciprocated by the family. Now. If we are the kind of readers inclined to cooperate with the text we should already go in alert mode: that meeting and that conversation may be normal, but they are definitely not essential: if they are there, there must be a reason. So, quietly, this piece of information is registered by our brain as an anomaly, a "rifle", and – along with the other rifles we will encounter during the movie – we will expect things to happen that relate to those anomalies, in one way or another.

The meeting with the neighbor is thus a preparation FP, and we expect delivery FD to occur at some point, and maybe also some reinforcements FR or variations FV. The movie goes on: as we know Truman becomes more and more suspicious that he is victim of a huge conspiracy, and things get to a climax when he tries to escape, is stopped in ways that look less and less credible to his eyes and finally has a heated confrontation with his wife, that only the intervention of his best friend prevents from escalating. At that point, the show production team, and particularly the main producer Christof try to re-establish a certain peace in Truman's mind, and devise a strategy that seems to be the only solution: the reappearance of his father, who had been allegedly killed in a sea storm in front of a terrified young Truman, years before. That death had served the psychological purpose to implant a phobia for water in Truman's mind, but now this unexpected comeback has apparently a sedating effect on Truman, who seems to go back to his old self: the unaware puppet of the big show, who conducts his manipulated life with joyful repetitiveness. One of the routine items that the movie adopts in order for us to see that everything is back to normal is exactly our FP: the meeting with the neighbors is re-proposed as FV, that is, with the same significance as FP, but with some different nuance. We indeed witness an additional and amusing moment, as some Asian fans of the show are seen practicing their English by clumsily repeating "Good afternoon, good evening and good night!". This small variation offers some comic relief, but also serves the purpose to show that this morning encounter is so regular that a non-American audience can even use it to practice the language. Plus, remarkably, it gives us a reminder of the importance of the line and on the fact that it seems to be one of the catch phrases of the show.

After a little while, indeed, Truman, who had actually faked his return to serenity in order to loosen the control over him, manages to escape exactly via the sea, on a small sailboat. From this point on, things become biblical and allegoric. Christof, who had created that whole dystopic "Eden" and very much feels like Truman is his property, tries desperately to stop him by launching one storm after another, like an infuriated god. Truman survives, and when the self-styled god gives up with the violent methods (hoping to play the winning card of persuasion), the "true-man" hits the fences of the TV studio – giant walls with realistic sky-looking trompe l'oeil. Truman starts following the perimeter of the walls, and while doing this he looks like he is walking on the water (which is at the same level as the paving of the studio). He finds stairs that go up (an ascension to the "real world") and a door – the quintessential symbol of transition (transition from diegesis to reality, among other things). While hesitating to open the door, Truman hears a voice from the sky (the production room is hidden inside the "moon" of the fictional city): it is Christof, of course, who gently tries to convince Truman not to leave the show and remain in the safe "heaven on earth" that he has constructed.

Christof: Truman. You can speak. I can hear you.

Truman: Who are you?

Christof: I am the creator... [small pause] of a television show that gives hope and joy and inspiration to millions.

    [That little pause after "creator" is a clever idea that reminds us that Christof really thinks he is a god of sorts].

> *Truman: And who am I?*
>
> *Christof: You're the star.*
>
> *Truman: Was nothing real?*
>
> *Christof: You were real. That's what made you so good to watch. Listen to me, Truman. There's no more truth out there than there is in the world I created for you. Same lies. The same deceit. But in my world, you have nothing to fear. I know you better than you know yourself.*
>
> *Truman: You never had a camera in my head!*
>
> *Christof: You're afraid. That's why you can't leave. It's okay, Truman. I understand. I have been watching you your whole life. I was watching when you were born. I was watching when you took your first step. I watched you on your first day of school. [smiles tenderly] The episode when you lost your first tooth [smiles tenderly] ... You can't leave, Truman. You belong here...With me.*

Having known Truman inside out for all these years, Christof knows very well how to play on his insecurities and weaknesses, and manages to install a doubt in his mind. Truman is now taking some time to think whether after all he should stay or leave, but silence is an intolerable occurrence in the bulimic world of TV shows, and this is where the improvised god makes his fatal mistake: he encourages Truman to talk, first gently, with the same protective, fatherly tone he had used this far, and then with much more irritation:

> *Christof: Talk to me... Say something. . . HELL, SAY SOMETHING, GODDAMMIT! You're on television! You're live to the whole world!*
>
> This is the final blow for Truman: at this point he has no more doubts on the fact that he means no more than a puppet to Christof (as well as to the whole audience), and he now knows that he will go through that door. Wanting to leave that huge farce in style, our hero recovers his signature line:

> *Truman: In case I don't see ya, good afternoon, good evening and goodnight. Hahaha! Yep!*

He makes an emphatic bow, like a veteran actor, puts a disgusted grin on his face as he gives the last look towards the source of Christof's voice, turns around and leaves. Here it is: the delivery FD has appeared – the rifle has fired, making a mighty blast, as it provides the emotional catharsis and dramatic apotheosis of the movie. And it also acquires a high "symbolic" value, therefore qualifying as symbolic foreshadowing as well, albeit a less complicated one than that displayed in *The Shining*.

One final note, before turning to our actual case study. Besides symbolic foreshadowing and Chekhov's rifle, there is a number of other foreshadowing forms which take different and colorful names, like "Dreaming of Things to Come", "Call-Forward", "Foreseeing My Death", "Prophecy", "Flashback", etc., which I however suggest to group into the single definition of timeshadowing. This refers to any form of foreshadowing that has to do with an explanation of the clue provided by time-passing: we may see a movie opening with a sequence from the future, where the clue is launched, and then the movie explains how we get there. Or: we may see a clue in the present, and only some revelation from the past (in the form of flashback, or even prequel) explains that clue. Timeshadowing is not so much a rifle that shoots, but rather a revelation of why the rifle was put there in the first place.

3 Case study: the multiple rifles of Roberto Benigni

The Italian Roberto Benigni is an actor and director who takes foreshadowing very seriously, and enjoys being very creative with it. Foreshadowing is to him not just a narrative expedient, but a prominent feature of his style and conception of "film", particularly his recurrent employment of *magic realism*, a form of narration that is realistic in most of its components, but introduces few, or even just one, specific elements of more or less supernatural nature.

Benigni's stories are quite diverse, but they tend to convey similar messages: that life is always worthy of being lived, that its training forces are love and poetry, that one should never give up hope and will to overcome adversities, and that – indeed – reality is "magic" in its own earthly way, and this magic, far from being supernatural, can be manifest when we simply learn to appreciate the "here and now" and the joys of our existence. In all this, Benigni employs his typical clumsy, exuberant and surreal mask reminiscent of the likes of Charlie Chaplin and Dario Fo. Benigni does not attack adversities, wars and prejudices: he rather *mocks* them. The sequence where the *Life Is Beautiful* protagonist makes fun of the *Race Manifesto* is a formidable example of how irony and satire can disintegrate a concept more effectively than a bomb. For a full-round analysis of his filmography and character construction, I shall recommend Celli 2001.

The other tool Benigni employs to draw his characters is exactly foreshadowing, and particularly Chekhov's rifles. In fact, he used so many of them that one is left wondering how few minutes his movies would last, had he decided to cut them all out from the final montage. If we take *Life Is Beautiful*, his most famous and celebrated work, as an example, we find literally dozens. The movie, released in 1997, is a bittersweet drama set in Italy during the Mussolini era, before and during the racial laws and mass deportations that generated the Shoah. Benigni is Guido Orefice, a clever but ordinary Jewish gentleman who falls in love with a Christian woman named Dora. With creativity and force of will, he manages to win her heart. The first half, which has distinctively comic tones, ends and gives room to a much more dramatic second half – another film altogether. This part opens five years later with a young boy called Giosué, fruit of Guido's and Dora's love. The racial laws have been enacted in Italy as well, and Guido is first harassed by the fascist police, then finally deported to a concentration camp, along with Giosué. Dora, who would normally be spared the horror due to her Christian confession, vigorously demands that she is also put on board the train by the Nazi soldiers, hoping to rejoice with her family. Sent to the female area of the camp, Dora will never have the chance to see her husband again. Guido, in the meanwhile, aware of the imminent tragedy and of the terrible trauma his son is about to experience, devises a monumental farce and tells Giosué that the train is taking them to a grand team contest where the winner's prize is a real military tank (Giosué's favorite toy). Motivated by this perspective, Giosué manages to swallow the horrific conditions of the camp as not real, and in order to gain points in the competition, obeys his father's instructions (which are, in reality, mere actions of survival). Defeated by the allied forces, the Nazis need to quickly evacuate the camp, and start killing the prisoners on sight. Guido manages to safely hide Giosué, but ultimately is himself killed. As the morning comes and the camp is almost deserted, Giosué comes out from his hideout, in time to see his prize appearing: a huge American tank is indeed exploring the area in search of prisoners to rescue. Giosué is noticed by the soldier, who invites him to jump on the tank and to drive it with him. On the way out of the camp, Giosué sees his mother, who had managed to survive, jumps off the tank and shouts "Mamma, we have won, we have won!", concluding the movie.

In a film that manages to make us laugh first and cry later with a complete turnover of the narration, there are plenty of items of foreshadowing. It may be useful to mention the functioning of some of the most relevant ones. First and foremost, the comically emphatic march by which Guido reassures his son Giosué when dealing with fascist and Nazi authorities. In its configuration FP, the march is displayed in the not-too-threatening moment when the Italian fascist police take Guido to the police station for the sole fact that he is Jewish: we first see Giosué scared by the brutal manners of the policemen, and then we see him smiling after the father's caricature of military marches. In its configuration FD, the march is performed in the concentration camp, during the movie's climax, when the Nazi soldiers capture Guido and take him to a spot where he will be executed. Giosué, hidden in a metal box, smiles again. Aware of FP (and Giosué's reaction to that situation), the appearance of FD makes us understand that Guido, despite his imminent death, has indeed managed to save his son's life once more – and for good. In this sense, this particular "rifle" takes the first place, as it is the one that characterizes the climax of the movie, much like the "good afternoon-good evening-good night" one from *The Truman Show*.

Secondly, we have one of the many leitmotifs of the dialogues – the famous quotation by Schopenauer that celebrates the force of one's will, by saying "I am what I want to be". As he is told this quotation the first time (by his buddy Ferruccio), Guido interprets it as a sort of magic formula that one can use to turn any event to his favor, and repeatedly tries to use it. In the first occasion (FP), Guido and Ferruccio are in bed trying to sleep, but they forgot to switch off the light: with hand gestures that mimic illusionists, he "orders" the light to go off, except that of course his patient friend Ferruccio gets up and presses the switch, *de facto* confirming Guido's expectations. In a configuration FV, Guido is again successful in his magic, by making the woman of his dreams Dora turn towards him one evening at the opera, despite her sitting in a distant seat in the theatre. With this second step we are given a reinforcement, but most of all a variation, since Dora turned her look unsolicited. It is a coincidence, of course, but we now understand that the magic formula is not just a gag, but will make sense at some point in a more meaningful manner (FD). And indeed, towards the end, as Giosué is already hiding in the above-mentioned metal box, a Nazi soldier's dog sniffs his presence and starts barking. Guido, who is watching from a distance, hidden himself, and who knows that if the soldier paid attention to his dog, that would mean the end of his son, begins chanting Schopenauer's magic formula, until, quite miraculously, the soldier pulls his dog away and proceeds further. Whispering "good old Ferruccio!" to himself, Guido understands that Giosué will still be safe. Besides being another significant rifle, this particular instance, brought to a symbolic level, is a good illustration of how Benigni interprets the poetics of "magic realism": love, faith, perseverance can act like magic, if we truly believe in them.

The third example is probably the best remembered catch phrase of the whole movie: that "Buongiorno Principessa!" ("Good day, Princess!") that Guido utters repeatedly to Dora, since the first time they meet. It is always employed as an expression of courtship in FP and in the various FR that we witness throughout the first half, but we hear it one more time, in the second dramatic part in a configuration FV, as Guido manages to serve as waiter to a party of German officers in the headquarters of the camp, finds a gramophone in one room and puts Offenbach's barcarolle "Belle nuit, ô nuit d'amour" (heard the very night they exchanged eye contact at the opera theatre), loud enough so it can be heard also outside. He then gets hold of the PA system, grabs the microphone and shouts "Buongiorno Principessa!" knowing that somewhere, in the female area of the camp, Dora will hear and recognize the call. After that, Giosué gets to talk too, and tells his mum that they are playing this game and that he might win a tank at the end. This gives comfort and hope to Dora, reassuring her that both her son and her husband are still alive, and that her son is actually experiencing a less confronting version of the tragedy.

The *barcarolle* itself is a touching example of symbolic foreshadowing. We hear it the first time during the opera sequence, when Guido and Dora look at each other with eyes that show that they are already falling in love. The movie devotes an unusually long time to show what happens on the stage and to make sure that we hear almost the whole piece. So, once more, our brain registers that something meaningful is associated with that sign. When Offenbach's famous duet is heard at the concentration camp, we already know that, even before the "Buongiorno Principessa!" calls, the music will have a special effect on Dora. But in fact we also see other prisoners receiving comfort by that single moment of beauty, within the misery of their situation. In that sense, the *barcarolle*, in both its configurations, becomes a moving metaphor of the power of art, a medium for desire and love first, and for comfort and compassion later. It is not accidental, I believe, that, for this purpose, Benigni chose the music of a German Jewish composer, that is, of someone who represented a most harmonious junction between the two cultures that Nazism had tragically made enemies.

Whoever has seen this movie will also remember many other Chekhov's rifles distributed all over the place: the repeated gag of the stolen hat, the riddles played with the German officer, the guy who shouts "Mary, throw the keys!" to his wife, and so forth.

In all these cases, we detect a set of important common characteristics in the rifles: they are real and concrete events; that they are slightly unusual, but not impossible or illogical; and also they do not create stylistic confusion. If, say, the rifle consists of the appearance of an alien, the author would not just be telling us that the "story" will take a certain direction, but they would also be dictating a turn of the genre altogether. The gag of Schopenauer's quotation, in Benigni's movie, leaves us in no doubt that Guido's "telepathic successes" have nothing to do with any supernatural power. Quite trivially, Guido is lucky on those occasions, and of course things would have gone exactly the same way even if he had not engaged in the illusionist's pantomime. All the rifles in *Life Is Beautiful* are perfectly possible and concrete, and could belong to any cinematographic genre. At the same time, though, they are bizarre and unusual enough to attract our attention.

Or: are they? Come to think of it, something like "Buongiorno Principessa!" is anything but bizarre. We all have sweet nicknames for our beloved ones: it may be "sweetheart", "honey", up to the "chocolate cream soldier" from *Arms and the Man.* Nothing is madly unusual about such nicknames, however eccentric they may be.

Let us take another of Benigni's movies, *The Tiger and the Snow* (2005), where we roughly have the same abundance of foreshadowing as in *Life Is Beautiful.*

While the setting is historically different (the American second attack to Iraq in the early twenty-first century), this movie is thematically kindred to *Life Is Beautiful*, sharing the common message of love and hope prevailing over hatred and difficulties. Benigni is Attilio de Giovanni, a university professor who teaches poetry in Italy. A poet himself and a romantic soul (he mentions that his vocation was inspired by a bird perching on his shoulder during his childhood), Attilio is in love with Vittoria, but his love is unrequited. He constantly dreams of marrying her, and a recurrent dream involves Tom Waits (played by himself) singing the song "You Can Never Hold Back Spring", while he and Vittoria get married. As the second Gulf War breaks out, Vittoria travels to Iraq with her friend, Fuad, a celebrated Iraqi poet, but she is seriously injured by a bomb and falls into a coma. Attilio runs to her side, and assists her every day in the middle of a raging war and all the difficulties related to providing decent medical assistance in times like that. Every time he leaves the hospital, Attilio kisses the sleeping Vittoria on the forehead, and in doing so, a necklace regularly slips out of his shirt, gently touching Vittoria's face. With the usual skills Benigni's characters display and promote – love, hope, creativity and perseverance – Attilio manages to heal Vittoria, but, due to a clumsy misunderstanding with the American soldiers, he is arrested before he can reach the hospital, missing the moment when she wakes up from the coma and returns home, unaware of who has saved her life. Weeks later, as they both are safely back in Italy, Attilio visits her, deciding not to reveal the truth. Vittoria, still convalescent, will however find out as Attilio bends to kiss her on the forehead and once more the necklace slips out from his shirt: Vittoria's brain had registered that regular sensation, and now she is able to recognize it. The movie ends with Vittoria sending a loving and grateful smile to Attilio, while a bird perches on her shoulder.

Just as with *Life Is Beautiful*, the key sequences that bring *The Tiger and the Snow* to both a narrative and conceptual resolution are based on foreshadowing. One, again of a symbolic type, is the moment when the bird perches on Vittoria's shoulder, a FD of the story that Attilio tells at the beginning of the movie, explaining what made him fall in love with poetry during his childhood. The fact that a relatively rare and certainly charming event like that happens to both of them is obviously a sign of the deep connection between them on the micro level, and between poetry and love on the macro level. Needless to say, this is another shining example of Benigni's fondness for magic realism.

The other one, a rifle, is of course the necklace worn by Attilio, and it is a rather obvious one. The first kiss to the comatose Vittoria is indeed shown with an unmistakable close-up that makes sure that we just cannot miss the particular movement of the necklace (FP). The kiss+necklace moment is repeated a few times in exactly the same configuration (so they are FR), while – as the movie approaches its end – we see a FV: after the necklace once again touches her face, Vittoria finally moves her eyes a fraction, giving us a visible clue that she is about to wake up from the coma. The rifle "shoots" during the last sequence of the movie, as Attilio visits the convalescent Vittoria at her house in Italy. Still unaware of her savior, Vittoria needs rest and dismisses Attilio. As she closes her eyes, Attilio kisses her on the forehead, once more touching her face with the necklace. Vittoria opens her eyes and finally understands who had been by her side and brought her back to life.

We may even object that there are too many reinforcements, making the narration a bit overstated. On the other hand, Benigni had a problem of scientific credibility, in this particular case: it is said that comatose patients register some stimuli, especially when repeated regularly: this is why friends and relatives are often invited to speak to them, regardless of the lack of visible reactions. Attilio's repeated kisses serve the function of stimulus *par excellence*, making it credible that Vittoria's brain would actually register that one in particular, and not others.

Now, despite being the good director that he is, Benigni is not necessarily one who challenges conventions in filmmaking. That means that we have to expect him to work on meanings in a classic way, that is – as we explained – "by subtraction". It is this subtractive process that prepares fertile soil for foreshadowing (or sideshadowing too, for that matter). The narrative catalyst in *The Tiger and the Snow* is of course Vittoria's journey to Bagdad, where a bomb explosion makes her fall into a coma. First step of the process: Attilio rushes to Iraq and finds her in a hospital, already pronounced as virtually dead. Her bed is removed to make place for cases that still seem to bear some hope of recovery. Second: Attilio, who has not lost hope, finds some room under the hospital's staircase, close to the entrance, and thereby he begins his heroic attempt to bring Vittoria back to life. Third: being close to the hospital's entrance makes Vittoria vulnerable to thieves and acts of looting. A thief breaks in and attempts to steal some of her possessions. Fourth: Attilio sets the thief on the run, managing to recover Vittoria's necklace, but decides that it is safer to wear it himself. Now we finally get to the kiss. In a classic filmic representation of a realistic situation, it is more than acceptable that Attilio kisses Vittoria's forehead: after all, he flew to Bagdad to assist her – a kiss is quite permissible. As readers, thus, we do not become suspicious by the kiss as such, but we see two things happening: 1) the kiss is repeated more than once, and, within a régime of "subtraction", repetition in a text is always suspicious; and most of all 2) there is that necklace dangling from Attilio's neck, and we see that every time the kiss as such is displayed.

In all this setting and preparation, the necklace is simply not necessary. The kiss is not any more explicit or complete without the necklace: Attilio had in fact already kissed Vittoria once before the robbery (without the necklace), thus the idea of the kiss (and therefore an attribution of "normality" to it) was already installed, making other kisses not so important. However, if we add something else, such as indeed a dangling necklace, then we are adding ambiguity to a scene and by consequence that ambiguity must or must not resolve, somehow. If we choose resolution it means we have officially created a rifle, that is, we have introduced a sign that was not necessary within the "minimal" (subtractive) representation/expression of a situation, and that is meant to anticipate some contents that shall be revealed afterwards.

The rifle is thus slightly excessive, in that it is not necessarily something strange, but something more, whose construction in itself is not pertinent to the communication of some contents. In other words, the contents would remain the same even if that sign was absent, just like Attilio's kisses would remain kisses even without a necklace.

The choice to explain all this is appropriate in that Benigni has made foreshadowing a central element of his storytelling and directing style altogether. His most interesting and charming ideas are often generated around the various items of foreshadowing he disseminates in his works. Foreshadowing can be extremely dull if one chooses to simply employ it rhetorically, as mere narrative expedient. Benigni, on the contrary, does not reduce it to its purely instrumental function: he puts it at the center of his filmic vision, almost a *raison d'être* of at least some salient moments. Benigni loves to make a rather refined and rich (and, again, symbolic) use of foreshadowing, one that does not just point the finger at an expected configuration FD, but that actually conveys more meanings that may even go beyond the story as such. We have partly seen this in the example of Offenbach's Barcarolle (in version FD), when we said that Benigni was not just describing a moment of comfort for Dora, but was actually conveying a denser metaphor of the power of art.

To conclude this case study, another poetic example of symbolic foreshadowing comes from *The Monster* (1994), the most mature of Benigni's comedy-only movies, before he engaged in the comedy+drama mixture that gave him international fame. Benigni's character, Loris, is erroneously suspected by the police to be a sex maniac serial killer. Police agent Jessica is hired to flirt with Loris in order to elicit his murderous erotic obsession. A plot like this is rather typical in classic comedy, where misunderstandings and identity mix are often used as source for comic situations. In the movie, police, criminologists and psychologists alike seem to have no doubt that Loris is indeed the man they are looking for: more than misunderstandings, thus, Benigni stages a comedy of prejudices, as all the specialists involved in the case express their opinions on the basis that Loris displays a slightly eccentric personality. In reality, he is a normal person, with his own problems, particularly an ongoing quarrel with the administration of the apartment block where he lives. For this reason, when going out, Loris tries to avoid the doorman by walking crouched down under his desk. Of course, when we see that, our brain registers a "rifle alert". While shown several times performing his escape plan, Loris often bumps into a polite neighbor who never fails to greet him. Absolutely not disturbed by Loris's peculiar way of walking, the neighbor has always a kind word or two for Loris and then proceeds on his way. His repeated appearance also smells like foreshadowing, so we are in the peculiar situation of two configurations (which we shall call FP1 and FP2) employed together, thus with the expectation of resolving into FD1 and FD2.

As the story develops, Loris will turn out to be innocent, and he will also find love in Jessica, who grows increasingly fond of a person who would not harm a fly, and who is also charmed by the many little oddities that he displays, including the escape strategy from the angry doorman. Jessica learns to crouch down too and silently pass beneath his booth. We have now a configuration FV1. As usual, our polite neighbor appears too, and, after greeting them, he also discreetly inquires about Jessica: "Your lady, I suppose? Pleased to meet you!". This is FV2.

Meanwhile, the real killer is captured and Jessica and Loris become a couple in all respects. Embarrassed by the presence of some construction workers nearby, they crouch down behind a car in order to – finally – kiss for the first time. FP1, thus, has turned into FD1: what was just a posture to escape the doorman now becomes a metaphor of Loris's and Jessica's love and understanding, a demonstration that they are really made for each other. While this happens, FP2 is also given a chance to resolve and become FD2. The car, indeed, happens to belong to the polite neighbor, who, rather apologetic for showing up at such an intimate moment, gently asks if he can get into his car. Loris and Jessica wave the neighbor off and disappear into the horizon while still walking in that clumsy posture. As he watches them heading out, the neighbor tenderly smiles, sincerely happy that two soul mates who share so much in common have found each other and can feel free to walk in that manner, hand in hand. On this note, the movie ends.

With such a poetic finale, the transformation of FP1 and FP2 into FD1 and FD2 has probably produced the nicest and most meaningful message of the whole movie. The neighbor indeed was not only a colorful note in the comic moments of Loris's escapes from the doorman: he was probably the most important character of the whole movie. He was the only person who, from start to end, had not surrendered to the temptation to "judge" Loris for his oddities, despite the fact that he was actually exposed to one of the protagonist's weirdest behaviors. The neighbor represents the noblest form of tolerance: he is the person who does not see diversity as evil, and who has chosen a discreet and polite cordiality as a life philosophy. He is the neighbor that we would all love to have, or aspire to be. The ending of the movie shows him not only as a "polite" person, but also as an empathic one, one who is sincerely happy when he sees other people in their happiness.

4 Conclusions

In an attempt to construct a prolegomenon to a semiotic theory of foreshadowing, the present essay has addressed the following topics:
1) A semiotic definition of foreshadowing (particularly the symbolic form and the so-called "Chekhov's rifle"), in relation to the need for "essentiality" in storytelling and what has been here called a "subtractive reality".
2) An articulation of foreshadowing in four operative steps: preparation, reinforcement, variation, delivery – probably the most innovative part of this article.
3) The case study of Roberto Benigni's filmography as concrete application of the reflections proposed.

References

Celli, Carlo. 2001. *The Divine Comic: The Cinema of Roberto Benigni*. London: Scarecrow Press.
Eco, Umberto. 1979. *Lector in fabula*. Milano: Bompiani.
Hockrow, Ross. 2015. *Out of Order: Storytelling Techniques for Video and Cinema Editors*. San Francisco: Peachpit Press.
Martinelli, Dario. 2020. *What You See Is What You Hear: Creativity and Communication in Audiovisual Texts*. Berlin/New York: Springer.
Niccol, Andrew. 1998. *The Truman Show*. New York: Newmarket Press.
Shchukin, Segius. 1911. "Из воспоминаний об А.П. Чехове" [From the memoirs of A.Chekhov]. *Русская мысль* [*Russian Thought*] 10. 37–61.

Evripides Zantides
Differences, similarities and changes of national identity signs in print advertisements. The advertising discourse as a mirror of locality and vice-versa

Abstract: The current paper aims to identify what kinds of nonverbal signs that conceptualise national identity are portrayed in print advertisements of the Republic of Cyprus since state independence from 1960 to 2010, as well as which ones are prevailing in an overall corpus of n=1860 advertisements. A methodological approach that utilizes quantitative content analysis and qualitative analysis, based on a semiotic interpretation of advertisements is implemented to withdraw results. While the findings of the study depict predominant cultural values and characteristics of the Cypriot national identity, a mapping of the nonverbal signs over time portrays which ones are affected or not, throughout the socio-political development of the island as a post-colonial, independent state. The paper shows that national credentials in print advertisements can, on the one hand, reflect values and characteristics of the people in a given culture, and on the other, portray differences, similarities and changes of locality in a reciprocal way over time.

Keywords: semiotics, national identity, advertising, typography, graphic design

1 Introduction and theoretical framework

The definition of national identity is a multidimensional issue inviting interdisciplinary approaches and ideas. From a primordialist perspective, modern nations are ethnic groups which share common ancestry, history, culture, religion and language. In contrast to primordialism, constructivism suggests that nations are not based on certain pre-ordained, characteristics but are cultural constructions resulting from social interaction. According to primordialism, ethnicity is actually constructed from social similarities and kinship (Bayar 2009), and can be considered fixed once it is created. As Bayar (2009: 1643) argues, it "is solidified by violent out-group conflict and/or mass literacy" and has an overwhelming

---
Evripides Zantides, Cyprus University of Technology

https://doi.org/10.1515/9783110662900-012

influence on people's behaviour in so far as humans consider their assumed kinship ties bonds of major importance and priority. Constructivism, on the other hand, proposes that individuals can have more than one ethnic identity, and that ethnic groups are in constant flux as ideas and interpretations of identity are redefined. As such, constructivist theories oppose the idea of natural, fixed national identities and prioritise a historical understanding of national identity as a symbolic artifice.

Embracing the constructivist perspective, our study explores the idea that national identity is grounded in socio-cultural practices and symbolic processes. In the early 1980s, Benedict Anderson (2006) suggested that nationess was a notion essentially constructed on the basis of a range of cultural artefacts. He went on to argue that the nation was an imagined political community conceived by its members as both a limited and sovereign entity not on the basis of some actual, concrete intercommunal experience but of a particular kind of imagination developed through shared cultural artefacts. Even in societies with small populations, individuals are connected to other people they have never seen before by ties that "were once imagined particularistically as indefinitely stretchable nets of kinship and clientship" (Anderson 2006: 6). In this imagined construction, the nation is also limited, as nations are bordered by other nations and therefore can't be imagined as single entities to which all of mankind could belong. Moreover, nations are imagined as sovereign because they wish to be free and stand out from other nations with different identities. In Anderson's own words, in "the allomorphism between each faith's ontological claims and territorial stretch, nations dream of being free" (Anderson 2006: 7). Lastly, the nation is imagined as a community because it depends on the ideals of fraternity and companionship that actually explain why its members are willing to die for the nation and even kill for it. On the basis of this conception, the core substance of national identity is a particular kind of imagination, and specifically, of the sense of belonging to a community of similars.

Anderson has coined the term *print-capitalism*, arguing that the proliferation and ramifications of the printing press in the middle of the 15th century contributed greatly to the 'imaginative construction' of the national community and the notion of nationess. The print-mediated stabilization of languages and the mass production of texts didn't only reproduce and disseminate sacred knowledge, as is commonly assumed, but also provided the foundation for the construction of nationalism as a distinctly modern mode of thinking. This was achieved in three different ways. First and most important, once texts were widely disseminated, people were able to exchange, communicate and understand ideas from other cultures, as well as keep up to date on what was being done in their fields by other 'fellow-readers to whom they were connected through print'. Second, print-capitalism unified languages, which had previously been subject to modification by 'the individualising and "unconsciously modernising" habits' of monks who specialised in the production of manuscripts. As a result, texts became accessible and comprehensible over longer periods of time. Third, certain emerging print languages dominated and absorbed dialects, whilst other languages 'lost caste' because they failed to succeed in print (Anderson 2006: 44). The emergence of imagined communities in the wake of capitalism, print technology and the proliferation of printed languages also introduced new collective practices and everyday habits. For example, as Anderson suggests, when a modern newspaper reader observes "exact replicas of his own paper being consumed by his subway, barbershop, or residential neighbours" (Anderson 2006: 35), this not only confirms the presence of an imagined political community that shares the same language and texts, it also shows that these narratives are strongly rooted in everyday life.

The importance of everyday practices for the cultivation of national identity and national belonging is especially highlighted by the concept of banal nationalism introduced by Michael Billig. Billig (1995) focuses on the informal and innocuous practices through which national sentiments are expressed and cultivated in everyday rituals, media messages and public displays. While his thinking encompasses the vast symbolic and official repertoire of nationhood, including national anthems, coins and currencies, stamps, national holidays, monuments, etc., he argues that "in the established nations, there is a continual 'flagging', or reminding, of nationhood", whereby discourses about national identity materialise in unnoticeable ways in everyday public life. A typical example of this kind of banal nationalism is "the flag hanging unnoticed on the public building", a constant, if unobserved, reminder of nationhood for citizens. Billig shares Anderson's idea that a nation has to be 'imagined' as a community, and draws our attention to signs of nationhood found in routine symbolic practices of social life and language in which "[t]o have a national identity is to possess ways of talking about nationhood". Having a national identity means feeling sentimentally, physically, socially and legally "situated within a homeland, which itself is situated within the world of nations" (Billig 1995: 8).

Another such reminder of shared nationhood is the use of targeted political language. Specifically, inconspicuous words like 'we', 'this', 'now' and 'here', when spoken to fellow citizens, become charged signifiers of ideas like 'us' and 'ours'. These are words of linguistics "in a routine 'deixis', which banally points out 'the' homeland." (Billig 1995: 94), and therefore act as a point of reference that stands collectively for all members of the community. The dissemination of this language is mainly carried out by the mass media, which bring such 'flaggings' into households through newspapers, computer monitors and television screens. Nationalist discourses that reinforce the idea of belonging to a common homeland are projected onto the ordinary lives of millions of people not only by politicians but also by sport writers and academics. Phrases like 'fellow citizens', 'united we stand here now', 'our national team plays tonight in its homeland' are some examples of this type of language. Following Anderson, Billig underlines the vital role played by newspapers in the reproduction of national identity, with their headline messages and verbal clichés evoked in a national context.

The importance of everyday practices in fostering nationess and a sense of national belonging is also the focus of the work of Edensor (2002), who examines how national identity is reproduced and transformed in popular culture and everyday life. Edensor agrees with Anderson's conception of the nation as an imagined community; however, he considers that Anderson's approach, by not paying attention to other ways in which the nation is imagined – for example, in public places and official ceremonies, official documents and ordinary habits – lacks comprehensiveness. In addition, he criticizes Billig for not including in the symbolic repertoire of 'banal nationalism' the habits and practices of everyday life and popular culture. Extending Billig's ideas, Edensor explores how national identity is expressed and experienced in ordinary, emblematic spaces of everyday life, such as the home and the popular performances of sport, or in everyday objects and commodities. Moreover, he explores nationhood as it is expressed in popular culture, such as in popular food and drinks, or in the mass media (i.e. television programmes, films, daily news, advertising, etc.) As Edensor suggests, "[a] sense of national identity . . . is not a once and for all thing, but is dynamic and dialogic, found in the constellations of a huge cultural matrix of images, ideas, spaces, things, discourses and practices" of everyday life. According to Edensor, national identity is cultivated in "the things we watch and read, the places we visit, the things we buy and the pictures we display" (Edensor 2002: 17). Both Billig and Edensor, then, extend Anderson's concept of the nation as an imagined construct built on typography and print media, museums and maps, monuments and censuses, by including the vast and multiform expanse of unofficial and informal material and symbolic everydayness. Moreover, whereas Anderson traces the emergence of nationalism to its cultural roots, both Billig and Edensor focus instead on its more adult stage, and specifically, on the cultural resources and social dynamics of its continuing collective appeal and symbolic sustenance. It is on the basis of this extended and dynamic framework of conceiving national identity that our research investigates a corpus of print advertisements taken over a fifty-year period from the Republic of Cyprus.

Goldman asserts (2011: 38) that advertisements depend on systems of meaning 'that already have currency with an audience' and provide an arena in which this meaning is transferred and rearranged. As such, their presence in newspapers and magazines allows them to be studied as artefacts of collective meaning in accordance with the constructionist schema outlined above. As pointed out earlier, for Anderson (2006), the invention of printing enabled people to rally around a common national identity through shared readings and imaginings. In a similar fashion, we claim that modern day advertisements reassert and reinforce their audience's sense of a shared national identity, of their belonging to a common imagined community. Our corpus of advertisements can be seen as a crucial addition to the symbolic repertory examined by Billig (1995) under 'banal nationalism', since it is a vital contribution to the innocuous and 'continuous flagging' of nationess and national belonging, a contribution that often takes the quite explicit form of the political linguistic deixis of 'we' (the consumers) and 'our' (products). While the nation is 'materialised' as a collective practice during this communication process, it is also consumed 'unnoticed', since the reading of the advertising messages is a routine, everyday habit. In this way, our investigation concurs with Edensor's approach, which, building on Billig's notion of banal nationalism, perceives the construction of national identity as grounded in the realm of the quotidian and fostered by the experience of daily mass media. Advertisements regularly display images of familiar locations and objects, cultural values and icons, traditional customs and practices, etc. in order to promote a wide variety of goods or services. As such, they can be considered cultural resources of major collective significance not only because of their pervasive presence in everyday life, but also because of the national symbolic content they convey.

Advertisements comprise the central focus of this paper and are explored as popular artefacts, grounded in the consumption routines and signifying practices of everyday life that consumers share as members of an imagined community, and at the same time 'flag' a sense of national identity and belonging through their verbal and nonverbal messages.

It is important to clarify that the present study refers to national identity as "a form of collective identification that serves the purpose of binding people together within a community, giving them a sense of membership of a cultural or ethnic group" as put forth by Bulmer & Buchanan-Oliver (2010: 200), and does not investigate socio-political conflicts, definitions and beliefs in relation to Greek-Cypriot or Turkish-Cypriot identities in the Republic of Cyprus. Instead, it focuses on how advertisements in Cyprus have exploited national identity signs and, thus, promoted 'collective identifications', since the country's independence in 1960 up until 2010. Content and semiotic analysis will also show how these ads were distributed over time and how they can be interpreted. While advertisements cultivate national values and a sense of national belonging in mass audiences, they also reflect the socio-political conditions of the time they were produced. As objects of popular culture, they are seen everywhere and are thus part of the everyday fabric of reality for entire populations.

1.1 Purpose of the study

The assignment of cultural meanings to commodities through advertisements, as articulated by the use of images and texts, does not necessarily relate to the attributes of the product but to other ideas or values. Within this system of signification, we are interested in exploring advertisements as carriers of cultural meanings, and specifically national identity signs in the context of the Republic of Cyprus.

Specifically, the current study aims to identify which signs of national identity are depicted and prevail in commercial print advertisements in the Republic of Cyprus over a period of 50 years, from its independence in 1960 to 2010. In this context, advertisements will be studied as "an indispensable interpretive key to understanding complex historical developments" (Goldman 2011: 193) as vehicles both of social consensus and social change. As a newly established, post-colonial state, Cyprus needed to define and solidify its national identity. Next to the official identity discourse of its flag, currency, coat of arms, anthem, monuments, etc., advertisements have contributed in their own way to the construction of 'Cypriotness' and Cypriot national identity through their visual and textual messages. Mapping the signs of national identity throughout the aforementioned period will reflect and shed light on historical developments in Cyprus, as well as reveal dominant national identity values and representations.

2 Methodology

2.1 Quantitative content analysis with semiotic analysis

Content analysis has been widely used in the social sciences in order to quantify patterns of content in both verbal and nonverbal communication, especially when studying large-scale data corpora. Content analysis has been particularly popular in media studies, because it offers an objective, systematic and generalizable description of communications content Kassarjian (1977), and thus allows for the verifiability and reproducibility of its results.

This method is employed in our study in order to answer a series of crucial questions. First of all, it will help us to determine the incidence and chronological distribution of the basic kinds of national identity signs used to conceptualise national identity from the establishment of the Republic of Cyprus as an independent state in 1960 until 2010. It will also allow us to make comparisons between different decades and identify changes in the prevalent modes of visual representation.

One of the most often noted limitations of content analysis is that by itself "[it] is seldom able to support statements about the significance, effects or interpreted meaning of a domain of representation" (Bell 2010: 13). Questions of significance and meaning, however, are central to comprehending the cultural implications of the interlacing of the symbolic text of advertisements with national identity signs. For this reason, we have chosen to complement our research methodology with semiotics, an analytical-interpretative approach widely acknowledged for its ability to provide in-depth qualitative and focused interpretations.

The use of a combined content analysis-semiotic approach to examine both the denotative and connotative meaning of advertisements is common in studies of advertisements. In general, the most productive approach to the investigation of advertising as a dimension of banal nationalism, which forms a major dimension of our study, "is the analysis of the complete 'set' of advertising messages themselves" (Leiss, Kline, and Jhally 1990: 197) and "[f]or such purposes, one must focus on messages themselves rather on the reactions of consumers to them. Two major methodologies have been employed in the study of advertising messages: semiology and content analysis." (Leiss, Kline, and Jhally 1990: 198). The mixed methods research design, which combines quantitative and qualitative approaches, seems to be appropriate and in alignment with the research questions of the current study. Quantitative content analysis combined with semiotic analysis is a synergistic methodological approach whereby the two investigative procedures complement each other.

2.2 National identity signs in Cypriot print advertisements as units of analysis

Being an empirical approach, content analysis depends directly on our basic theoretical assumptions. It is the latter that determine the content units, i.e. the specific thematic categories, that need to be studied, for "content analysis can only measure what one deems important to measure" (Leiss et al. 1990: 219). Depending on the perspective from which the researcher approaches a corpus of advertisements, different kinds of research questions may be articulated.

In the case of our study, the basic assumption is that national identity characteristics that make up the sense of 'Cypriotness' are not natural, fixed or hereditary, but culturally constructed and transformed over time through the workings of state institutions, like education, or official occasions, such as ceremonies and collective rituals, but also through non-official social activities or contexts, such as sport, media, leisure or entertainment practices, and other daily social routines.

Even though advertisements typically consist of both verbal and nonverbal messages, the current research focuses on the visual (or nonverbal) representations and considers the verbal messages only secondarily. The focus of our research, in other words, is on the whole wide range of national iconography, i.e. all the various visual signs that comprise the visual representations of a certain nation. National iconography is typically quite diverse, including symbols like the flag and the national flora and fauna, the shape of a country, iconic personalities, buildings and artworks, etc. In order to proceed in a systematic fashion, we need to identify specific categories of national identity signs, and define their particular cultural function and significance.

2.3 Constructing categories for the coding procedure

In order to achieve an objective, systematic and quantitative description of the manifest content Kassarjian (1977) of national identity signs in our corpus, a working tool needs to be implemented where the concept of 'national identity signs' is not only split into representative categories for empirical research, but these categories need to be mutually exclusive (i.e. non-overlapping) and exhaustive. Moreover, different coders working with the same corpus should independently converge on the same descriptive categorisations, using the same set of criteria for all the data under examination. Similarly, they should not be 'reading between the lines', but measuring only 'the manifest or "surface" content of the message under study' (Leiss et al. 1990: 218–220). For this reason, a series of categories common to similar investigations, as well as other studies, were selected to provide a coding scheme for the current research, in addition to other possible emerging categories from the corpus itself.

In more detail, the different categories that were implemented to measure the national identity signs in other relevant studies, (Avraham and Daugherty 2012; Avraham and First 2003; Hogan 1999; Pritchard and Morgan 2001), were carefully examined. In combination with a preliminary study of the national identity signs in the images of our corpus, the following categories were drawn and summarised to define our coding scheme:

Geography: national locations, distinctive landscapes and landmarks or distinctive flora and fauna;

Leisure practices: popular pursuits, everyday practices, patterns or trends;

Cultural heritage: monuments, traditional occupations or costumes, historical, contemporary or cultural-icon figures, handicrafts like embroidery, weaving, tapestry, woodwork, ceramics, metalwork, basket-weaving, leather making, mosaics;
Social relationships and social values: images of family (nuclear and extended), co-workers, friendship and mates.
National identity symbols and official state symbols: signs and shapes of the visual national identity repertoire and general governmental national iconic references.

In his attempt to map 'Britishness' in everyday practices, Edensor (2002) defines the term 'Andscape' through a series of emerging constellations depicted in hundreds of photographs representing "Britain". By asking various people to respond to printed leaflets displayed in various local shops, stores and public spaces like Marks and Spencer, and in libraries, he questioned 'what one thing represents something good about to Britain and why?'. Consequently, a 'National Portrait' representing 'profusion and multiplicity' was created. Building on emerging constellations, he suggests that dimensions of national identity can also be reflected in representations of "'things', 'food and drink', 'geographies', 'people', 'animals and plants', 'popular cultural forms', 'technology and innovation', and 'cultural practices'," (Edensor 2002: 175). Drawing from Edensor's more inclusive range of categories, we have also included in the coding scheme the following:

Objects: culturally distinctive popular items used in everyday practices and daily habits.
Food/Drinks: food and drinks not as mere products, but objects invested with national cultural relevance, their consumption being perceived as a national custom, i.e. as an ethno-cultural practice, as the performance of an ethno-cultural protocol, e.g. Cypriot coffee, halloumi cheese, etc.

Finally, to measure the frequency of the advertisements that include a national identity sign in their images, as well as to count how many such signs are depicted every time in the corpus, two further variables were included in our coding frame:

Presence of national identity signs: presence or not of a national identity sign; and
How many national identity signs: number of national signs depicted in the advertisement.

2.4 Implementing the coding system

Given that the main purpose of the content analysis of images is to discern the 'dominant content', the coding scheme must include clearly defined categories which are mutually exclusive (i.e. non-overlapping) and exhaustive. For each advertisement, only the nonverbal signs were coded, while all other kinds of verbal signs, like trademarks, headlines or slogans, were not taken into consideration. Furthermore, for an advertisement that contained more than one national identity sign in the same category, the coding was made on the basis of the dominant element. In other words, since we were interested for 'one principal feature of representation' (Bell 2010: 17) we coded the presence of one value in each category/variable. To study in more detail how certain categories were constructed and visually portrayed, a series of values were assigned separately for each category under the scope of what Bell (2010: 16) suggests are elements of the same logical kind and constitute values of a particular variable. As the study investigates the construction of Cypriot nationhood, these values should be based on portrayals that belong to the official national identity visual repertoire, the historical or traditional landmarks, customs, popular practices and behaviours encountered in the Republic of Cyprus. The map of the island behind a product advertised, for example, would be an example of a visual value under the category of *national identity symbols*. All values assigned to the aforementioned categories are described below and will be implemented in the coding procedure as follows:

The first category, *Geography,* is divided into eleven values for: (1.0) coding the *Absence* of landscape and landmark references; (1.1) coded representations of a highly touristic landmark location known as *Petra tou Romiou,* which according to the Cyprus Tourism Organisation is ". . . Aphrodite's mythical birthplace . . . an interesting geological formation of huge rocks along one of the most beautiful coastlines on the island" (Cyprus Tourism Organisation, *Birthplace of Aphrodite – Petra Tou Romiou*); (1.2) coding the mountains of *Troodos* as a landscape location with national association; (1.3) coded seashores, and more particularly the *Beach*, a popular place almost all year round in Cyprus and a visual reference also portrayed often in tourism advertisements; (1.4) coding the *Absence* of fauna references from the ones described below; (1.5) coding *Mouflon* as 'an indispensable part of our natural heritage and one of the symbols of Cyprus' (P.I.O.); (1.6) coding the Cyprus *Donkey* "as the donkey breed of the Mediterranean island of Cyprus" (ibid); (1.7) coding the *Absence* of any flora references from the ones described below; (1.8) coding any representations of *Olive trees*; (1.9) coding the presence of *Citrus fruits* and (1.10) coding nonverbal signs of *Grapes*. All flora values selected are based on the fact that "when Cyprus achieved independence in 1960, the backbone of its economy was agriculture . . . while after the 1974 invasion the south retained nearly all of the island's grape growing areas and deciduous fruit orchards" (Library of Congress). As specific values are rooted historically in Cypriot agriculture, they also become part of popular culture and are often associated with national references too.

The second category, *Leisure practices,* is divided into five values for: (2.0) coding the *Absence* of any leisure practices; (2.1) coding image representations of people *Swimming;* (2.2) coding any nonverbal references of people *Hanging out with friends;* (2.3) coding any visual representations of people *Singing,* and similarly (2.4) coding any *Dancing* representations. All values used in this category are based on daily practices and routines; however, singing and dancing are exceptional cases which mainly happen at traditional weddings and are not expected to be found regularly in the corpus.

The third category, *Cultural heritage,* is divided into five values for: (3.0) coding the *Absence* of Cultural heritage references; (3.1) coding *Historic monuments,* for example, ancient theatres or temples; (3.2) coding *Traditional occupations and industry* where people practice folklore, traditional professions like weaving, potmaking or making traditional products and silverware; (3.3) coded the presence of *Aphrodite* as the goddess of love and symbol of Cyprus who, based on legend, 'was born in Cyprus' (Cyprus Tourism Organisation, *Birthplace of Aphrodite – Petra Tou Romiou*), and (3.4) coding the presence of people in *Traditional uniforms* or any other objects covered with traditional textiles and embroidery.

The fourth category, *Social relationships and social values,* is divided into four values for: (4.0) coding the *Absence* of social relationships and social values; (4.1) coding any nonverbal signs of *Hospitality,* thus representations of people offering food or drinks or inviting other people (or even the viewer of the advertisement) to be treated or 'offered with a welcoming' or giving 'personal treatment to a guest'; (4.2) coding representations of *Extroversion,* as in people expressing themselves with outward gestures of smiling or laughing, socialising with other persons, talking with others, shaking hands, celebrating joyfully, or in general portraying people with an open heart, and (4.3) coding *Family Values* where representations of family relationships were evident in the images, like parents interacting with children.

The fifth category, *Objects,* is divided into four values for: (5.0) coding the *Absence* of Object references; (5.1) for including the coding of *Sun-umbrellas;* (5.2) the coding of *Sun-beds,* both as objects for protection and relaxation often seen in images by the beach, especially in tourism advertisements; (5.3) the coding of *Beach-ball,* as an object signifying entertainment, and (5.4) the coding of *Objects made out of clay* in different shapes and functions, without necessarily having any folklore relevance. All objects in this category are not coded as products but as contributing signs of the overall composition of the advertisement.

The sixth category, *Food and drinks*, is distributed into six values for: (6.0) coding the *Absence* of Food and drinks references; (6.1) coding the consumption of *Traditional products* of unique character mainly made in Cyprus like 'halloumi'-cheese or 'soutzoukos', a traditional, chewy sweet made from grape juice; (6.2) coding in the same way any *Traditional foods* like 'trahanas', a soup made out of groats, or other local dishes like 'afelia', a traditional Cypriot pork dish; (6.3) coding the consumption of *Fish*; (6.4) coding the representation of sharing *Coffee*, either hot or cold 'frappe', and (6.5) coding any presence of enjoying *Wine or Spirits*, including any traditional Cypriot drinks (e.g. 'Commandaria' fortified wine or 'zivania' brandy).

The seventh category, *National identity symbols*, is divided into four values for: (7.0) coding the *Absence* of National identity symbols references; (7.1) coding any representations of the *Cypriot Flag*; (7.2) coding the presence of the *Coat of Arms of Cyprus*, and (7.3) coding any *Shape or Map of Cyprus*. This category has the strongest association with national affiliations because of the official relevance of the assigned values with the Republic of Cyprus.

The eighth category, *Product category*, is divided into thirteen values to accommodate the type of product that is promoted in the advertisements as follows: (8.0) *Recreation and travel*; (8.1) *Auto and related products*; (8.2) *Jewellery*; (8.3) *High-tech devices*, (8.4) *Apparel*; (8.5) *Cosmetics*; (8.6) *Movies and entertainment*; (8.7) *Food and drinks*; (8.8) *Household items, (e.g. cleaning detergents)*; (8.9) *Personal hygiene*; (8.10) *Financial services*; (8.11) *Home appliances*, and (8.12) *Absence* of the aforementioned categories.

The ninth category, *Presence of National Identity signs*, was used for coding whether one of the aforementioned values was present in the advertisements or not, in order to measure the presence of national signs in the corpus. (9.0) coded the *Absence* of national identity signs in the advertisement, while (9.1) was used for coding the *Presence* of national identity signs in the advertisement.

The last category, (10.0) *How many national identity signs*, was numerical and was applied to measure the number of national signs depicted in an advertisement in order to count the total number of national signs in the corpus, as some advertisements had more than one value portrayed each time.

In general, the values adopted under the categories *Geography, Leisure practices, Cultural heritage, Social relationships and social values, Objects, Food/Drinks, National identity symbols* and *How many national identity signs* will allow a more detailed response to our purpose that investigates how national identity signs are portrayed in print advertisements in Cyprus since independence in 1960 to 2010.

Finally, all advertisements were coded into *Decades,* and the corpus divided into five decade clusters, the first being from 1960 to 1969, the second from 1970 to 1979, the third from 1980 to 1989, the fourth from 1990 to 1999, and the fifth from 2000 to 2010. This categorisation, together with the aforementioned values would, on the one hand, provide an answer to how the presence of national identity signs change from 1960 to 2010, and, on the other, let us identify whether there is a relationship between the historical landmarks of the Republic of Cyprus and the way national identity signs are portrayed in the corpus.

An example of how the coding of national identity signs would look is shown in the example of the Cyprus Airways advertisement in Figure 11.1. Under the category *Leisure practices*, the value of 'Hanging out with friends' in the square photograph at the bottom left can be observed, while at the same time, the value of 'Hospitality' is also evident under the category of *Social relationships and social values* in the gesture of the air hostess offering beverages to the passenger.

2.5 Sampling

To study the signs which conceptualise national identity in images of the print advertisements produced in the Republic of Cyprus over a period of fifty years (1960–2010), a convenience sample of print advertisements was used for content analysis. The advertisements were drawn from the Press and Information Office (P.I.O.) in Cyprus, the archives of different newspapers and collectors, magazine publishing houses and advertising agencies. They were chosen according to the following criteria: 1) the writing of verbal messages were in the Greek language, 2) their main target audience was the citizens of the Republic of Cyprus, and, 3) the products or services advertised were sold locally and not abroad.

All duplicates were excluded from the corpus. An effort was made to have a balanced number of advertisements for every decade as shown in Table 11.1 above. We need to remember that the presence of advertisements in print media throughout the decades proceeded in parallel with the development of local industry needs, the production of new local press media and magazines, as well as with the opening of advertising agencies. Our final corpus consists of a total of 1860 unique advertisements. Advertisements with long presence in the local market are of special interest as they portray national signs that change over time and have a story to tell. Such examples are advertisements for companies like KEO beer, Cyprus Airways which in addition was a governmental company, and Laiko Kafekopteio (People's Coffee).

Figure 11.1: Cyprus Airways hospitality adult, 17 May 2003, *Phileleftheros* newspaper.

Table 11.1: Frequency and Percentage distribution of advertisements across decades (1960–2010).

	Frequency	Percentage
1960–1969	287	15,4
1970–1979	345	18,5
1980–1989	387	20,8
1990–1999	375	20,2
2000–2010	466	25,1
Total	1860	100,0

The sample was limited to those advertisements which included at least one national identity sign resulting in a sample of 452 advertisements as shown in Table 11.2. Out of these 452 advertisements, there were $n=550$ national identity signs as shown in Table 11.3 above; specifically, 372 ads had one national sign, 62 had 2, and 18 had 3, resulting in 550.

Table 11.2: Frequency and Percentage distribution of advertisements with national identity presence.

	Frequency	Percentage
Absence	1408	75,7
Presence	452	24,3
Total	1860	100,0

Table 11.3: Frequency and Percentage distribution of advertisements and number of nonverbal national identity signs.

	Frequency	Percentage
0	1408	75,7
1	372	20,0
2	62	3,3
3	18	1,0
Total	1860	100,0

2.6 Reliability agreement

Two coders participated in the coding process; the first coder was the author of this work, while the second coder was trained by the first author to apply the coding scheme of the content analysis. The coding was done according to a number of criteria as already described in sections 2.3 and 2.4 of the coding scheme for the quantitative content analysis. The coders went through a pilot coding of 100 advertisements (see, for example, Plakoyiannaki and Zotos 2009) prior to the main study so as to experience the coding scheme worked. During the pilot study, the coders reached a general consensus regarding the application of the coding scheme. The coding of the whole corpus was conducted by the first coder. For a reliability check, the second coder independently coded randomly selected advertisements from the overall corpus (n=186), or about 10%, as suggested by Rogers and Escudero (2015: 97). An intercoder Kappa reliability test was conducted, as described by Peat (2002). The results revealed a percentage of agreement in Cohen's kappa coefficient between the first and second coder. For example, the measure of value agreement between the first and second coder for the presence of national identity signs was 98,4% (K=0,959, p<0,001). According to Peat (2002), this value represents a strong agreement, almost perfect in this specific case. It is also observed that in all cases there is a statistically significant agreement at level above the required 70%, with most values reaching perfect agreement, and therefore deemed appropriate.

3 Results

As already described, for the classification of the advertisements, as well as for their distribution into specific categories and values, a reliability agreement between two coders was conducted, whilst both IBM SPPS 19 and Excel 2010 were used for the data analysis.

Table 11.4 (above) presents the frequency and percentage distribution of advertisements with national identity presence across five decades. The chi square test indicates that there are statistically significant differences between the decades (X^2_4=70,955, p<0,001). In other words, the decades (time) and advertisements with national identity presence are not independent. We observe that the proportion of advertisements with national identity signs reaches its highest level (33,8%) in the first decade after independence (1960–1969). In the subsequent two decades there is a marked downward trend (17,4% during 1970–79 and 11,9% during 1980–89). The trend is reversed in the final two decades, since the proportion of advertisements with national identity signs increases significantly during 1990–1999 (26,9%) and furthermore during 2000–2010 (31,8%).

These findings suggests that, immediately after independence, a significant range of companies and organisations redefined their advertising strategy on

Table 11.4: Frequency and percentage distribution of advertisements with national identity presence across decades *n=1860*.

Time	Presence of National Identity								
	Absence			Presence			Total		
	f	Row %	Column %	f	Row %	Column %	f	Row %	Column %
1960 1969	190	66,2%	13,5%	97	33,8%	21,5%	287	100,0%	15,4%
1970 1979	285	82,6%	20,2%	60	17,4%	13,3%	345	100,0%	18,5%
1980 1989	341	88,1%	24,2%	46	11,9%	10,2%	387	100,0%	20,8%
1990 1999	274	73,1%	19,5%	101	26,9%	22,3%	375	100,0%	20,2%
2000 2010	318	68,2%	22,6%	148	31,8%	32,7%	466	100,0%	25,1%
Total	1408	75,7%	100,0%	452	24,3%	100,0%	1860	100,0%	100,0%

the basis of Cyprus' newly established statehood, embracing a nationalistic discourse so as to take advantage of the atmosphere of popular enthusiasm and pride. These advertising practices are commonly studied under the rubric of 'commercial nationalism', whereby advertisers develop narratives which assign national meaning to consumer products or services in order to promote them (Castelló and Mihelj 2017). The subsequent decrease, during the 1970s and 1980s, is attributed to the humanitarian, economic, but also to the identity, crisis the 14-year-old state experienced due to the traumatic events of July 1974, i.e. the anti-Makarios coup d'état engineered by the Greek military junta and the consequent Turkish invasion in July 1974. It is only after the economic boom in the 1980s and 1990s that we observe an increase in the presence of national identity signs in the advertisements once again. This trend continues up until Cyprus' entry into the European Union in 2004 and finally reaches a level in the fifth decade (2000–2010) that is almost equal to the initial high of the post-independence 1960s. Just as in the 1960s, 2000–2010 marks a turning-point for the Republic of Cyprus, whereby Cypriot citizens had the opportunity to reassert their identity, this time in the totally new context of international competitiveness and tourist attractiveness. It was a climate propitious for the rekindling of consumer nationalism; this time, however, in a globalised, European context more geared to competitive advantage and nation branding.

The frequency and percentage distribution for the seven categories is presented in Table 11.5. Responses are sorted in descending order for the seven categories for

Table 11.5: Frequency and percentage distribution of national identity signs into categories n=550.

Category	Frequency	Percentage
Social relations and values	211	38,4%
Geography	131	23,8%
Leisure Practices	93	16,9%
National identity symbols	37	6,7%
Cultural Heritage	34	6,2%
Objects	22	4,0%
Food and drink	22	4,0%
Total	550	100,0%

a total of 550 items. It can be seen that the majority of advertisements are related to *social relations and values* (38,4%), with *geography* (23,8%) and *leisure practices* (16,9%) being the second and third most frequently substantial categories. It must be noted that the percentage of advertisements classified in the categories *national identity symbols, cultural heritage, objects* and *food and drinks* is less than 7% for each case.

We also observe that the categories of *social relations and values* and *leisure practices* decrease from the first decade (1960–1969) to the second (1970–1979), and then increase again during the following decades. The categories of *cultural heritage* and *objects* follow a more or less similar path; however, *cultural heritage* never regains the significance it held in the 1960s. Contrary to these categories, the categories of *geography and national identity symbols* both see an increase between the first decade (1960–1969) and the second decade (1970–1979), and then decline in the decades that follow. The category of *food and drinks*, on the other hand, has a quite stable presence across the five decades.

4 Discussion

Due to word limitation, we will discuss selectively the findings of values that prevailed from the content analysis. Looking analytically at the first category of *social relations and values*, we observe that *family values* were portrayed a total of 156 times in the corpus, accounting for 73.9% of this category's presence. Our data also revealed that in this category, *extroversion* comprises 20.4% of the category, followed by 5.7% for *hospitality. Family values* were widespread not only in print advertisements but also in much of the audio-visual advertising in Cyprus.

In another study, for example, Pavlou (1992) investigates radio commercials of the Cyprus Broadcasting Corporation and argues that the 'priority of family' is among the most important values portrayed. He also mentions that "[t]he family is the nucleus of Cypriot society and the parents' mission in life is to make their children's life [sic] as easy as possible" (Pavlou 1992: 10), a practice that still holds true today. Pavlou argues that focusing on family is a collective and socially acceptable priority among Cypriot people as a result of the austere poverty they lived through during the British colonial period. Specifically, as "most parents were poor as children, they now want their children to live as comfortably as possible" (Pavlou 1992: 10), a mentality that most Cypriot parents still share today, regardless of the economic advances the country has made in recent decades. Consequently, family ties are strong and are an essential part of life in Cypriot society. This long tradition, together with the importance and priority given to *family values* in local culture, explains why this particular value stands out from the rest.

As observed in a series of advertisements, visual signs depicting family relations and contexts are central to advertisements during the whole of the period studied. Parents and children are portrayed as major participants in the visual narratives and have a central role in the consumption of the advertised product or service. It is noteworthy that as years go by, the emotional bond among family members looks stronger. The children, for example, are placed in closer proximity to their parents, suggesting stronger emotional ties. In advertisements from the 1960s, the children are often placed on a separate level from their parents, whereas in more recent advertisements, children are fully integrated into the family unit, occupying, moreover, a central place in the foreground. This proximity and foregrounding reflects how the family unit has become a more egalitarian and child-centred emotional group at the expense of independent family roles. One of the possible explanations for these representations is the waning of patriarchy over the years, and a loosening of the strict roles ascribed to children traditionally i.e. to obey, to behave ('know their place'), and to assist either the breadwinning father or the homemaking mother. Headlines like 'Your dreams are no longer . . . dreams' in a *Universal Life Insurance Company* advert reinforce this trend. The value of family is, therefore, not only prominent in nonverbal signs, but in verbal messages as well. In a *Lambrianidi Brothers* food advert, we observe how the bold headline '. . . from our family to yours . . .' emphasises the dominant value of the family code. Written in huge, black, lowercase, bold letters, the headline is a strong, dynamic statement that aims to draw our attention both to the co-presence of two generations in the family depicted, and to the associative meaning of the company as a 'family' that addresses its products to actual families. The concept of tradition, of the passing from one family to the next, runs through both these associative paths, highlighting the family core values as the lynchpin of collective life, both on a personal and a collective (social, business) level. The salience of men, as the carriers of these values, attests to the deeply patriarchal construction of national identity at play here.

The second most prevalent category found in the corpus is *geographic references*, accounting for 23.8% of the presence of national signs in print advertising. According to Edensor (2002), the relationship between national identity and space is vital, and is attested by frequent references to borders, areas and sites of national symbolism, gatherings or household places. After state independence, *geographic references* appear with a low percentage of 15,4% and increase rapidly to a maximum of 39.4%. Then they decline in the 1970s, have a slight increase in the 1980s, and then decline again to 18.7% in the 2000s. This can be explained easily as the 1974 Turkish invasion disoriented the land, flora and fauna and caused confusion as to where the Cypriot landscape begins and ends.

Our data also revealed that in the *geographic location* category, *beach* comprises 90.8% of the category, followed by 4.6% for both *Troodos Mountains* and the *birthplace of Aphrodite (Petra tou Romiou)*. The location of the *beach* is not an official governmental landscape, therefore proving that ". . . the national is evident not only in widely recognised grand landscapes and famous sites, but also in the mundane spaces of everyday life" (Edensor 2002: 37). Cyprus is the third largest island in the Mediterranean Sea and the *beach*, as a location, is inevitably rooted in its culture, associated with nostalgia, childhood memories, relaxation, *family values*, being with friends and having fun. From a tourist perspective, research indicates that Cyprus is promoted primarily as a safe, welcoming, summer, sun-sea-and-sand destination (Sharpley 2001). The location becomes so familiar that while it "constitutes a sense of being in place in most locations within the nation" (Sharpley 2001: 51), it is repeatedly found on postcards, national logos, fashion magazines, national television series, album covers, and in other popular media. Under *geographic references*, however, we need to remember that "[t]hese mundane signifiers are also accompanied by the recognisable forms of flora and fauna which recur throughout most environments" (Edensor 2002: 52).

The *beach* has been present in the advertisements of KEO, an emblematic local beer brand, since the early 1960s. While its association with summer sun and pastime makes the *beach* a favourite motif in tourism, spirits and refreshments adverts, the location is often used to advertise other commodities as well, such as financial services or clothing. Its potential as a semiotic resource is exploited in diverse ways. The Cyprus Tourism Organisation, for example, promotes Cyprus' beaches as a uniquely intense sensual experience: "[f]ew feelings can compare to that of sinking your toes into warm sand . . . of the sun kissing your skin, and your senses taking in the fresh, salty breeze and the endless views of glittering blue waters" (Cyprus Tourism Organisation, *Sun & Sea*). In the advertisement for a Cypriot wine), we find the *beach* used as a place to hold a wedding party. The festive but also erotically charged atmosphere is underscored by the windy, rough sea in the background. Manipulated to look like a romantic painting, with spontaneous brush strokes and an oil effect, the veil of the bride is blown away, revealing a natural sensuality that pairs nicely with the accompanying text. Contrary to this, in the *Bank of Cyprus* advert, we see an elderly man posing for a typical summer vacation souvenir photo, with a bright, golden sand *beach* and a very calm sea as background. The salient presence of an aged person evokes a specific set-up that inspires 'wishful thinking', i.e. it's never too late to do what you always dreamt, enjoying a great summer vacation and surfing. The 1960s Volkswagen van, symbol of hippie culture, heightens the sense of an unyielding youthful spirit, whereas the matching turquoise colour of the board, van, sky, sea and towels multiplies the promise of a sea-side relaxation and paradisiac location. In all cases, the *beach*-experience is highlighted as the characteristic mode of experiencing Cyprus, which thus assumes the character of an endless seaside space, full of light, warmth and the promise of joy, play and sensuality. Although from a geographical point of view it is an ambiguous and fuzzy border that looks outwards into the horizon rather than inwards towards the homeland, the *beach* is in effect transformed into a privileged rhetorical topos for partaking the essence of the national psyche.

The third most popular category of national identity signs in the corpus is *leisure practices*, which comprises 16.9% of all the national symbols found in advertisements. Specifically, 88.2% of ads in this category portray the value of *hanging out with friends*, followed by two quite smaller groups related to *swimming* (7.5%) and *dancing* (4.3%). What is interesting about this category is the fact that the engagement of collective practices *like hanging out with friends* is evident in "a host of popular sites where less formal, directed rituals occur . . . referring to more quotidian homely spaces where national festivities are carried out" (Edensor 2002: 78), such as the living room, the Sunday lunch gathering in the kitchen, the *beach* or the mountains, in the domestic garden or on the veranda, on the streets shopping, and in taverns and coffee shops. Habitual performances and leisure practices of everyday life "persist despite the changes and disruptions. Perhaps if they didn't, shared forms of identity would break down more easily" (Edensor 2002: 88). Socialising and *hanging out with friends* is found in the corpus since 1960, and friendship for Cypriots is found to be almost as essential as *family values* and social bonding. A possible explanation for this is the relatively small population of the island, which creates an effortless familiarity between people. Friends are found in the relatives of other friends or in parents who know each other from before, friends since high school, the army, work, and so on. In general, the society of Cyprus has all the features and norms of a small island; therefore, imagined communities are not so imagined. In fact, they often know each other. Another reason for this strong social bonding among friends and relatives is the constant threat under which the island has lived throughout its history. The recent British rule (until 1960) and the Turkish invasion in 1974, whose forces continues to occupy the island, inevitably brought people together in collective actions for purposes of survival and recovery. In a *KEAN* refreshment advertisement, friendship and caring not only begin at an early age but can by catalysed by sharing an orange drink, another nonverbal sign of national affiliation, as discussed above. The meaning of the specific advert, as articulated explicitly by the linguistic message and the informal handwritten headline, 'The refreshment of Cyprus!' under the logo, regards the promotion of the specific brand as a symbol of national identity, which would make buying *KEAN* products an act of patriotism. In the second example, a group of friends are served juice in a living room, revealing the home environment as a place where social interaction takes place. Moreover, the accompanying headline 'Our friends think that we produce orange and lemon juice in our garden', implies the idea of collectiveness and an imagined community with shared practices. Another everyday Cypriot habit is enjoying good company and having shots of the local white spirit zivania around a table, as shown in, a *Cytamobile* phone advertisement. What is noteworthy in this advert is the coexistence of the older generation with the younger one, under the elegantly confident headline 'We chat with friends'. In what seems to be the garden of a traditional house, we observe an extroverted, joyful scene of socialisation among friends. In the advert, friendship is a timeless notion that has no age barriers.

The fourth category, *national identity symbols*, turned up in 6.7% of ads using national identity signs. While this category begins with a 4.9% presence in the first decade after state independence (1960–1969) and increases to a maximum of 21.2% in the 70s, it then declines in the following decades, reaching a low of 2.2%. Again, the 1974 war brought confusion to a clear definition of official *national identity symbols*, which in addition to Cyprus's entry into the European Union in 2004, and for the same reasons, have almost caused the disappearance of this category from the corpus.

Breaking that down, we find 67.6% fell under the *shape/map of Cyprus* sub-category, 29.7% under the *coat of arms of Cyprus* sub-category, and 2.7% under the *Cypriot flag* sub-category. In the category *national identity symbols*, nonverbal signs tend to be more conventional. They are not as vernacular, as is the case of the other categories we have examined so far, because they are formally defined and acknowledged as the official national language. Since the *flag* of

Cyprus adopts the *map of Cyprus*, as mentioned above, we can argue that the depiction of the island's map/shape is directly associated with the concept of nationalism. For Avraham and First (2003), among all the national symbols used by an imagined community, the flag is the most profound. It is an ancient symbol that is used in modern times "as a mechanism of ontological security ... needed to sustain unifying factors" (Avraham and First 2003: 287).

Most of the time, the *Cypriot flag* is presented along with the European Union and Greek flags, hanging outside cultural and governmental buildings, public schools and army camps, behind politicians, national spokespeople, and during other official ceremonies. Although a national symbol is important as a carrier of patriotism, it also functions as a "metonymic image of banal nationalism ... hanging unnoticed on the public building" (Billig 1995: 8). Along the same lines, the presence of the *shape/map of Cyprus* in the corpus in a number of advertisements in different versions and sizes is at once evident and 'unnoticeable' to viewers. The use of national symbols in advertisements is also a practice that invites consumers to "become involved in consumer nationalism by engaging in acts of consumption of such nationalised products" (Castelló and Mihelj 2017: 6).

Figure 11.2: CYPRIOTmania, 12 November 1972, *Haravgi* newspaper.

In the example of Figure 11.2, titled 'CYPRIOTmania' a product label hanging from a bold headline saying 'CYPRIOTmania' depicts a male figure carrying the *map of Cyprus*. While the main text encourages readers to prefer Cypriot products because of better prices, a circular mark-stamp at the right bottom of the advert asks 'Why xenomania? CYPRIOTmania' to boycott any preferences for foreign products. It is, therefore, obvious that the map, a highly visible and iconic national symbol, is used to cultivate the notion of locality and pride in local products among citizen-consumers and, by extension, pride in the Republic of Cyprus itself. As already mentioned, the shape of the island is also the major graphic element of the flag of Cyprus, and as such is a sacrosanct symbol loaded with national significance and respect. For this reason, and to raise the morale of the people of Cyprus, the map appears most frequently in advertisements right after the 1974 war.

Figure 11.3: Buy Cypriot products 1, 5 October 1975, *Phileleftheros* newspaper.

The two advertising examples in Figures 11.3 and 11.4 are part of an official campaign that was conducted after Turkey's invasion of Cyprus during the summer of 1974, with the purpose of boosting the national economy. In fact, this kind of advertisement falls under the type of political consumer nationalism that fosters ". . . the consumption of particular national products with the aim of affecting a nation's economy" (Castelló and Mihelj 2017: 9) by using emotional appeals to promote the financially destroyed Republic of Cyprus. As we can observe in both cases, there is an illustration of a hand holding the *map of Cyprus* in its palm. This 'lend a hand' drawing is a visual wave for help where the rest of the body is not visible as if it were drowning. While it holds the island to support it, it is also

Figure 11.4: Buy Cypriot products 2, 19 October 1975, *Phileleftheros* newspaper.

a radio signal transmitting an SOS, which takes the form of radio waves transmitted from the centre of the island. The headline in emphasised capital letters 'BUY CYPRIOT PRODUCTS and you're helping your country and honouring your labour' becomes a slogan that will accompany the symbol of the hand whenever it is used, as we can see in Figure 11.4, the second poster in the series. Interestingly, even though it's only a year after the war and 40% of the island is under Turkish occupation, the map is visually portrayed as a whole, filled with local industrial and agricultural products. The visual denial of the island's partitioning reflects the then-widespread prospect of its imminent reunification. The lowercase text above the island states 'Cyprus, our homeland, needs you, your job, your money', reinforcing the image below and vice-versa. The wording of the text, 'our homeland' implies an imagined community to which the readers of the advertisement belong and which can call upon them 'for Cyprus when needed'. In the wake of this campaign, various companies redefined their advertising messages so as to echo its buycott emphases, i.e. emphasised locality and collective-patriotic sentiments. Such is the case with the next advertisement for *Carlsberg* beer, which uses the same tactic, and adopts a combined commercial and political nationalist agenda to increase the company's sales as well as help the national economy.

Figure 11.5: Drink Carlsberg beer, 25 July 1976, *Simerini* newspaper.

In the headline of the Carlsberg beer advertisement (Figure 11.5), the Danish company produced locally too, states that it is 'A Cypriot quality beer that employs hundreds of our compatriots', and asks its readers to 'Drink Carlsberg'. As in the previous example, the phrase 'our compatriots' in the headline invites consumers to join an imagined collective family, while triggering a strong national feeling. The advertisement reproduces the 'Help your country, honour your labour' slogan, which appears next to the hand/radio wave. The proximity of the open hand that holds a map of Cyprus and the sub-heading slogan enhances the implied politically consumer message that buycotts the product for the economic help of the country.

The fifth category in the corpus was *cultural heritage*, which accounts for 6.2% of all the national signs used in advertising in the period. In this category, the visual signs of *Aphrodite, historic monuments, traditional uniforms* and *traditional occupations & industry* featured in 44.1%, 29.4%, 17.6% and 8.8% of the advertisements, respectively. It comes as no surprise that representations of *Aphrodite* dominate this category. According to Karageorghis (n.d.), Aphrodite was mentioned as 'Kypris' and 'Golden Aphrodite' by Homer as far back as the 8th century BC and "[C]yprus has always been considered as the Island of Aphrodite and no other place in the world can boast of being the birthplace of the goddess of love."

As a visual sign, *Aphrodite* has been 'flagging' her daily presence in many official and non-official places, products and services. The official website of the Cyprus Tourism Organisation (CTO) has included her in countless advertisements to promote Cyprus as 'the island of Aphrodite, Goddess of Beauty and Love', and as such, she is also represented as an iconic figure on the logo of the organisation.

Other examples that indicate Aphrodite's presence on official documents include the 20-pound note before Cyprus' entry into the European Union and on various stamps. Her presence in the mass media and in everyday culture is very common, too. While her name is broadcast in the daily news as the gas site *Aphrodite* off the south coast of the island of Cyprus, her figure features in a wide variety of contexts and uses. Just as the *map of Cyprus*, *Aphrodite* is a sign of Cypriotness deeply rooted in Cypriot culture. According to Papadakis (2006: 238), Aphrodite has been used as a cultural symbol by the British, and both Greek- and Turkish-Cypriots, "revealing insights into the island's politics, as they encompass issues of colonialism, nationalism, historiography, gender and migration". Her typical associations as a symbol of beauty and love, or as a national reference, are invoked according to the marketing approach and style of the advertised product or service.

The sixth category examined was *objects*, which appeared in 4% of the corpus. In this category, *sun umbrella* featured in 59.1% of the advertisements, *sunbed* 31.8%, and both *beach-ball* and *objects made out of clay* 4.5%, respectively. The placement of *objects* at home, in public spaces or in the work environment has numerous cultural dimensions that are relevant to the manifestation and perpetuation of national identity. *Objects* organise individual and collective narratives around them that anchor people to a place. Because of their historical and geographical context, *objects* are also invested with powerful connotations of shared stories that allude to a common past. The common consumption of artefacts triggers nostalgia and collective memories. This is how *objects* enter the shared national biography. For example, if a red phone box is a sign of Britishness, or a tulip a symbol of Dutchness, then a *sun umbrella* was found to be the most popular cultural symbol of Cypriotness. The *sun umbrella* is an object that is found in almost every Cypriot house and is used on the typical Sunday family visit to the *beach* during the summer or while having afternoon *coffee* on the veranda or in the home garden.

In a *KEO* Beer advert and the *Lanitis Juice* adverts, the *sun umbrella* is as important an element as the *beach* itself, or as the *citrus fruits* and *family values* discussed earlier. An object that offers protection to the family, it also becomes part of the visual identity of national life.

Finally, the seventh and last category examined, *food and drinks*, featured in 4% of advertisements in the corpus (see Table 11.9). The values in this category were *coffee, traditional food, traditional products, fish* and *wine*, accounting for 27.3%, 13.6%, 9.1% and 4.5% of sign representations, respectively. The consumption of *food and drink* is not only a physical need, it is a social habit, and as something shared by fellow citizens, it can also be seen as a practice that instils a sense of national belonging.

In the case of a *Jacobs Coffee* advertisement, we can observe how *coffee* is a cultural sign that, according to its headline, has a 'real coffee aroma that awakens the senses'. The 'aroma' is written in lowercase brown letters running in a wavy line as if leaving the coffee cup and imitating the smell of coffee. A smaller, black and white photograph of a happy couple in the background seems to interpret what the smell of *coffee* brings to the mind of the female coffee drinker.

5 Conclusions

The findings of the quantitative content analysis identified the kinds of national identity signs that were found in the images of commercial print advertisements in the Republic of Cyprus from state independence in 1960 to 2010, as well as which of those were most prevalent in the overall corpus. They have also been influenced in a similar way by socio-political changes like the establishment of the Republic of Cyprus in 1960, the 1974 Turkish invasion and the country's entry in the European Union in 2004. While their advertising presence is high after state independence, it reaches its low in the 1970s, and ascends in the decades that follow. Such findings suggest that Cypriot citizens redefine their identity during socio-political turning points so that to compete the challenges of a new state independent status, the European family and the international competitiveness of globalisation and tourist attractiveness.

Social relations and values, particularly *family values*, as well as *leisure practices*, specifically *hanging out with friends*, were the two most popular national identity signs in the corpus. Contrary to this situation are the national identity signs categorised under *geographic references* and *national identity symbols* whose low presence after state independence reaches a high in the 1970s, and then decreases in the decades that follow. This shows that human-centred values are more likely to be portrayed after critical historical events rather than more impersonal values, such as *geographic references* and *national identity symbols*.

'Claiming nationality' is commonly observed in many advertisements for products or services right after the end of British rule in August 1960. A political consumer nationalism alongside 'buycotting' techniques are found in the discourse of governmental advertisements after the 1974 Turkish invasion, whereby the promotion and consumption of domestic products are used to support the domestic industry and create more local jobs. This nationalisation of commercial products, which was aimed at increasing sales as well, is a marketing strategy that falls under the broader umbrella of economic nationalism.

This paper is part of a wider study on Semiotics of national identity in advertising communication, and the case of Cyprus from 1960 to 2010. The overall research incorporates tools for in-depth semiotic analysis of verbal and nonverbal signs on print advertisements. By doing so, it identifies what additional cultural meanings are found in advertisements of official institutions, such as the Cyprus Tourism Organisation and Cyprus Airways, as well as compares them to the advertisements of two local non-state companies, KEO Beer and Laiko Kafekopteio (People's coffee). In addition, it provides a critical analysis and comparison between quantitative and qualitative findings, as well as an investigation and interpretation of how the presence of national identity signs changes over time, in relation to the product categories that are depicted during the aforementioned period.

References

Anderson, Benedict. 2006. *Imagined Communities*. New-York, London: Verso.
Avraham, Eli & Anat First. 2003. 'I Buy American': The American Image as Reflected in Israeli Advertising. *Journal of Communication* 53(2). 282–299. https://doi.org/10.1111/j.1460-2466.2003.tb02591.x.
Avraham, Eli & Daniel Daugherty. 2012. Step into the Real Texas: Associating and claiming state narrative in advertising and tourism brochures. *Tourism Management* 33(6). 1385–1397. https://doi.org/10.1016/J.TOURMAN.2011.12.022.
Bayar, Murat. 2009. Reconsidering primordialism: An alternative approach to the study of ethnicity. *Ethnic and Racial Studies* 32(9). 1639–1657. https://doi.org/10.1080/01419870902763878.
Bell, Philip. 2010. Content Analysis of Visual Images. In Theo van Leeuwen & Carey Jewitt (eds.), *Hanbook of Visual Analysis*, 11–34. London: Sage.
Billig, Michael. 1995. *Banal nationalism*. London: Sage.
Bulmer, Sandy & Margo Buchanan-Oliver. 2010. Experiences of brands and national identity. *Australasian Marketing Journal* 18(4). 199–205. https://doi.org/10.1016/j.ausmj.2010.07.002.
Castelló, Enric & Mihelj Sabina. 2017. Selling and Consuming the Nation: Understanding Consumer Nationalism. *Journal of Consumer Culture* 18(4). 558–576. https://doi.org/10.1177/1469540517690570.
Cyprus Tourism Organisation, CTO. *Birthplace of Aphrodite – Petra Tou Romiou*. http://www.visitcyprus.com/index.php/en/discovercyprus/culture-religion/sites-monuments/item/250-birthplace-of-aphrodite-petra-tou-romiou. Accessed 8 Aug. 2017.
—. *Sun & Sea*. http://www.visitcyprus.com/index.php/en/discovercyprus/sun-sea. Accessed 26 Feb. 2018.

Edensor, Tim. 2002. *National Identity, Popular Culture and Everyday Life*. Oxford: Berg.
Goldman, Robert. 2011. *Reading Ads Socially*. New-York: Routledge.
Hogan, Jackie. 1999. The construction of gendered national identities in the television advertisements of Japan and Australia. *Media, Culture & Society 21*(6). 743–758. https://doi.org/10.1177/016344399021006003.
Karageorghis, Jacqueline. *Aphrodite Cultural Route*. http://www.visitcyprus.com/index.php/en/discovercyprus/culture-religion/cultural-routes/item/288-aphrodite-cultural-route. Accessed 10 Mar. 2018.
Kassarjian, Harold H. 1977. Content Analysis in Consumer Research. *Source Journal of Consumer Research* 4(1). 8–18, http://www.jstor.org/stable/2488631.
Papadakis, Yiannis. 2006. Aphrodite delights. *Postcolonial Studies* 9(3). 237–250, https://doi.org/10.1080/13688790600824963.
Pavlou, Y. Pavlos. 1992. *The use of the Cypriot-Greek dialect in the commercials of Cyprus Broadcasting Corporation*. Retrieved from https://files.eric.ed.gov/fulltext/ED371617.pdf.
Plakoyiannaki, Emmanuella & Yorgos Zotos. 2009. Female role stereotypes in print advertising: Identifying associations with magazine and product categories. *European Journal of Marketing, Vol. 43*(11/12). 1411–1434.
William, Leiss, Stephen, Kline & Jhally Sut. 1990. *Social Communication in Advertising: Persons, Products & Images of Well-Being*. 2nd edn. London & New York: Routledge.
Library of Congress. *Agriculture*. http://countrystudies.us/cyprus/37.htm. Accessed 11 Dec. 2017.
Peat, Jennifer. 2002. *Health Science Research: A Handbook of Quantitative Methods*. London: Sage.
Pritchard, Annette & Nigel J. Morgan. 2001. Culture, identity and tourism representation: marketing Cymru or Wales? *Tourism Management* 22(2). 167–179.
P.I.O. *Cyprus Mouflon*. http://www.aboutcyprus.org.cy/. Accessed 11 Dec. 2017.
Rogers, L. Edna & Valentin Escudero (eds.). 2015. *Relational Communication: An Interactional Perspective To the Study of Process and Form*. 1st edn. London & New York: Routledge.
Sharpley, Richard. 2001. *Tourism in Cyprus: Challenges and Opportunities*. doi:10.1080/14616680010008711.

Elena Negrea-Busuioc, Diana Luiza Simion
What's in a nickname? Form and function of sports' team nicknames

Abstract: The aim of this paper is to examine the morphological and semantic structure of nicknames assigned to 50 of the most valuable sports teams in the world, as ranked by Forbes in 2017. Furthermore, the paper seeks to discuss the function of team nicknames as mechanisms for community identification of both team members and fans. We argue that the nicknames assigned to sports teams fulfill a referential, descriptive function rather than an evaluative one, and that the nicknaming mapping practices reveal the dynamic nature of sports culture. Sports team nicknames are a valuable symbolic resource of the team ethos that define and reshape the social imagery associated with the team. Furthermore, they serve as a mirror in which the whole team community experience and history are reflected.

Keywords: nicknames, nicknaming practices, sports teams, sports brand community

1 Introduction

Assigning nicknames to athletes and teams is a common practice in many sports, especially in popular team sports known to bring together wide and active fan communities, such as soccer, baseball, American football or basketball. Nicknaming is an important identity component of the sports ethos, acting as a symbolic resource that athletes, teams and fans turn to when defining and differentiating themselves from others. Nevertheless, scholarly literature on nicknaming practices in sports has focused mainly on inventorying and categorizing nicknames of athletes (Skipper 1984; Wilson and Skipper 1990; Kennedy and Zamuner 2006) rather than teams. In this paper, we seek to examine the structure and semantic function of nicknames assigned to the world's 50 most valuable sports teams, as ranked by Forbes (2017), arguing their social meaning and branding potential.

The analysis of the corpus revealed a nickname-mapping pattern, comprising more nicknames used either separately or concomitantly, rather than a single nickname associated with one team. According to their structure, the analyzed

---

Elena Negrea-Busuioc, Diana Luiza Simion, National University of Political Studies and Public Administration (SNSPA), Bucharest, Romania

https://doi.org/10.1515/9783110662900-013

nicknames are formed by shortening (e.g. *Seattle Seahawks > The Hawks*), abbreviating (e.g. *Minnesota Vikings > The Vikies*) or indexing (e.g. *Boston Celtics > The C's*) the official name of the team. Based on their origin and meaning, the nicknames were classified into categories related to color (e.g. *Manchester City > Blues*), toponyms (e.g. *New York Mets > The Metropolitans*), team behavior (e.g. *Baltimore Ravens > Death on Wings*) and animal imagery (e.g. *St Louis Cardinals > The Redbirds*). All these build on brand identity elements and their symbolic value, whether they refer to the origin, brand heritage, history or visual identity (i.e. mascots, crest) of the team. Along with the consistency of the nicknaming strategies, we can also notice the power they have in activating a broader imagery build around brand team identity. Thus, we aim to assess the potential of nicknames not only as a descriptive function applied to brand identity, but also as a symbolic and expressive identity function, fostering a strong sense of belonging to the team community.

This chapter begins with a brief presentation of previous research on nicknames and nicknaming practices, followed by an account of the symbolic power of nicknames in building a sense of belonging within sports communities. We argue that nicknaming practices in sports teams are a valuable and powerful tool for enhancing team identification and regulating group membership. In the fourth section, a corpus-based analysis of the form, structure and function of the nicknames assigned to the world's 50 most valuable sports teams (as ranked by Forbes, 2017) will be performed. Based on the nicknaming patterns identified and analyzed, we then discuss the role of nicknames for the team's brand identity and question the practical implications of nicknaming for sports marketing and communication.

2 Structure and function of nicknames

The name is an essential tool for identification and classification that people rely on to make sense of the social world (Levi-Strauss 1966). Unlike the institutionalized procedure that governs names and naming, nicknaming practices are optional, transient, fluid and, more importantly, sensitive to the cultural context in which they appear (McDowell 1981; Holland 1990; de Klerk and Bosch 1997; Kennedy and Zamuner 2006).

Since they are optional linguistic categories (Wierzbicka 1992), nicknames are more likely to depend on changes and transformations in the culture and the society in which they appear. Nicknames are given, "unofficial" names whose choice is strongly influenced by context and social relationships. Research has shown that nicknaming practices are frequently used in gangs (Zaitzow 1998), in politics (Adams 2009), the army (Potter 2007), in school (Crozier and Dimmock 1999; Chevalier 2004; Starks and Taylor-Leech 2011), in music (Skipper 1986a; Skipper and Leslie 1988) in sports (Skipper 1984; Kennedy and Zamuner 2006), in online communities and gaming (Alderman 2009; Hagström 2012) or in mobile phone user communities (Persson 2013).

Many nicknames show high degree of semantic transparency as they tend to relate to their bearer's physical characteristics (e.g. *Skinny, Shorty, Beanpole, Big Ears, Fat Cat, Blondie*) or to their personal habits and mental traits (e.g. *Couch Potato, Rat, Brainiac, Geek, Nerd*). Some nicknames give clues about the ethnic background of their bearers (e.g. *Jewboy, Aussie)*, while others may be based on word and rhyme play (e.g. *Leonard > Len, Smith >Smittie, G.W. Bush > Dubya*) or hypocoristic versions of bearers' personal or family name (e.g. *Ashleigh > Ash, Jennifer Lopez >J Lo*). Nevertheless, as previous studies have shown, a wide range of nicknames is derived based on an indirect way of nomination, and, irrespective of their positive or negative connotation, their structure is suited to the context of their usage (Kennedy and Zamuner 2006).

In addition to their identity and classificatory functions, nicknames usually fulfill a richer array of functions and roles that available literature, especially based on sociological and anthropological research, has identified and described. They are thus seen as a dynamic process and output of the interaction between language and culture, requiring an interdisciplinary research approach (Garayeva, Akhmetzyanov and Khismatullina 2016: 103–109). For example, nicknames are used to build and consolidate group membership and cohesion. In many cultures and societies, nicknames have a boundary-maintaining function (Gilmore 1982), serving as "instruments of social solidarity" (Holland 1990: 258). In an interview-based analysis of the prevalence of nicknames among coal miners in West and Southwest Virginia, Skipper (1986b) discovered that nicknaming is a common practice among the interviewed coal miners and that it is indicative of work group solidarity, which is crucial for survival in a coal mine (Skipper, 1986b: 145). Even though many of the assigned nicknames were not at all complimentary (many of the nicknames reported by the interviewees had negative or sexual connotations, e.g. *Flat Chest, Big Balls*), miners used them (but only inside the mine) because being nicknamed is an important and honored tradition. Thus, nicknames among coal miners become a symbol of work group acceptance and membership.

There are, however, examples of nicknaming practices that reinforce individuality and autonomy, rather than solidarity. In a study of community nicknaming in Spain, Gilmore (1982) remarks that nicknames are used to deride and minimize, they infantilize and trivialize by distorting the name of the bearer. According to Gilmore (1982: 687), nicknaming in the Mediterranean area are a form of verbal aggression. In the Fuenmayor community in Southwest Andalusia, nicknames do not promote or reinforce solidarity and group coherence; on the contrary, they have the opposite effect, symbolizing rejection by the community and stigmatization.

Nicknames are also used as a form of social control. Chevalier (2004) shows that nicknames can single out individuals both in favorable, positive (e.g. praising the bearer for his/ her achievements) and negative contexts (e.g. highlighting a mistake made by the bearer or disapproving the bearer's behavior). Adults, parents in particular, use nicknames to reinforce norm on their children by targeting their unacceptable behavior. For example, a mother invented the nickname *O Great Salami* to point out the bossiness of one of her children (Chevalier 2004: 130). This nickname is derived from an affectionate nickname of the child, namely *Saus*, which is short for *Sausage* (salami being a hyponym of sausage).

Another important function of nicknames is to evoke affection and friendship (Lawson and Roeder 1986; Wierzbicka 1986). Affectionate abbreviatory names (usually formed by adding a diminutive suffix to the bearer's first name) are often coined and used in intimate and family contexts. Wierzbicka (1986; 1992) shows that the use of nickname (e.g. *Jacqueline* > Jackie) or an affectionate nickname (e.g. *Jacqueline> Jacks*) expresses positive warmth and may imply affectionate in-group closeness.

Sometimes, nicknames are used to indicate social hierarchy and power relations within a group (Holland 1990). Nicknaming may be a seen as an aggressive form of gaining dominance over others in competitive contexts. Nicknames can also be used to reinforce stereotypes and to influence perceptions of others (de Klerk and Bosch 1997), because of the semantic loading of some nicknames (e.g. *Jewboy, Japs, Nazis*).

3 Nicknames as an identity marker in sports communities

Group identification and differentiation are key aspects that make a field resourceful in terms of nicknaming practices. Given that "questions of identity and identification are of critical importance both for the routine functioning of sports and for some of the problems recurrently generated in connection with them" (Dunning 1999: 3), it is not surprising that nicknames are an important identification mechanism supporting the 'we-ness' feeling within sport communities.

Nicknaming practices are quite popular in sport-related contexts. Sports narrative is marked by the frequent use of nicknames carrying a wide range of social and semantic functions (Skipper 1984; Skipper 1990; Holland 1990; Kennedy and Zamuner 2006). According to Mashiri (1999: 99), most of the nicknames are used and popularized by the teammates, the fans and the soccer commentators. Moreover, they tend to be intensely used in competitive situation, when differentiating from other teams and/or fans and, thus, activating the "us" vs "them" scenario.

Skipper (1986a; 1990) noticed that shifting patterns in public nicknaming – sports included – seemed to be informed by changes in the society in which they occurred. In a series of studies focusing on nicknaming practices for football and baseball players, the author remarks that a large proportion of place nicknames comes from the South, probably reflecting a sense of Southern identity and solidarity (Skipper 1990). Furthermore, the high frequency of nicknames assigned to baseball players is indicative of the closeness that the public felt with the respective athletes. In the sports field, one can notice an interesting phenomenon of coexistence and blending between individual and team nicknaming. However, the two aspects have been examined separately in some of the previous studies on this topic. On the one hand, when it comes to athletes' nicknames, it seems that the main nicknaming patterns and strategies are based on: *behavior traits (i.e.* Marvin Hagler > *Marvelous*, Marshawn Terrell Lynch > *Beast Mode*, Carles Puyol > *Lionheart)*, which are frequently used to warn against an unacceptable mode of behavior or dispute any allegation leveled against an individual member of the team (Babane and Chauke 2015); *physical aspect* (i.e. Maurice Jones-Drew> **Pocket Hercules,** Bartolo Colon> **Big Sexy,** Johnny Dickshot > **Ugly)**; and *field of play* – nicknames that are directly linked to the sports itself and athlete-role "in action", routed in moments, sport performances or playing style that individualize (and, at the same time, distinguish them from others) athletes in the field (i.e. **Michael Jordan > Air Jordan,** Usain Bolt > **Lightning Bolt,** Chauncey Billups > **Mr. Big Shot).** The latter is the most resourceful nicknaming practice in terms of branding potential, serving both as a reflection of the performance expectations and as the athlete's "signature". Moreover, in addition to their symbolic or referential function, many these nicknames are based on humorous analogies or puns, thus fulfilling an entertainment function as well.

On the other hand, when speaking about team nicknames, these can be considered part of the wider brand identity repertoire (as developed by the team itself or its fan community), but also evaluative labels assigned by an outgroup name-giver, and carrying pejorative, racist or ethnic overtones (Jensen 1994; Williams 2006). Nevertheless, nicknames assigned to a team across time and in different social and cultural contexts could be considered as capsule history of that team. Varying nicknames across time select and amplify some features of the team, its sport or managerial performance. Whether they overlap for a while, become dominant or are highly ephemeral and die after only one occurrence, nicknames of a team are very contextual and dynamic in usage.

In this chapter, we focus on the nicknaming practices for sports team by means of a three-layered analysis comprising the morphological and semantic structure (i.e. source domain, ways in which nicknames are built, their literal meaning), the team nickname dynamic discursive repertoire or mappings (i.e. nickname "family"), and the role and potential the nicknames have in terms of brand identity communication.

4 Nicknaming practices for Forbes world's best 50 sports teams. A corpus-based identification and classification of nicknames

The corpus and research questions

Our analysis of the nicknames for teams is a corpus-based investigation of the current and past nicknames assigned to the world's most valuable sports teams in 2017, according to Forbes annual franchise valuations. The corpus comprises 174 nicknames for well-known teams from American football and baseball (accounting for 76% of the corpus), basketball and soccer. As expected by the distribution of sports performed by the analyzed team, the number of team outside the U.S. included in the ranking is very low; only 7 out of the 50 teams originate from Europe and all seven are soccer teams: four from the UK (Manchester United, Manchester City, Arsenal and Chelsea) two from Spain (Barcelona and Real Madrid) and one from Germany (Bayern Munich). The nicknames attributed to the Spanish and German teams, respectively, were collected in Spanish and German (English translations for each identified nickname were provided).

The aim of the corpus-based analysis was to identify and classify the nicknames that the sports teams examined have had across time and to reveal the nicknaming practices that have shaped the team identity. To this purpose, the following research questions were formulated:

RQ1: How are sports teams nicknames morphologically built?
RQ2: What are the sematic associations evoked by the team nicknaming practices?
RQ3: What do the nicknaming practices patterns of sports teams reveal about sports community and brand identity?

To answer these questions, we draw on the results of previous research on nicknaming practices in sports (Skipper 1990; Kennedy and Zamuner 2006). Particularly, we build on the morphological model for professional hockey and baseball athletes nicknames assignment developed by Kennedy and Zamuner (2006). The model includes both Homeric and hypocoristic nicknames used as referring expressions in sports. On the one side, Homeric nicknames are mainly meant to be descriptive, serving a referential function; they often involve loose wordplay on the giver (athlete's) formal name, metaphorical imagery, toponyms or reference to personal traits (e.g. *Ted Lindsay > Terrible Ted*; *Helmuts Balderis > The Riga Express*; *Claire Alexander > The Milkman*, Kennedy and Zamuner 2006: 402–403). Hypocoristic nicknames, on the other side, function as both referential and addressing expressions, adding an evaluative dimension to the nickname's referential function; they often involve truncation (clipping) and creative suffixation (e.g. *Donald Brashear > Brash; Chris Dingham > Dinger; Mark Potvin > Potsie*, Kennedy and Zamuner 2006: 404–405). Finally, to capture the complex phenomenon of nicknaming, we propose a four category-based analysis of semantic associations manifest in sports team nicknaming practices.

Nickname identification and inventory

For each of the 50 sports teams included in the ranking, a thorough search on the internet was performed to identify and inventory the nicknames assigned to the team over time. For the majority of the 50 teams in the list, the history of the team available on its website was used to identify the nicknames; where team website was not found, the additional resources were used (e.g. the website of the national X sports association, Wikipedia, online clubs' communities). None of the identified nicknames were discarded from the analysis, with the exception of duplicates. Thus, a total of 174 nicknames for the 50 teams in the Forbes list were used. We illustrate below some examples of sports teams and their nicknames included in the corpus.

(1) Sports: American football
   Team: The New York Giants
   Nicknames: *G-Men, Jints, Big Blue, Big Blue Wrecking Crew* (1986), *Crunch Bunch* ('80s), *Midgets, Dwarfs, Ants, Gigantics*

(2) Sports: Basketball
   Team: Boston Celtics
   Nicknames: *Celtics, The Green, The C's, C-Green Smash Machine*

(3) Sports: Soccer
   Team: Arsenal
   Nickname: *The Gunners*

Nickname classification

The 174 nicknames were further classified according to their morphological and semantic characteristics into seven categories in total (three morphological categories and four semantic categories). First, we discuss the morphological nicknaming practices identified in the analyzed corpus. The hypocoristic nicknames identified were mainly constructed by clipping, with some fewer examples involving suffixation.

1. Hypocoristic nicknames based on clipping, with the resulting truncation consisting of one syllable plus the definite article
   Examples:
   (1) Seattle Seahawks > *The Hawks*
   (2) St Louis Cardinals > *The Cards*
   (3) New England Patriots > *The Pats*
2. Hypocoristic nicknames based on clipping plus suffixation
   Examples:
   (4) Minnesota Vikings > *The Vikies*
   (5) New York Mets > *The Metsies*
   (6) Dallas Cowboys > *Dalloos*

Apart these two categories in which most of the identified nicknames were included, the analysis also revealed two hybrid categories based on clipping, which comprised significantly fewer nicknames.

3. Hypocoristic nicknames based on clipping, with the resulting truncation consisting of an initial plus/ minus definite article
   Example:
   (7) Boston Celtics > *The C's*
4. Hypocoristic nicknames based on clipping, with the resulting truncation consisting of an initial plus modifier
   Example:
   (8) Dallas Cowboys > *Big D*
   (9) New York Giants > *G-Men*

Furthermore, as far as the semantic association evoked by nicknaming practices used to refer to the analyzed sports teams, as mentioned earlier, we have classified the identified nicknames according to four categories, as follows:

1. Homeric nicknames based on color
   Examples:
   (1) Manchester United > *The Red Devils*
   (2) Real Madrid > *Los Blancos (The Whites)*
   (3) Boston Celtics > *The Green*
   (4) Bayern Munich > *Die Roten (The Reds)*

2. Homeric nicknames based on animal imagery
   Examples:
   (5) St Louis Cardinals > *The Birds, The Redbirds*
   (6) New York Giants > *Ants*
   (7) Atlanta Falcons > *The Dirty Birds*
   (8) Philadelphia Eagles > *The Eagles*
   (9) Chicago Bulls > *The Bulls*
3. Homeric nicknames based on place-names, toponyms
   Examples:
   (10) Los Angeles Clippers > *Lob City*
   (11) New York Mets > *The Metropolitans*
   (12) Pittsburg Steelets > *Blitzburgh*
   (13) San Francisco 49ers > *The Village People, Earthquakes*
4. Homeric nicknames based on team performance and behavior
   Examples:
   (14) Arsenal > *The Gunners*
   (15) New York Yankees > *The Bombers*
   (16) Washington Redskins > *The Deadskins*
   (17) Dallas Cowboys > *Doomsday Defense*

What is interesting to notice is the use of the definite article "the" whose role as an identifier and determiner does not serve a referring function, but rather a delimitation or a categorization function: the reference is made not to Bombers, but to "*The* Bombers"; not just Eagles, but "*The Eagles*". Together with the implicit use of capital letters to convert common nouns into proper nouns *(i.e. The Bulls*, not *the bulls; The Dirty Birds*, not *the dirty birds)*, this practice goes beyond the referential function of nicknaming, serving as an identification and categorization mechanism, too. Moreover, it involves the idea of a community built around the team, on more informal grounds. This type of brand community is even more inclusive, leveling down the differences between its member, since it is no more about Manchester United players, Manchester United fans, Manchester United coach etc., but about The Red Devils, no matter their role or position within the brand community.

In addition to the mere name of the club, one of the key sources used as input for nicknaming is the crest, both in terms of colors and symbols (e.g. Arsenal > *The Gunners;* Philadelphia Eagles > *The Eagles;* Manchester United > *The Red Devils)*. While the first three semantic categories rely on a simple and rather explicit nicknaming process (i.e. color, animal imagery and toponymy), the last one – team performance/ behavior – goes for less explicit analogies or metaphors, which require a higher level of team-related knowledge, as they refer to specific performances or playing style. This, in turn, makes the resulting nicknaming more resourceful in terms of brand narratives, as they entail a story explaining the nickname assignment; nicknames in this category are nor as self-explanatory as the ones based on color, animal or place name references.

One final aspect that we want to address in our analysis is what we called the "nicknaming map" built around a team. We rarely speak about a one-to-one correspondence between team name and team nickname. We can find distinct nicknames given to the same team, as we have previously seen. The meaning of this nicknaming map goes further than a simple random enumeration of nicknames. First, based on the frequency and spread of their use, we can distinguish between dominant, secondary and occasionally nicknames. While some can be very generic, wildly accepted and well know, others are more local or context-dependent, or even "dead" nicknames (e.g. Chicago Bulls > *The Bulls, Broncos, Orange Crush*) linked to different moments in the team's history but serving as a function of time rather than as a function of usage (Kennedy and Zamuner 2006). Secondly, based on their naming pattern, they reflect different aspects of the team imagery, coexisting and complementing each other (e.g. Boston Celtics > *Celtics, The Greens, The C's, C-Green Smash Machine*). Finally, the naming map could be seen as a proxy to a team's identity card, as it brings together core-elements of the brand identity (i.e. color, symbols, name), as well as elements from its history and sport spirit. This, in turn, makes it a resourceful brand community anchor that both internal and external stakeholders can relate and identify with.

5 Conclusions

Sports nicknames serve as important tools for identification and classification, being deeply embedded culturally and carrying a wide range of social and semantic functions. This chapter went from identifying the morphological and semantic nicknaming patterns to a more inclusive and dynamic overview of the nicknaming map patterns and its branding potential. A first aspect that we noticed was that, when it comes to teams' nicknaming, there is a dominant Homeric approach, serving rather a referential, descriptive function than an evaluative one. Team nicknames are mainly leveraging the social imagery embedded in the name, crest (i.e. colors, symbols) and origin (toponyms), thus acting as powerful identity resources. This comes to confirm previous studies (Kennedy and Zamuner 2006) on nicknaming patterns. However, this study adds to the previous work in this field insights into the way in which nicknaming practices contribute to building a sense of belonging to brand community.

The Bulls, The Blues, The Redbirds, The C's, The Dolphins speak as much about the brand identity of a club, as they speak about an entire community that is built around it. Nicknaming allows for a more informal and inclusive identification process, a form of appropriation though the "given" nickname that leverage the power of the sport brand but creates a sense of familiarity and intimacy among the community members. Whether it is based on symbolic analogies or on pun playfulness, team nicknames emphasize and consolidate the idea of community, of a "we-ness" surrounding the team. Therefore, from the use of the capitalization of common nouns, to the use of the definite article "the" and the plural form, the nicknaming patterns highlight the uniqueness and distinctiveness of the collective identity they refer to.

Furthermore, the structure and dynamics of the nicknaming reveal the richness in declinations and combinations of both the morphological and semantic associative patterns. Nicknaming expressivity is closely related to the semantic associations evoked, as they engage symbolic resource form the team ethos (i.e. history, place, colours, etc.). Therefore, based on its symbolic meaning and community-identification value, nicknaming constitutes a resourceful communication asset, allowing for both relevant, as well as creative ways of addressing, connecting and mobilizing sports team communities around a team-related message or issue. As the role of brand communities has increased and has been redefined by the new online spaces and practices, further research in this direction could contribute valuable insights to both sports and marketing specialists.

References

Adams, Michael. 2009. Power, politeness, and the pragmatics of nicknames. *Names* 57(2). 81–91. https://doi.org/10.1179/175622709X436369.

Alderman, Derek H. 2009. Virtual place-naming, internet domains, and the politics of misdirection: The case of www.martinlutherking.org. In Jani Vuolteenaho & Lawrence D. Berg (eds.), *Critical Toponymies: The Contested Politics of Place Naming*, 267–283.

Babane, Thembhani Morris & Thomas M. Chauke. 2015. An Analysis of Factors Involved in Nicknaming of Some Soccer Players in South Africa. *Anthropologist*, 20(3). 780–787. https://doi.org/10.1080/09720073.2015.11891785.

Badenhausen, Kurt. 2017. Full List: The World's 50 Most Valuable Sports Teams 2017. Forbes. com, Jul 12, 2017. https://www.forbes.com/sites/kurtbadenhausen/2017/07/12/full-list-the-worlds-50-most-valuable-sports-teams-2017/#2561318b4a05.

Chevalier, Sarah. 2004. Nicknames in Australia. *Bulletin VALS-ASLA* 80. 125–137.

Crozier, W. Ray & Patricia S. Dimmock. 1999. Name-calling and nicknames in a sample of primary school children. *British Journal of Educational Psychology* 69(4). 505–516. https://doi.org/10.1348/000709999157860.

De Klerk, Vivian & Barbara Bosch. 1997. Nicknames of English adolescents in South Africa. *Names* 45(2). 101–118. https://doi.org/10.1179/nam.1997.45.2.101.

Dunning, Eric. 1999. Sport matters: Sociological studies of sport violence and civilization.

Dunning, Eric. 2013. *Sport matters: Sociological studies of sport, violence and civilisation*. London and N.Y.: Routledge.

Garayeva, Almira K., Ildar G. Akhmetzyanov & Lutsia G. Khismatullina. 2016. The Significance of Learning Nicknames of Public Figures in Modern English and American Language Models of the World. *International Journal of Environmental and Science Education* 11(4). 10337–10345.

Gilmore, David D. 1982. Some notes on community nicknaming in Spain. *Man* 17 (4),new series. 686–700. https://doi.org/10.2307/2802040.

Hagström, Charlotte. 2012. Naming me, naming you. Personal names, online signatures and cultural meaning. *Oslo studies in language* 4(2). 81–93. https://doi.org/10.5617/osla.312.

Holland, Theodore J. 1990. The many faces of nicknames. *Names* 38(4). 255–272. https://doi.org/10.1179/nam.1990.38.4.255.

Jensen, Robert. 1994. Banning 'redskins' from the sports page: The ethics and politics of native American nicknames. *Journal of Mass Media Ethics* 9(1). 16–25. https://doi.org/10.1207/s15327728jmme0901_2.

Kennedy, Robert & Tania Zamuner. 2006. Nicknames and the lexicon of sports. *American Speech* 81(4). 387–422. https://doi.org/10.1215/00031283-2006-026.

Lawson, Edwin D. & Lynn M. Roeder. 1986. Women's full first names, short names, and affectionate names: A semantic differential analysis. *Names* 34(2). 175–184. https://doi.org/10.1179/nam.1986.34.2.175.

Levi-Strauss, Claude. 1966. *The savage mind*. Chicago: University of Chicago Press.

Mashiri, Pedzisai. 1999. Terms of address in Shona: A sociolinguisticapproach. *Zambezia*, XXVI (i). 93–110.

McDowell, John H. 1981. Toward a semiotics of nicknaming the Kamsá example. *Journal of American folklore* 94(371). 1–18. https://doi.org/10.2307/540773.

Persson, Marcus. 2013. "Loli: I Love It, I Live with It": Exploring the Practice of Nicknaming Mobile Phones. *Human IT: Journal for Information Technology Studies as a Human Science* 12(2). 76–107.

Potter, Terrence M. 2007. USMA Nicknames: Naming by the Rules1. *Names* 55(4). 445–454. https://doi.org/10.1179/nam.2007.55.4.445.

Skipper, James K. 1984. The sociological significance of nicknames: The case of baseball players. *Journal of Sport Behavior* 7(1). 28–38.

Skipper Jr, James K. 1986a. Nicknames, folk heroes and Jazz musicians. *Popular Music & Society* 10(4). 51–62.

Skipper, James K. 1986b. Nicknames, coal miners and group solidarity. *Names* 34(2). 134–145.

Skipper, James K. 1990. Placenames Used as Nicknames: A Study of Major League Baseball Players. *Names* 38(1–2). 1–20.

Skipper, James K. & Paul, L. Leslie. 1988. Women, nicknames, and blues singers. *Names* 36 (3–4). 193–202.

Starks, Donna & Kerry Taylor-Leech. 2011. Research project on nicknames and adolescent identities. *New Zealand Studies in Applied Linguistics* 17(2). 87–97.

Wierzbicka, Anna. 1986. Does language reflect culture? Evidence from Australian English. *Language in Society* 15(3). 349–373. http://www.jstor.org/stable/4167767.

Wierzbicka, Anna. 1992. *Semantics, culture, and cognition: Universal human concepts in culture-specific configurations*. Oxford: Oxford University Press.

Williams, Dana M. 2006. Patriarchy and the 'Fighting Sioux': A gendered look at racial college sports nicknames. *Race Ethnicity and Education* 9(4). 325–340. https://doi.org/10.1080/13613320600956621.

Wilson, Brenda S. & James K. Skipper. 1990. Nicknames and women professional baseball players. *Names* 38(4). 305–322. https://doi.org/10.1179/nam.1990.38.4.305.

Zaitzow, B. H. 1998. Nickname usage by gang members. *Journal of Gang Research* 5(3). 29–40.

Constantin Popescu
Architecture and painting codes in *the Annunciation*. Oltenia (XVIIIth–XIXth centuries)

Abstract: The rural churches in the northern Oltenia built in the XVIIIth and XIXth centuries by small nobles, priests and peasant communities move off architecturally and iconographically from the medieval models. There are neither materials, nor people able to keep their beauty. Nor financial means. The painters, many of them peasants, do not minutely understand the coherence of the iconographic programs, therefore deviate from them. This paper examines the interpretations given to the *Annunciation* in some churches of the Vâlcea County, on their *eastern exterior walls* (*pertaining to the sanctuaries*). The constructors and painters, most of them peasants, tried different solutions to harmonise codes of architecture and fresco painting, sometimes – the *Annunciation* proves it – with notable results.

Keywords: rural churches, religious architecture, Oltenian Annunciations, iconic images, visual communication

The churches erected at the end of the XVIIIth and the beginning of the XIXth centuries in the villages and small towns of Oltenia – part of Romania bordered by the Carpathians at the north and west, the Danube at the south and the river Olt at the east – have not inspired the researchers an attention comparable with the interest stirred up by the Romanian medieval monuments, especially those of Northern Moldavia. The founders –members of the petty nobility, priests in the inferior ranks of the Church hierarchy, communities of free peasants –, the materials that had been used, of rather poor quality, the constructors and the painters, with skills and knowledge inferior to those of the masters working with the former princely shops, the condition often very bad of the monuments – the church of Copăceni, Vâlcea county, one of the most beautiful, leans upon poles, and birds enter the tower by the broken glasses – have not invited to consistent scientific concerns, yet well deserved by the monuments.

These churches offer matter for various kinds of research; they can stress the ways modernity, which found its way so late in Romania, started its mani-

---

Constantin Popescu, Faculty of Letters, University of Bucharest, Romania

https://doi.org/10.1515/9783110662900-014

festations in the rural world; they can witness a patriotic feeling, oriented toward independence (Wallachia and Moldavia were under Ottoman dominance) and unity (the three Romanian Principalities were separated); they can indicate outstanding aspects of art history – the idea, widespread, that these buildings of worship and their frescoes, mainly, would be marginal zones, of rusticized and lacking depth expression, yields to the idea that the monuments here concerned are "one of the most important chapters of the Romanian art history" (Paleolog 1984: 6).

This paper examines the interpretations given to the *Annunciation* in some churches of the Vâlcea County, on their *eastern exterior walls (pertaining to the sanctuaries)* (I write Annunciation with roman fonts when it concerns a moment in the earthly life of Jesus, and *Annunciation* with italics for a theme/scene of the religious painting). In the Romanian orthodox churches, this theme can be found:

a) in the interior: in the pronaos (among the scenes of the Akathistos Hymn), rarely above the opening between the pronaos and the naos (Moldovița, in Moldova – medieval church; Urșani, in northern Oltenia); in the naos, rarely or very rarely, in the cupola's pendentives (Moldavian medieval churches of Arbore and Pătrăuți), on the feet of the triumphal arch (Polovragi, in Oltenia), on one of the pillars that support the cupola of a greek-cross church (the medieval Sf.Nicolae Domnesc, in Curtea de Argeș), under the crossing arches of the cupola's tambour (Moldavian medieval church, at the Dragomirna monastery), in an apse (Sf.Ștefan, church at the monastery of Hurezi); on the iconostasis (among the twelve Great Feasts and on the Beautiful Gates);

b) in the exterior: on the western wall (the icon of the sacred character or episode the church is dedicated to), on the southern wall (among the scenes of the Akathistos Hymn, at the Moldavian medieval churches), on the eastern wall (of the sanctuary), as a rectangular panel or under *arcatures*.

I am interested in a very precise spot; I do not know, in other regions of Romania, churches to present this theme here. I conceive sanctuaries from my area of concern as prisms with ten or 14 sides, half of them being visible in the exterior (the others can be mentally reconstructed, as situated in the interior of the church); in the superior part of the external walls, under the roof, some churches deploy arcatures (elements of decoration); in the case of the sanctuaries, the feet of the arches do not correspond to the prism edges (Figure 13.1), with two exceptions, at Urșani and Ceaușu); under an arch a single character is represented – with very rare exceptions; in some cases, the edge separating two sides of the prism cuts the character – at Măldărești, for example (Figure 13.2b).

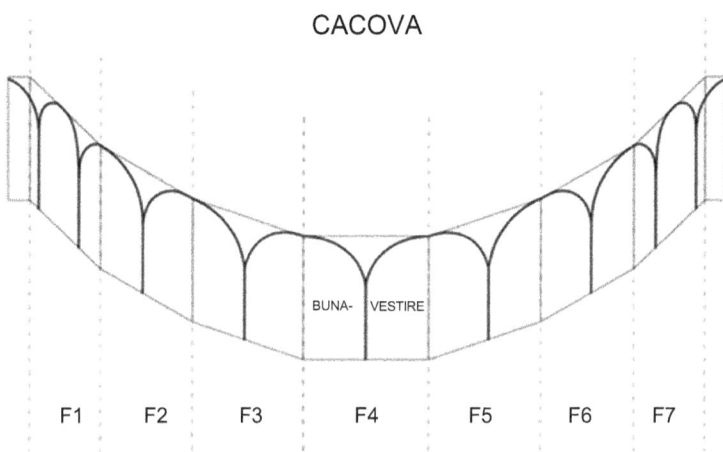

Figure 13.1: Sketch of the sanctuary, Cacova, commune of Stoenești.

What happens when to be represented is the theme of the *Annunciation*, which has two characters: the Virgin and Gabriel? Simplifying things, the constructor and the painter have three solutions: a) to squeeze the two characters under the central arch, which has the same height and width as the others; b) to place the two characters under two contiguous arches, with the same height and width as the others, and to link the characters in a way or another through the common foot of these arches; c) to extend the width of the central arch in order to make it host both characters.

It is easy to see that, whatever the solution, the constructor and painter have to produce a hierarchy between two codes: of architecture (which establishes the dimensions of the prism sides and, in the main, asks for a constant rhythm of the arcature) and of painting (which, at least for the Christian painting of Byzantine influence/inspiration, produces as solid isotopies as possible). What does isotopy means in the case of the *Annunciation*?

We can reformulate this question: what makes the theme of the *Annunciation* immediately recognizable? Hélène Papastavrou examined its components: the Father, the clouds (which hide him to the earthly gazes), the Holy Spirit, the beams on which he descends, Gabriel, the sceptre or the lyly he carries, the Virgin, the wool ball or the book she reads from, the throne, the table, the tree, the vase with flowers, the urban landscape in front of which the characters stand (Papastavrou 2007: 49–96, 226–323).

The throne reminds us that Mary is the Queen of Heaven, but the Virgin is sometimes called Throne of Christ. From the wool thread the Virgin was doing the veil of the Holy of Holies; the thread is a symbol of the Incarnation, the veil is the physical body of the Logos. The book is a symbol of the incarnated Logos. The lyly means chastity, excellence. Tradition has called "flower" both Mary and Jesus; in some *Annunciations* (Pătrăuți and Sucevița, in Moldova) a vase with flowers can be seen, Mary bearer of Jesus (Mary can be a golden vase too, bearing the manna, Jesus). The tree is associated to the tree of Jesse, the tree of life, the tree of the Cross, the Church; it is a Byzantine metaphor of Mary. Sometimes, in the urban landscape behind the Virgin there is a ciborium, reminding a church; or, Mary is the Church.

How many figures, how many appearances for the Virgin in the *Annunciation*? Mary herself, the throne, the lyly offered by Gabriel, the ciborium, the tree, the vase . . . Where does this strengthening of the scene's semantic coherence end up? It is just this *plurifiguration* of Mary in the *Annunciation* that allows the reduction of the scene: the Virgin and Gabriel. For Hélène Papastavrou (2007: 49), the essential components are Mary, Gabriel and the Holy Spirit; for Jean Paris (1997: 25), the Virgin and Gabriel. I am tempted to follow him, as nowhere else in the orthodox iconography one can find a winged character and a feminine character together; therefore, even with only two components the scene is easy to recognize. Moreover, in different spaces and moments, painters have figured the theme (it is true, rarely) through two elements: the Virgin and Gabriel.

The number of the elements in the *Annunciation* on the Oltenian churches' sanctuaries varies a lot. In all of the other places in the churches where the *Annunciation* is figured, the scenes have a constantly high number of elements (in these places, the interest for the representations' narrative dimension is decisive). The big variation in the number of components depends on the hierarchising solutions concerning the two codes: the extension of the central arch determines a higher number of elements for the scene; the squeezing of the two characters under an arch or their representation under two contiguous arches determines the reduction of the elements' number. In establishing the number of the selected elements, the neighborhoods too play an important role: under each arch there is a character (prophet, sibyl, philosopher, apostle . . .). Here, the concern to produce an illusion of reality is normally limited to the skyline – which makes the beholders believe that the characters stand on the ground – and a small, because distant, tree.

The options in point of compositional patterns are lower in number in the East than in the West when it comes to the *Annunciation*. Some western ones are commented on by Arasse, who is interested in the ways Renaissance painters, by resorting to the rules of perspective, tried to give artistic expression to the descent of eternity into time, of unmeasureness in measure, of what cannot be explained into words, of what cannot be represented into visible form: the archangel and the Virgin are placed on a line parallel with the composition plane; perpendicular to it there is a central linear zone with architectural elements running in depth and inviting the beholder's gaze to follow; somewhere in this zone he will discover the spirit's descent in matter (Arasse 2010: 25–28) (the dimensions of the closed door, at the end of Veneziano's perspective in *the Annunciation*, Fitzwilliam Museum, Cambridge, are increased to symbolise the miracle; the closed door is a figure of Mary, *Ezekiel*, 44, 2); the two can be placed on the same line parallel to the composition plane with Mary under an open small architectural construction and Gabriel outside, separated by a column contiguous to the composition axis (Piero della Francesca, around 1455, at San Francisco, Arezzo); or situated in different planes, with the archangel rushing in more or less acute angles in the Virgin's plane in order to accomplish the mission (Venetian scheme of the XVIth century, adopted by the baroque painters, and of high dynamism).

This last scheme implies the beholder. For Jacques Fontanille, an *Annunciation* of Lorenzo Lotto's (Villa Colloredo Mels, Recanati) in which Mary hastens toward the beholder looking into his/her eyes, while Gabriel lags behind in the other subfield of the painting and a cat tries to escape unharmed in another direction creates a "paradoxical situation, in which the divine message is received by one who was not its addressee" (Fontanille 1989: 88–90), in other words by the beholder.

In Romania at least, the by far most frequent compositional scheme for the *Annunciation* places Gabriel and the Virgin in a plane parallel to the plane of the painted panel and the wall of the urban landscape in the second plane; usually, strictly behind the characters and behind the horizontal line of a wall there are edifices with queer components.

> Massive architectures, viewed from three quarters and placed in the two (lateral – C.P.) parts of the composition suggest a relatively deep space. Their position has a double aim: to show the beholder as many possible angles in order to deepen the space and to focus his / her attention on the main characters. (Velmans 2011: 58)

All the *Annunciations* with urban landscape from the Oltenia of the concerned period emphasize the characters by means of an architectural frame. The other components are placed around the characters (the tree is between them, behind the wall, the vase upon the wall). This organisation makes very comfortable the suppression of the elements.

Before examining in a more detailed way the compositional problems for the *Annunciation* – the selection of a certain number of elements, their organisation under the arches –, problems that, as we have seen, depend on the relationships between the decorative surfaces – the prism sides and the extension of the arches –, let's consider how the presence of the *Annunciation* on the sanctuary external wall, above its window, could be explained. It is an important aspect, as it is linked to the mental structuration of the meanings implied by the code: "When a code apportions the elements of a conveying system to the elements of a conveyed system, the former becomes the expression of the latter and the latter becomes the content of the former" (Eco 1979: 48).

There are two levels, of different depth, of structuring the contents: a) that of the role the moment of the Annunciation plays in the ideological tissue (validated by tradition) of the Christian religion; b) that of the role founders, constructors, and painters intended to give to the scene in the iconographic programs of their churches. I'll examine them successively.

The sense of God's love for mankind is the deification of its members (the cancelling of their death). The Annunciation is a miracle which opens to Jesus the way toward the sacrifice meant to redempt them. The Incarnation, the birth, the Passion, the death, the Resurrection, stages during this way, are interdependent in the minute detail; if a single one is eliminated, the whole loses its meaning (Constas 2017: 161). There are some plastic arguments for this idea: Wrapped up as a corpse prepared for burial, the child in the manger reminds visually the corpse in the tomb; in this way, the end of the divine story is already present in its beginning (Constas 2017: 146). In an interpretation of the Mourning (church Sf.Pantelimon of Nerenzi), "the corpse is slighly raised by the mother and placed under her feet slighly separated, with the superior part of the corps leaned upon the womb, therefore suggesting the birth" (Velmans 2011: 98).

With the Annunciation starts the dissolving of the curse coming upon us by the disobedience and the deviation of our first mother, Eve. (Maximus the Confessor 1998: 19) The Son incarnated from the creative energy of the Holy Spirit, and not from seed. His trans-mundanity can be perceived from his birth without injury and pain (Maximus the Confessor 1998: 22), from the perfection of his body in the very moment of his conception (Vlachos 2004: 310–311). At the end of the kenotic way – *kenosis* = self-emptying, humbling (*Phil.*, 2, 7–8) – there is the open door of Heaven.

The second level. The Marian cult has had numerous periods of flourishment; Velmans comments several of them (end of the XIIth and beginning of the XIIIth centuries, then XIVth century, in the Orient) (Velmans 2011: 96, 101, 173–175). In a paper on the sibyls, a very important theme of the Romanian religious painting in the period that interests me, Teodora Voinescu mentions the "high prestige" that the Marian cult had in the Romanian territories, "from the end of the XVIIth century, especially in the monastic entourages" (Voinescu 1970: 205).

The author mentions hand-written translations, very popular during the whole XVIIIth century, of texts from the preceding century, as *The Holy Mother of God's miracles* and the *New Sky* (legends about Mary) of Ioannikie Galeatovski,

Ukrainian, or *Amartolon sotiria* (Redemption of sinners) of Agapie Landos, Cretan (Voinescu 1970: 201–205).

In 1728 a church is erected in Bucharest: the Church with Saints or the Church with Sibyls; the characters are painted under the exterior arcature. It is an isolated case in the Capital, but it inspires representations of Sibyls at the chapel of the Diocese of Râmnic, in Râmnicu-Vâlcea (1750–64); the practice then spreads in the whole Oltenia. Add the enduring idea of the Middle Ages to establish parallels between the Old and the New Testament, still alive (at Coasta, commune of Păușești-Măglași, Vâlcea county, the arcature deploys apostles on the southern wall and prophets on the northern). People with lives accelerated by the society dynamics greedily absorb fragments of hagiographies, wisdom literature, writings on Christian ethics, etc.; it follows that cultural agglomerates are born, whose elements, as if about to claim and acquire iconographic expression, were sometimes able to put the Annunciation in such a high spot.

Voinescu's idea, that the representations under the arcature of mainly philosophers and sibyls flow toward the sanctuary axis in order to celebrate the *Annunciation*, is reprised by Andrei Paleolog who, invoking sibyls and prophets, tries to redirect the theme's content toward earthly horizons: it is a matter not only of "immemorial waiting which finds its resources in a permanently refreshed hope of redemption and miracle" (Paleolog 1984: 31), or, in other words, a matter of redemption, but also of "*a waiting full of hope, continuously nourished by the vision of the national freedom and of the Romanian union.* In an iconographic context, to preach a Romanian homeland sovereign, united, and independent was still considered as pertaining to a 'miracle'" (Paleolog 1984: 42). The Romanian Provincialities too went through the turmoils of the national mouvements in Europe; the idea of getting out from under the Turkish rule fuelled people's imagination to an even larger extent, the more so as the Sublime Porte, since 1821, had no longer appointed Greeks from Fener as princes of Wallachia and Moldavia (he had done that from the beginning of the second decade of the XVIIIth century).

As long as the symbolic network (system) of the Christianity is so tight that the hypothesis to not find a link between two symbolic elements seems to be excluded, I think that we can record another explanation for the presence of the *Annunciation* on the eastern external wall of the churches. They are oriented toward East, when the sun raises it sees their sanctuaries. And Jesus says: "I am the Light of the world" (*John*, 8, 12; 9,5) and "whoever follows me will not walk in darkness, but will have the light of life." (*John*, 8, 12) And again, talking to his disciples: "You are the light of the world (. . .) let your light shine before others, so that they may see your good works and give glory to your Father who is in heaven." (*Matthew*, 5, 14–16)

The Christianity developed its solar symbolism on the ground of the "orientalisation of the Roman empire" (Baudry 2009: 85–88): Jesus is born at the winter solstice, when the days start to grow; the Annunciation is celebrated nine months before Nativity, Easter is celebrated in a Sunday (day of God, *Dominicus dei*, day of Sun), in their prayers the believers are turned toward east . . . It is hard to establish if founders and believers in the XVIIIth and XIXth centuries' Oltenia knew the efforts made during the centuries to give coherence to *solarity* within their own religion, but we cannot do them an injustice by depriving them of the joy to find expressions in architecture and painting to those Bible verses, that they surely knew, where Jesus was the Sun.

We can ask ourselves if the people could venerate their God showing him his own physical birth. To venerated beings, offerings are made to show them why they are venerated: the objects symbolize their excellence. The offering intends to consolidate the link between the venerated and those who venerate them by the very reason of veneration. If the lyly is a symbol of Mary's chastity and Mary receives a lyly, can we then admit that the Annunciation is a prediction and a symbol of redemption, and that Jesus having it before his eyes receives the proof of mankind's gratitude for what he does for them? Such an interpretation tallies with the hope for redemption (deification) included in the Annunciation episode.

*

Let us consider now, after we have seen the strata of the conveyed system, the ways constructors and painters use in order to solve the tensions produced in the conveying system by the clash between the architecture and painting codes. The *Annunciations* placed under the arcatures can be grouped following the hierarchies, visible, between these codes.

In Negreni and Măldărești, Gabriel and the Virgin are placed under contiguous arches, with height and width equal to those of the other arches from the arcature. In Măldărești (Figures 13.2a, 13.2b), the painter established a visual link between the two representations: to so many vertical lines (the feet of the arches, the bodies themselves) he added two horizontal lines that meet at a point of the foot the two arches reserved to the scene have in common: the forearms, one in continuation of the other (the angel's, Mary's); in fact, we can speak of one horizontal line, which cuts the common foot of the arches.

In Negreni (here the frescoes have been restored), the painter does not seem to have had problems concerning hierarchies of codes; he was satisfied that Gabriel's wings are visible and that he wrote Buna and Vestire (Annunciation = Bunavestire, in Romanian) near the characters; their forearms, oriented downward, follow different directions.

Architecture and painting codes in *the Annunciation* — 297

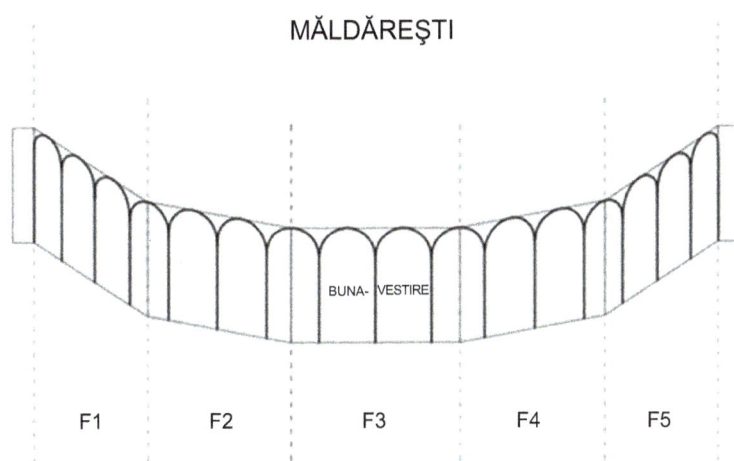

Figure 13.2a: Sketch of the sanctuary, Măldărești.

Figure 13.2b: Măldărești.

In Cacova, commune of Stoenești, the arches are wider (Figure 13.1); as the edges of the seven sides cut them in the middle, the painter could place two characters under each arch and separate them by the edge; it follows that a play of planes is generated, with slightly different orientations. At the left side of the scene there is Zephaniah (Sofonie, in Romanian), who prophetized: "The great day of the Lord is near – near and coming quickly." (*Zeph.*, 1, 14) and, respectively, Balaam, who, prevented by Michael to damn the Jews, blesses them and says

Figure 13.3: Cacova, commune of Stoeneşti, Vâlcea County.

(*Num.*, 24, 17): "I see him, but not now; I behold him, but not near. A star will come out of Jacob." (Figure 13.3)

In his half-space under the arch, Balaam rises his hand to indicate a star, that the painter placed on the edge, linking the halves of that space. Messiah announced by the prophets is evoked as a new star.

The painter unified the *Annunciation* placed under two arches by the urban landscape: the wall behind Gabriel continues in the zone reserved to Mary under the contiguous arch; above the point where their two curves meet there is the dove (Holy Spirit).

In Cacova the painter used a big amount of scene constituents; this was not the case in Câinenii Mici, where constructor and painter chose another solution.

The sanctuary has seven sides too, but fifteen arches (eight, in Cacova) (Figure 13.4a). In the sides 2–7, the succession is half-arch + arch + half-arch; this is why, in sides 4–5, Gabriel, at the right side of Mary, is cut by an edge. The builder unified the scene by eliminating the arch foot that separated the Virgin and Gabriel. The rest of the foot became a false console (Figure 13.4b).

Things are simpler in Copăceni and Urşani, where Gabriel and the Virgin stand under the same arch. In Copăceni, where the sanctuary has seven sides and fifteen arches, the squeezing of the two characters under the same arch diminished their dimensions; we should expect, as this is the central arch, to find under it dimensions able to stress the importance of both characters and position. At the left and at the right of the central arch, Solomon and David are cut by edges.

Architecture and painting codes in *the Annunciation* — **299**

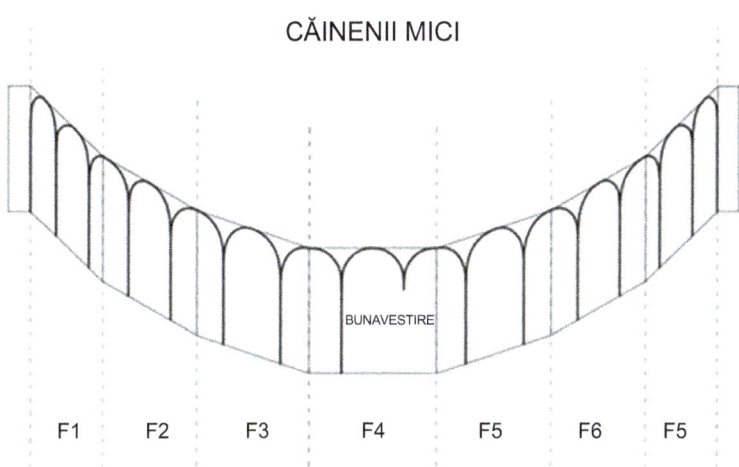

Figure 13.4a: Sketch of the sanctuary, Câinenii Mici.

Figure 13.4b: Câinenii Mici.

In Urşani, the feet of the arches correspond to the sides' edges (Figure 13.5a). Moreover, there are as many arches as sides. The big width of the arches does not claim the diminishing of the characters. In addition, there are several arches under which two characters stand. In Urşani there are two lylies: the Virgin already has one in her right hand, but Gabriel too has one, in his left hand. And he seems to move it away from the Virgin (Figure 13.5b).

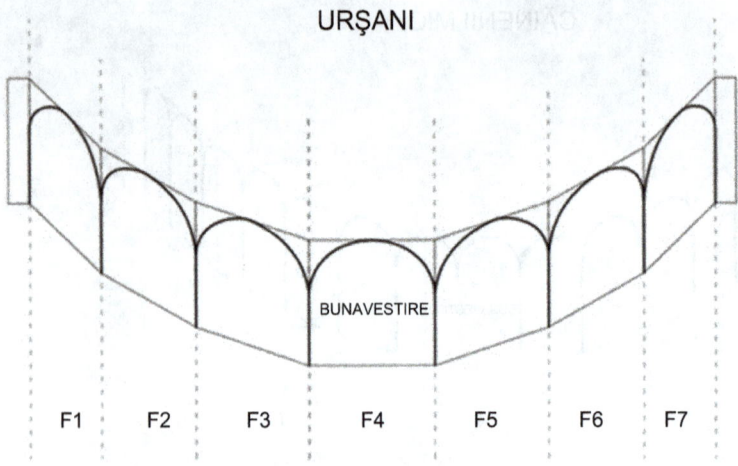

Figure 13.5a: Sketch of the sanctuary, Urşani.

Figure 13.5b: Urşani.

In Olari and Ceaușu (Figure 13.6a), two villages attached to the little town of Hurezi, the central arches are extended. The sanctuaries have five large sides, they are deep; the arches, of little width, are numerous. At the church of the former village, the painter had a lot of space to figure many scene components: the Father, the clouds, the archangel, the lyly, the table, the book, the urban landscape; a cimented area after an unhappy restauration covers the place where beams and dove could have been.

In Ceaușu, the *Annunciation* deserves all our attention (Figure 13.6b). The space under the arch is cut in the axis by a vertical that separates heavenly and earthly parts of the scene: Holy Spirit, beams, clouds, archangel, and, respectively, Mary, book, table, and curtain – sometimes, the building behind Mary becomes a curtain (Papastavrou 2007: 91). This vertical does not overlap an edge. It is not easy to explain this effort of classification. But its result stresses the relationship, *the new relationship*, between the scene elements. Here lies the beauty of the scene: if in the interpretations of the *Annunciation* the beams stop usually in Mary's ear, rarely in her womb, now they stop in the flower Gabriel offers to Mary *across* the line that separates the space under the arch.

Why *in the flower*? The tradition calls Mary *flower*. The flower bears a fruit, allusion to the Incarnation. The slipping from Mary to a figure of her's in such a moment (the Annunciation), with the modification of a such strongly codified fragment (beams stopping in the Virgin), deserves meditation. Where to place the modification, in a continuum from involuntary expressivity and deeply motivated intention?

*

In the variants of the external wall of the sanctuary, *the Annunciation* from the northern Oltenia (XVIIIth and XIXth centuries) shows the builders' and painters' waverings in the refreshing of orthodox paintings' canons. The last one was due to Constantin Brâncoveanu, prince of Wallachia, and the catolicon of the monastery of Hurezi, UNESCO monument, was the main reference; the church has two tiers separated by a string-course and on them deploys arches (the superior tier) and rectangular panels (the inferior tier); at the seven sides of the sanctuary, panels and arches have the same rhythm. The new sensibilities (*circa* hundred years separate Brâncoveanu's death – he is beheaded in Istanbul in 1714 – and the erection of the churches which concern us) and the scarcity of material means require changes in the canons. The inferior panels disappear, the rhythm of the arcature does not correspond to the edges of the polygonal prism, pronaos and naos are separated by a balustrade of plain wall, different characters are represented under the blind arches, characters and scenes with a high affective charge are pictured on the external walls (death, for example), etc.

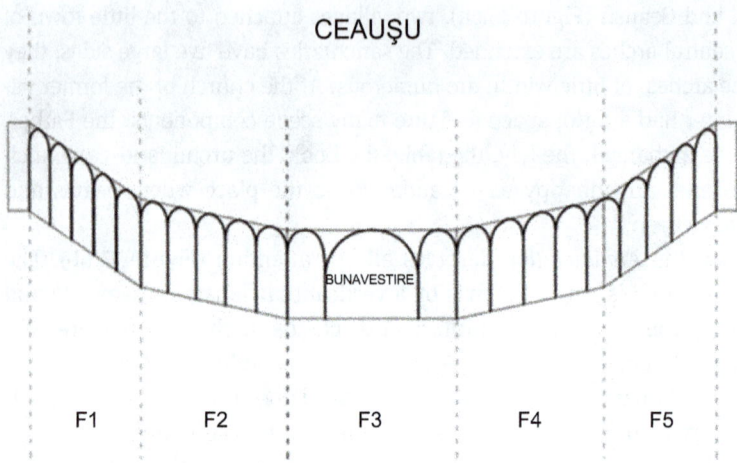

Figure 13.6a: Sketch of the sanctuary, Ceaușu.

Figure 13.6b: Ceaușu.

In the case of the *Annunciation*, builders and painters had to find a harmony between the arcature and the scene's narrative complexity. The conveyed system related to painting combines strata with different ages and relevances: the miracle of Incarnation and the interpretations inspired by a bounce of the marian cult. The corresponding conveying system allows the use of a variable number of the scene constituents, this variation depending, among other factors, on the plurifiguration of Mary in the scene (ciborium, throne, vase, tree . . .) and on the pressure of the iconographic neighborhoods, very simple (characters on monochrome background). The architectural conveying system consists of blind arches and their rhythm, the lateral sides of the prism (the sanctuary), the (non)concordance between the arches' feet and the sides' edges. It is meant to produce, by the crossing of the areas bordered by edges and arches, a kind of perceptive *allegresse* (the architectural conveyed system) which stresses the sanctuary's importance.

It is the wish to harmonise these systems which generates waverings, effect of a too weak inventory of solutions for a too complex mission. Nevertheless, the *Annunciations* of Câineni, Cacova, and, mainly, Ceauşu move us by their naive full of devotion expressiveness.

References

Arasse, Daniel. 2010. *L'Annonciation italienne. Une histoire de perspective*. Paris: Hazan.
Baudry, Gérard-Henri. 2009. *Les symboles du christianisme ancien. Ier-VIIe siècle*. Paris: Editions du Cerf.
Constas, Maximos. 2017. *Arta de a vedea. Paradox şi percepţie în iconografia ortodoxă* [The art of seeing. Paradox and perception in orthodox tradition], rom. transl., Iaşi: Doxologia.
Eco, Umberto. 1979. *A theory of semiotics*. Bloomington: Indiana University Press.
Fontanille, Jacques. 1989. *Les espaces subjectifs. Introduction à la sémiotique de l'observateur*. Paris: Hachette.
Maximus the Confessor. 1998. *Viaţa Maicii Domnului* [Life of the Virgin], rom. transl. Sibiu: Deisis.
Paleolog, Andrei. 1984. *Pictura exterioară din Ţara Românească (secolele XVIII-XIX)* [The external painting in Wallachia (XVIIIth-XIXth centuries)], Bucharest, Meridiane.
Papastavrou, Hélène. 2007. *Recherche iconographique dans l'art byzantin et occidental du XIe au XVe siècle*, Venise, Bibliothèque de l'Institut hellénique d'études byzantines et postbyzantines de Venise.
Paris, Jean. 1997. *L'Annonciation*. Paris: Editions du Regard.
Velmans, Tania. 2011. *L'image byzantine ou la transfiguration du réel*. Paris: Hazan.
Vlachos, Hierotheos. 2004. *Predici la marile sărbători* [Sermons at the Great Feasts], trad. rom., Galaţi: Ed. Cartea Ortodoxă – Ed. Egumeniţa.
Voinescu, Teodora. 1970. Un aspect prea puţin cercetat în pictura exterioară din Ţara Românească: motivul sibilelor [A too less considered aspect in the external painting of Wallachia: the theme of the Sibyls]. *Studii şi cercetări de istoria artei*, Seria Artă plastică, tomul 17, nr.2. 195–210.

Index

abduction 34, 37, 68
abductive inference 34
ability to perceive similarities and differences 4
Absolute responsibility 45, 79
accumulation 103, 105–107, 155
advertising 126, 198–199, 206, 245, 248–249, 251, 257, 260, 264–265, 269, 271, 273–274
affect 152, 163–164, 168
alterity 17–22, 49, 51
Annunciation 2, 8, 292–298, 300, 303–304
Apostle Matthew 3
Aristotle 34, 42, 98, 152
Arthur Koestler 112
artificial intelligence 6, 26
assemblative logic 38, 51
asymmetry 17, 50, 137
Audiovisual communication 227
Augusto Ponzio 4–5, 7, 29
authentic dialogue 4
axiology 74, 87, 184

Bakhtin 13, 21–22, 46, 48–51, 53–55, 64, 74, 82, 84, 87
banal nationalism 247, 249
Benedetto Croce 32
biopolitics 83
biosemiotics 32, 80–81
borders 130, 138, 185, 265
brand identity 278, 286

capacity for listening 66, 72–73, 80, 82–83
capitalization of trust 158, 161
Charles Morris 14, 30, 42, 60, 64, 74
Charles S. Peirce 30, 33, 62, 67, 87
chronotope 54
churches 2, 8, 291–292, 294, 296–297, 303
Cicero 113
code 5, 13, 51–52, 87, 99, 182, 199, 265, 296
cognition 4, 7, 37, 104, 122, 163–165, 168, 193
cognitive dissonance 166, 168
cognitive economy 190

cognitive linguistics 30–32, 35
cognitive semiotics 2, 72, 80
communication campaigns 3, 163, 166, 168–169, 173, 175
communicative competence 31
conceptual competence 31
Constantin Popescu 2, 8
creativity 31, 34, 41, 44, 68–69, 81, 86, 137, 183, 236, 239
cultural anthropology 126
cultural artifacts 2
cultural diversity 5, 161
cultural globalization 5
Cultural heritage 253, 255–256
cultural identity 5, 122–124, 142
cultural meme 192
cultural narrative identity 122, 129
cultural semiotics 7, 122, 181
cultural values 127, 137, 249
Culture 82, 122, 130, 194
cybernetics 122

Dante 54
deduction 37
democracy 25, 198
Descartes 32
diagrams 33
dialogism 25, 30, 44, 49, 52, 74, 84
dialogue 2, 5, 13–14, 25–26, 47, 49–52, 54, 81, 84, 86, 111, 207
Diana Luiza Simion 8
dictionary 67, 112, 116, 118
difference 3, 5, 14, 16, 18–19, 22, 26, 30, 37–38, 40, 43, 45–48, 51, 65, 107, 116, 132, 167
digital culture 7, 126
discursive functions 205, 211–212, 214–215, 218–219
Dostoevsky 17, 49
double bind 134
dynamic systems 7, 137

Edmund Husserl 79, 88, 99
ego 7, 151, 153, 160

ELAN 205, 210–211, 213, 219
Elena Negrea-Busuioc 8
Emmanuel Levinas 15–16, 18–20, 49, 65, 75
encyclopaedia 7, 112, 116–118
ethnic discrimination 170
ethosemiotics 72
Evripides Zantides 5
exogram 106–107
extended mind 97, 106, 108, 117
Extralocalization 48
Extremism 7, 181, 184, 200

fashion 155, 186, 188, 194–200, 249, 252, 265
fear of difference 5
Ferruccio Rossi-Landi 26, 64
fiction 43, 112, 142, 152, 185, 228
final interpretant 46
firstness 39, 46
foreshadowing 8, 228–229, 231, 234–236, 238–241, 243
fresco painting 8
functional theory 204–205, 211

Galen 7, 17, 34, 73, 83, 87
gesture 209–211, 214–219, 228, 257
gestures 210–212, 216–219, 237, 255
Giambattista Vico 30
glasnost 195–197
global semiotics 46, 59–64, 66, 72, 79–80, 82, 85, 87
globalization 5, 83, 85, 121, 129–131, 208
Göran Sonesson 2, 7
graphic design 245
great experience 21

hermeneutics 182–184
heterarchy 7, 121, 132–133, 138, 143
heterarchy of values 7, 121, 133, 143
Hippocrates 7, 17, 73, 87
humanism of otherness 13, 17–18, 89
hypothesis 20, 37, 112, 297

icon 33–35, 253, 292
iconic 33, 42, 44, 252–253, 269, 272
iconicity 30, 33, 39, 46, 68

iconosphere of difference 3
icons 42, 118, 249
identity 3, 5, 7, 13–14, 15, 16, 17, 18, 19, 20, 21, 22, 23, 24, 38, 45–54, 68, 121–124, 126–128, 130–132, 134–143, 151, 154–156, 189–190, 197, 207, 242, 246–254, 256–257, 259–263, 266–267, 272–274, 277–282, 286–287
ideologies 7, 46, 52, 81, 131, 185, 189
impression of similarity 6
induction 4, 37
International Conference *Semiosis in Communication* 1
Ioana Bird 3, 7

Jacques Derrida 1, 20, 48
Jakob von Uexküll 81
Jean-Jacques Rousseau 1, 35
John Locke 30, 87
Jonathan Friedman 5
Jordan Zlatev 3
Jurij Lotman 7

Kristian Bankov 7

linguistic competence 31
logic of association 39
Loredana Ivan 3, 7
Luis Emilio Bruni 5, 7
lying 1, 7, 43, 151–160

magic realism 235, 237, 239
Marcel Danesi 30
Martin Heidegger 15
Marx 23–24, 87, 191
Maurice Merleau-Ponty 79, 104
mediation 46, 71, 123, 184
medical semeiotics 17, 59–60, 62, 73, 83, 85, 87
mental representations 105
metalanguage 192, 195
Metaphor 30, 35–36
metaphors 33–36, 39, 98–99, 118, 131, 285
metasemiosis 41, 59, 61, 63–64, 66–67
Michael Tomasello 97
modeling processes 33

modeling theory 31, 40
modelling 30, 46, 64–66, 81, 87
Mondher Kilani 3
mood 164–165
moral emotions 7, 163, 165–166, 169, 173, 175–176
multicultural 5, 54
multiculturalism 131, 135
multimodality 204–205, 211, 219
MUMIN coding scheme 211
mystique of dialogue 4

Narrative 122–123, 128
narrative identity 7, 122–124, 127–128, 131, 137–139, 142
national identity 5, 8, 245–254, 256–257, 259–263, 265–268, 272–274
nationalism 8, 165, 246–249, 251, 261, 268–269, 272, 274
network society 6
new humanism 19
nicknames 1, 8, 238, 277–287
nicknaming practices 277–282, 284, 286
non-functionality 7, 13–14, 17–21, 23, 52

Oltenia 8, 291–292, 295, 297–298, 303
Orthodox painting 2
otherness 3–4, 5, 7, 14, 19, 21, 25, 30, 38–39, 44–48, 51–53, 67, 69, 74, 77, 86–87

Patrizia Violi 112
Paul Cobley 15
perception 18, 22, 100, 104, 113, 156, 170, 203–204, 230
perceptual similarities 4
personalisation 204
persuasion 164–165, 205, 233
persuasive potential 205, 212, 216–217, 219
phenomenology 2, 7, 79, 88–89, 99, 104, 109, 114, 117, 153–154
play of *differences* 1
political correctness 135–136
populism 181
positioning 8, 204–212, 220
positioning strategies 208, 211

positioning theory 204, 206–207, 209, 211
Post-truth 156
Pragmatism 67
Prague School 99
primary modeling 31, 40, 43, 186

reality effect 204
relational meanings 211, 219
relevance 97, 109, 112–115, 117, 227, 253, 256
Renaissance 83, 294
responsive listening 73, 84
responsive understanding 4, 14, 52, 84
rhizome 112
Roberto Benigni 8, 227, 235, 243
Roland Barthes 37, 84
Roma communities 1, 7, 168–169, 171, 173
Roman Jakobson 30, 63, 98

schemes of interpretation 109–111, 113
sedimentation 2, 97, 104–108, 110, 113, 117
semantic relations 210–211
semioethics 1, 7, 13, 17, 46–47, 59, 62–64, 67, 72–74, 77, 80, 82, 85, 87, 89
semiosis 2, 17, 41–42, 44, 46, 54, 60–67, 69, 71–74, 76, 80, 83, 85, 87, 89, 99–100, 117, 137, 185
semiosphere 5, 7, 32, 78, 130, 132, 135, 137–138, 141–142, 184–190, 192, 195, 197, 200
semiotic animal 53, 59, 61, 63–64, 66–67, 71, 77
semiotic behavior 218, 220
semiotic modes 210–211
semiotic potential 217–218
semiotic practice 206
semiotic resources 4, 138, 205, 210–212, 219–220
semiotic value of the gesture 216–217
sense of similarity 4
similarity 1, 4–5, 6, 30, 33–34, 37, 105, 175–176
small experience 21, 26
social control 280
social semiotics 69, 99
Socrates 25
Søren Kierkegaard 49

Spielberg 14
sports brand 277
sports teams 8, 277–278, 282–284
stereotypes 170–175, 280
Storytelling 227
structuralism 87, 137, 192
Susan Petrilli 4–5, 7, 13, 29
symbols 42, 168–169, 252–254, 256, 262–263, 266–268, 273, 285–286

Tartu School 98, 109, 117
texts 43, 87, 104, 116, 168, 173, 175, 188–190, 195, 228, 230, 246, 250, 296
textual analysis 7, 168, 173
the dichotomy *differences – similarities* 2
The Divine Comedy 54
the *imaginary of otherness* 3
the interplay of differences and similarities 1–6, 8
the *invention of the Other* 3
The Merchant of Venice 4, 37–38
the Prague School 7, 99, 103, 109, 117

The problem of alterity 15
the problem of *otherness* 3
Thomas Sebeok 30
typography 245, 248

Umwelt 81
understanding of otherness 4
Ur-Fascism 5

Vâlcea county 2, 291, 297
Values 122, 128–129, 152, 255
veil 198–199, 266, 293
Victor Ieronim Stoichiță 3
visual signs 169, 252, 264, 271
visual stereotypes 172

Walter Benjamin 23, 44
Web 2.0 160

Yuri Lotman 103, 126, 137